ULTRASOUND ANNUAL 1982

Ultrasound Annual 1982

Editor

Roger C. Sanders, M.D.

Associate Professor
Department of Urology and
The Russell H. Morgan Department of Diagnostic Radiology
The Johns Hopkins Medical Institutions
Baltimore, Maryland

Raven Press ■ New York

Raven Press, 1140 Avenue of the Americas, New York, New York 10036

Made in the United States of America

International Standard Book Number 0-89004-861-4

Preface

Diagnostic ultrasound represents a field in which the development of new equipment has led to rapid changes in diagnostic information. This improvement in imaging quality has revealed subtle structural detail and movement not previously available. Since information has been changing so rapidly, textbooks have not been able to keep up; they are usually two years out of date by the time they are published. Technical changes in the instrumentation can have a major impact on the performance of procedures and, indeed, can lead to the development of entirely new fields as is the case with Doppler B-scanning and small parts scanners in the investigation of the carotid artery. As in any other field, it is important that the sonologist (or physician who devotes most of his time to ultrasound) remain abreast of "state-of-the-art" ultrasonic methodology. This volume is intended to further this goal.

Reviewed are ten areas of rapid development in ultrasound. To clarify traditionally problematic areas of diagnosis such as the pancreas, gallbladder, renal medical disorders, and obstetrical measurement an overall composite of a dispersed group of observations has been assembled. New areas to which ultrasound just recently has been applied are neonatal intracranial ultrasound, the use of ultrasound in the carotid artery, breast ultrasound, and fetal cardiac ultrasound. The practical uses of ultrasound as an aid in performing puncture techniques and as an aid in the operating room are also reviewed.

Each annual edition will be timed to appear on the occasion of the American Institute of Ultrasound in Medicine meeting. Each volume will consist of ten relatively exhaustive reviews on topics in which major developments have occurred or in which a comprehensive overview of a complex situation is necessary. Most of the material will be on clinical aspects of ultrasound but future annual volumes will devote some material to basic sciences or instrumentation. Many of the authors are well-known experts in their fields; others are individuals who have something new and original to contribute but whose work has previously been published in smaller segments. A drawing together of this dispersed material in one comprehensive review is of great value.

This book is designed for the physician with a major interest in ultrasound and for the sonographer (truly a physician's assistant) in an attempt to provide an up-to-date assessment of areas in which rapid changes in ultrasonic practice have occurred and in which confusion exists. Obstetricians with an interest in ultrasound will find the chapters on breast, obstetrical measurements, neonatal intracranial ultrasound, and fetal echocardiography helpful. Surgeons will find the information on puncture techniques, the use of ultrasound in the carotid artery, and operative real-time ultrasound of special interest.

Roger C. Sanders

Contents

Contributors

Richard Chang. *The Russell H. Morgan Department of Radiology and Radiological Science, The Johns Hopkins Medical Institutions, Baltimore, Maryland 21205*

Julio C. U. Coelho. *Department of Surgery, Abraham Lincoln School of Medicine, University of Illinois, Chicago, Illinois 60680*

David Graham. *Departments of Gynecology and Obstetrics and The Russell H. Morgan Department of Diagnostic Radiology, The Johns Hopkins Medical Institutions, Baltimore, Maryland 21205*

Michael C. Hill. *Department of Radiology, George Washington University, Washington, D.C. 20037*

Hedvig Hricak. *Department of Radiology, University of California, San Francisco, San Francisco, California 94143*

Charles S. Kleinman. *Department of Pediatrics, Yale University School of Medicine, New Haven, Connecticut 06510*

William Robert Lees. *Department of Ultrasound and CT Scanning, The Middlesex Hospital, London, United Kingdom*

Edward Robert Lipsit. *Department of Radiology, George Washington University Medical Center and Department of Radiology, Columbia Hospital for Women, Washington, D.C. 20037*

Junji Machi. *Department of Surgery, Abraham Lincoln School of Medicine, University of Illinois, Chicago, Illinois 60680*

Roger C. Sanders. *Department of Urology and The Russell H. Morgan Department of Diagnostic Radiology, The Johns Hopkins Medical Institutions, Baltimore, Maryland 21205*

Bernard Sigel. *Department of Surgery, Abraham Lincoln School of Medicine, University of Illinois College of Medicine, Chicago, Illinois 60680*

Thomas L. Slovis. *Departments of Radiology and Pediatrics, Wayne State University School of Medicine and Children's Hospital of Michigan, Detroit, Michigan 48202*

Hsu-Chong Yeh. *Department of Radiology of the Mt. Sinai Hospital and School of Medicine of the City University of New York, New York, New York 10029*

Ultrasound Annual 1982,
edited by Roger C. Sanders.
Raven Press, New York © 1982.

Pancreatic Sonography:
An Update

Michael C. Hill

*Department of Radiology, The George Washington University,
Washington, D.C. 20037*

Before sonography and computed tomography (CT) were available our ability to diagnose diseases of the pancreas was limited. Radiographically one could identify a soft tissue mass, calcifications, or air in the area of the pancreas. With barium studies one looked for displacement of the stomach, duodenum, or loops of bowel. All of these findings, however, were decidedly insensitive in detecting pancreatic disease. Sonography and CT have changed all of this. We can now identify minimal changes in both the appearance and the contour of the gland. This has allowed us to detect disease processes at an earlier stage, and follow-up studies have provided greater insight into the correlation between the clinical findings and the morphological changes of the gland. This is especially true in acute pancreatitis. It is unfortunate that, although we can now detect pancreatic carcinoma at an earlier stage, it appears the long-term survival of these patients has not at present been altered significantly.

The role of sonography in the diagnosis of pancreatic disease improved by leaps and bounds with the development of gray-scale imaging. Its diagnostic role, however, has been modified by CT. There are many reports in the literature comparing sonography and CT not only with each other but also with more invasive techniques (2,10,16,28,47,58–60,71,94). Despite the fact that overall CT is a little more accurate than sonography, the latter, because of its noninvasive nature, lack of ionizing radiation, and lower cost per examination, has not been replaced by CT for evaluation of the pancreas as predicted by some a number of years ago. When considering which diagnostic modality should be used in any given patient with suspected pancreatic disease, the most important factor to consider is the patient's body habitus. Obviously, the thin, emaciated individual is better evaluated by sonography than by CT, whereas the reverse is true of the obese patient. The majority of one's patients, however, lie between these two extremes; therefore in our institution we screen most patients with an initial sonogram, except those patients suspected of having the complications of acute pancreatitis, i.e., phlegmon, abscess, hemorrhage (94).

If the study in those patients initially screened by sonography is normal, then CT is performed only when there is a high index of suspicion of pancreatic disease. If the examination is suboptimal or if there is limited visualization of the gland, and this occurs most often in the area of the tail, then CT is recommended.

NORMAL SONOGRAPHIC ANATOMY

The pancreas is both an exocrine and an endocrine gland that lies in the upper portion of the retroperitoneum draped across the aorta and inferior vena cava. The exocrine tissues produce inactive proenzymes (trypsinogen, chymotrypsinogen, procarboxypeptidase, lipase, etc.) that are activated in the small bowel and are necessary for the digestion and absorption of carbohydrates, proteins, and fats (30). The endocrine portion of the gland is represented by the islets of Langerhans, which are groups of cells scattered throughout the pancreas. The main hormones they produce are insulin (in the beta cells) and glucagon (in the alpha cells) (30). Diseases that impair function evoke signs or symptoms only when far advanced because there is such a large reserve of tissue (82).

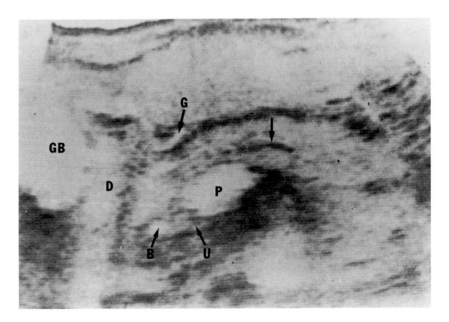

FIG. 1. A: Transverse sonogram of the pancreas. The pancreatic duct can be identified in the body (*arrrow*). P, Portal vein; B, common bile duct; U, uncinate process; D, duodenum; GB, gallbladder.

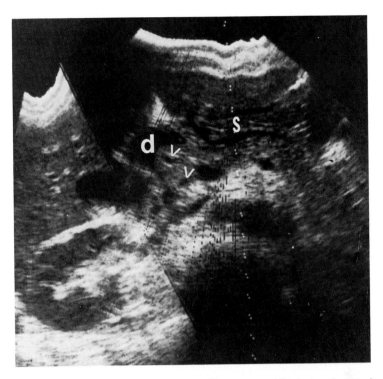

FIG. 1. B: Transverse sonogram of the pancreas. The common bile duct and gastroduodenal artery (*arrowheads*) can be seen in the head of the pancreas. The posterior wall of the stomach (S) could be mistaken for the pancreatic duct. D, Duodenum.

Anatomic Landmarks

Anatomically the pancreas is divided into a head, neck, body, and tail (109). The neck lies anterior to the superior mesenteric vein and beginning of the portal vein and separates the head from the body (Fig. 1). The head of the pancreas lies anterior to the inferior vena cava and right crus of the diaphragm, nestled in the C loop of the duodenum. The uncinate process, which varies in size, is the lower medial portion of the head of the pancreas that lies posterior to the superior mesenteric vein and rarely posterior to the superior mesenteric artery as it proceeds to the small bowel mesentery by crossing anterior to the third portion of the duodenum (Fig. 1). Infrequently, the head may lie anterior to the abdominal aorta and rarely it may lie to the left of it. In such cases, it still maintains its relationship to the duodenum, surrounding vascular structures (portal vein, superior mesenteric vein, hepatic, splenic, and gastroduodenal arteries) and the common bile duct (52,84). The craniocaudad relationship of the head can also vary. It is usually at the level of the first three lumbar vertebrae; however, it has been described as low as the sacral promontary (52,84).

The body of the pancreas lies anterior to the aorta just below the origin of the celiac trunk and anterior and superior to the superior mesenteric artery (Fig. 2). It is separated from the aorta by connective tissue, small lymph nodes, and a variable amount of fat. In classic anatomic texts, the pancreatic tail is described as being that portion of the gland abutting the spleen within the two layers of the splenorenal ligament (109). In conventional sonographic and CT anatomy, however, the tail is defined as all of the pancreas that lies to the left of a vertical line drawn along the left lateral margin of the lumbar vertebral body. From right to left it is related posteriorly to the left crus of the diaphragm and the left adrenal gland, and is separated from the left kidney by the anterior renal fascia (Gerota's fascia). In some patients (10%) the tail may be cranial to the upper pole of the left kidney (72). Along the anterior aspect of the pancreas, where the body joins the tail, there can be a slight outward bulge of parenchyma. This "promontary" is usually seen in patients in whom the tail angles acutely posteriorly and can be mistaken for a mass. The position of the tail of the pancreas is variable and has to be taken into account during

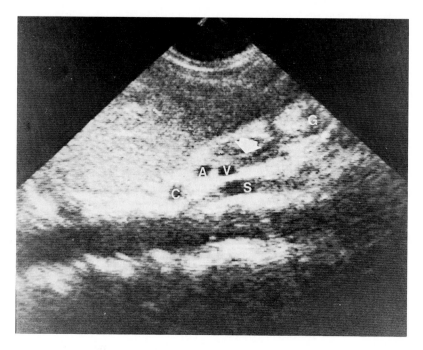

FIG. 2. Midline longitudinal sonogram of the abdomen. The body of the pancreas and pancreatic duct (*arrow*) can be seen. C, Celiac artery; S, superior mesenteric artery; A, splenic artery; V, splenic vein; G, stomach.

transverse scanning of the gland. The tail may be at a higher level (41%) or at the same level (51%) as the body and infrequently (2%) it is at a lower level (72). When the left kidney is absent the tail is more medial and dorsal.

The main pancreatic duct (duct of Wirsung) runs from left to right, starting in the tail and traversing the body until it reaches the head, where it turns inferiorly and posteriorly in the pancreatic parenchyma. It joins the common bile duct just prior to where it enters into the posteromedial wall of the second portion of the duodenum at the ampulla of Vater. In 82 to 84% of normal patients the duct can be demonstrated sonographically in the body of the pancreas (Figs. 1 and 2). It is situated in the pancreatic parenchyma slightly anterior and inferior to the splenic vein. Less frequently it is seen in the head (32%) or tail (12%) (73,74). It is identified as a linear echogenic line or as a fluid-distended tube with well-defined echogenic walls and internal diameter of no greater than 2 mm (9,74). One should be careful not to mistake the echo-free posterior wall of the antrum of the stomach for the duct. Confusion can also arise in the area of the tail of the pancreas where the splenic artery can run in the pancreatic parenchyma or where there is a double splenic vein (74,85). These anatomic variations will be recognized as such by tracing the vascular anatomy.

The accessory pancreatic duct (duct of Santorini) runs transversely in the upper anterior portion of the pancreatic head at a higher level than the main pancreatic duct. It enters the duodenum separately via the minor papilla approximately 2 cm above and slightly ventral to the ampulla of Vater (109). It is uncommonly identified sonographically (74).

The portal vein is formed behind the neck of the pancreas by the junction of the superior mesenteric and splenic veins. The splenic vein runs from the splenic hilus along the posterior superior aspect of the gland. The superior mesenteric vein runs posterior to the lower portion of the neck of the pancreas and anterior to the uncinate process (Figs. 2 and 3). The portal vein courses superiorly at varying degrees of obliquity in the free edge of the lesser omentum to enter into the liver at the porta hepatis. On rare occasions, as a result of an embryological anomaly, the portal vein may run anterior to the duodenum (pre-duodenal portal vein). In such cases it has a more anterior location than the superior mesenteric artery (67).

The splenic artery arises from the celiac axis and runs along the superior margin of the gland slightly anterior and superior to its fellow vein (Fig. 2). As it approaches the lateral portion of the tail it may run anterior to the pancreatic parenchyma, depending upon the degree of the tortuosity of the artery and the position of the splenic hilus.

The common hepatic artery also arises from the celiac axis and is easily identified in most patients (92%) (81). It courses along the superior margin of the first portion of the duodenum and divides into the proper hepatic artery and gastroduodenal artery, usually where it crosses onto the front of the portal vein. The proper hepatic artery is usually seen (75%) as it proceeds superiorly along the anterior aspect of the portal vein with the common bile duct lateral to it (Fig. 3). The gastroduodenal artery is less frequently seen (30%) (81) (Fig 1B). It travels a short distance along the anterior aspect of the head, just to the right of the neck, before it divides into its superior pancreaticoduodenal branches (anterior and posterior). These branches can be identified in the lateral margin of the pancreatic parenchyma close to the duodenum. They join with their counterparts (inferior pancreaticoduodenal arteries) which arise from the superior mesenteric artery. In some patients (14%) the right hepatic artery arises from the superior mesenteric artery (replaced right hepatic artery). It can be demonstrated posterior to the medial portion of the splenic vein and runs along the posterior aspect of the portal vein to enter the hepatic parenchyma (46) (Fig. 4). As might be expected, the size of the hepatic artery in such patients is smaller than usual.

The superior mesenteric artery arises from the aorta behind the lower portion of the body of the pancreas (Fig. 2). It crosses anterior to the third portion of the duodenum to enter the small bowel mesentery. The superior mesenteric vein lies along its right side; however, it can run directly anterior to the artery. The artery itself runs directly anterior to the aorta, but in older

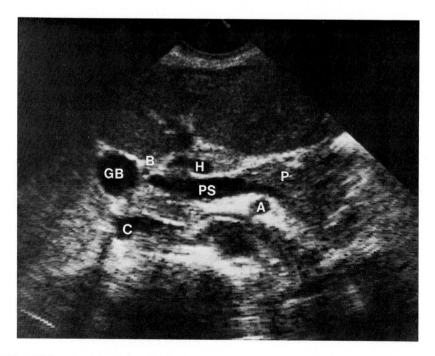

FIG. 3. Oblique sonogram along the long axis of the portal-splenic vein (PS). The proper hepatic artery (H) is along its anterior aspect with the common bile duct (B) lateral to it. GB, Gallbladder; C, inferior vena cava; A, superior mesenteric artery; P, pancreas.

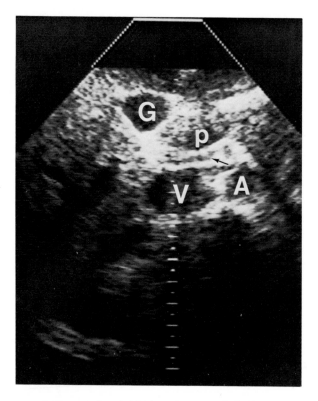

FIG. 4. Transverse sonogram, replaced right hepatic artery. G, Gallbladder; P, portal vein confluence; V, vena cava; A, aorta; *arrow* indicates replaced right hepatic artery.

people with atherosclerotic vessel disease it may be tortuous and run to the left of the midline. Depending on its size, a small portion of the uncinate process can occasionally lie between the superior mesenteric artery and the aorta.

The normal common bile duct traverses along the anterior aspect of the portal vein to the right of the proper hepatic artery (Fig. 3). Where the portal vein crosses anterior to the inferior vena cava, the duct passes off the front of the portal vein and goes behind the first portion of the duodenum to course inferiorly and somewhat posteriorly in the parenchyma of the pancreatic head (Fig. 5), where it lies close to the second portion of the duodenum. It joins the pancreatic duct close to the ampulla of Vater. In the head of the pancreas the internal diameter of the duct should measure no more than 4 mm.

The upper anterior portion of the pancreatic head and the anterior aspect of the neck, body, and tail are related to the lesser sac that separates the pancreas from the stomach. The echo patterns of the stomach and duodenum vary and depend on whether they are empty or contain fluid, food, or air.

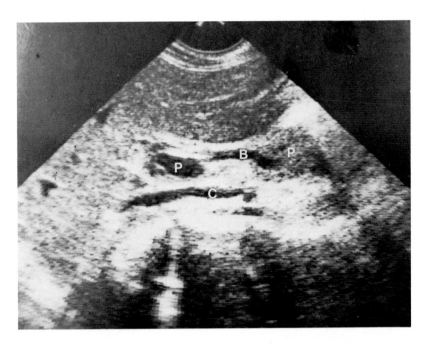

FIG. 5. Oblique sonogram along the long axis of the portal vein (P). The common bile duct (B) descends from the front of the portal vein to enter into the head of the pancreas (P). C, Inferior vena cava.

When empty a central area of echogenicity is seen surrounded on both sides by a thin echo-free wall. The size of the central area of echogenicity varies with the content and degree of distension of the bowel (Fig. 2). When water alone is present, the lumen will be echo-free after the echogenic microbubbles of air have dissipated. Food is identified as echogenic particles of different sizes in a variable amount of fluid. Air, on the other hand, totally reflects the sound waves and no diagnostic information deep to the bowel lumen is elicited (Fig. 4).

Pancreatic Outline and Echogenicity

The pancreas has no capsule but is surrounded by a thin layer of connective tissue. Thin septa arise from this layer and pass into the substance of the gland between the many small lobules. The lobules contain a varying amount of fat between them and each one is drained by a branch of the pancreatic duct (66). The pancreatic outline has a smooth appearance that becomes somewhat lobulated by fatty infiltration (66).

The echogenicity of the normal pancreatic parenchyma is homogeneous and equal to (48%) or greater than (52%) that of the adjacent portion of liver (27). The degree of pancreatic echogenicity is determined to a great extent by the amount of fat present between the lobules of the gland and to a lesser extent by the interlobular fibrous tissue (66). In adults, the degree of echogenicity is greater than in children, as fatty infiltration occurs with aging; by the age of 60 the pancreas is almost always echogenic (102) (Fig. 6A). Fatty infiltration may also be found with obesity (Fig. 6B); however, in such cases the process is reversible with weight reduction.

The echogenicity of the gland in fatty infiltration may be such that it blends with the surrounding retroperitoneal fat so that its outline cannot be distinguished (Fig. 6B). It is only by observing the normal vasculature that one can ascertain that the area of the pancreas is definitely being scanned. Although the degree of echogenicity of the pancreas can vary in different patients, it is important to remember that in any one patient the level of echogenicity is homogenous throughout the gland. It is only by taking this into account that small intraparenchymal pancreatic masses that do not distort the contour of the gland will be detected (99).

It is obvious that the pancreatic echogenicity should not be compared with the echogenicity of a diseased liver i.e., cirrhosis, fatty infiltration, and metastasis. If there is doubt about the normality of the liver, comparison with the kidney, which is normally less echogenic than the liver, is helpful. If the liver is used as a comparison standard, an adjoining portion of liver at the same depth from the patient's skin as the pancreas should be used. If the

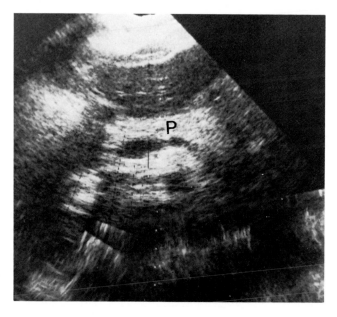

FIG. 6. A: Transverse sonogram of the upper abdomen. Fatty infiltration of the pancreas in elderly normal patient. Smooth-bordered evenly echogenic pancreas (P).

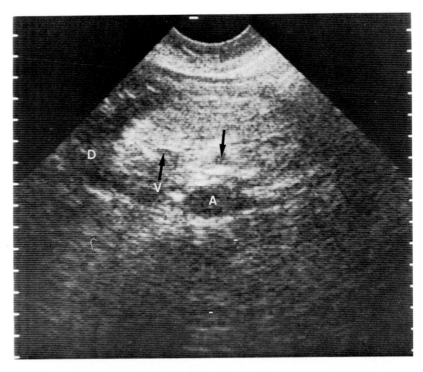

FIG. 6. B: Transverse sonogram of the upper abdomen in an obese patient. The pancreatic outline cannot be identified; however, the superior mesenteric artery (*arrow*) and vein (V) can be seen, together with the aorta (A) and duodenum (D). The pancreas was normal on CT.

water-filled stomach is used as an acoustic window to visualize the pancreas, then the echo pattern of the gland cannot be compared with the liver at the same level (18). In such cases, the pancreas has to appear more echogenic than the liver, since there is less attenuation of the sound beam through the fluid-filled stomach than through a comparable thickness of liver (Fig. 7).

Size of the Normal Pancreas

There have been many reports regarding the size of the normal pancreas, and the multiplicity of reports indicates that there is a general disagreement among different authors. The relative size of the pancreas in children is larger than in adults. The size of the pancreas does decrease with age; however, there has been no definitive sonographic study of this finding. Normal antero-posterior measurements given for adults include 2.0 to 3.0 cm for the head, 0.7 to 1.0 cm for the neck, 1.2 to 2.2 cm for the body, and 0.7 to 2.8 cm for the tail (21,37,87,104). Since the pancreas is a curved organ, the maximum antero-

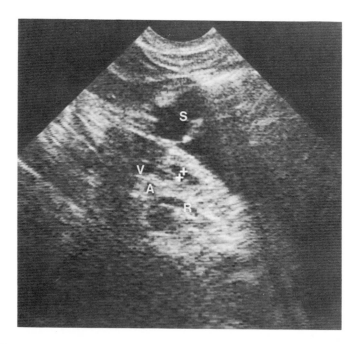

FIG. 7. Transverse sonogram of the pancreas through the water-filled stomach (S) demonstrating the tail with a dilated pancreatic duct (+). The echogencity of the tail is greater than that of the body as a result of the good through transmission through the water filled stomach. A, Superior mesenteric artery; V, splenic vein; R, left renal vein.

posterior diameter should be obtained at right angles to the true axis of the gland at that point (104). The diameter can vary with the phase of respiration; it is thicker in inspiration. The difference in thickness of the body between inspiration and expiration can be as great as 8 mm (39). Higashi has stated that in most patients the pancreas appears thicker while it is scanned longitudinally as opposed to transversely; however, this difference is not felt to be significant (21,39).

The normal craniocaudad dimensions of the gland are given as 2.4 to 4.8 cm for the head, 2.4 to 3.6 cm for the body, and 1.6 to 2.4 cm for the tail (87). The maximum long length given for the head is too small, as it can certainly extend for at least 6 cm.

· I agree with Johnson and Sarti that absolute measurements of pancreatic size can be misleading, as pancreatic pathology could be underdiagnosed in the 75-year-old asthenic female and overdiagnosed in the 20-year-old football player (47,87). What is more important is the symmetry and contour of the gland. The neck and body should measure less than the head, as should the tail outside of a couple of percent of normal exceptions.

Pancreatic Pseudomasses

Pseudomasses of the pancreas are usually due to fluid-filled loops of bowel, usually the stomach, duodenum, or proximal small bowel. The problem of these pseudotumors can usually be resolved by looking for peristalsis with real-time sonography or by having the patient drink a glass of water and noting microbubbles of air in the stomach (35).

A feces-filled splenic flexure can simulate a pancreatic tail neoplasm, especially during scanning in the prone position (6). A dilated afferent loop following Billroth II surgery and a large duodenal diverticulum can mimic cystic pancreatic masses. In such cases, correlation with the clinical history and barium studies should be done, and when necessary follow-up scans should be performed within 12 to 24 hr. Some of the pseudotumors seen on CT, including fatty pseudotumors and ectatic or aneurysmal vessels, can be resolved by sonography (14).

Examination Technique

In the past, a diagnostic examination of the pancreas could be obtained in only 70 to 90% of patients (2,58). Visualization of the tail (37%) was less than that of the head and body (70 to 90%), as it was obscured in the supine position by gas in the stomach. The number of inadequate examinations has been decreased by using the stomach filled with water as an acoustic window and by changing the patient's position to visualize different portions of the gland (7). By means of this technique, diagnostic sonograms can now be obtained in up to 99% of patients (65).

Patients are instructed to fast for 6 to 8 hr prior to the examination. To prevent an unsatisfactory study as a consequence of bowel gas, some recommend the use of Simethicone, 80 mg, four times a day for 2 days (98). Its effect at best, however, is somewhat marginal and has not received widespread acceptance. The presence of barium in the intestinal tract may prevent adequate visualization of the pancreas, and thus when a sonographic examination is to be performed it should be done prior to the barium studies.

The examination should be performed with digital gray-scale ultrasound equipment with either a single transducer and articulated arm or a high-resolution real-time sector scanner. The study can be performed more rapidly with the latter instrument, as the appropriate vascular anatomy can be more quickly delineated. If a unit with an articulate arm is used, then single-sweep sector scans should be performed because compound scanning will obliterate the fine detailed anatomy (Fig. 1A and B). The choice of transducers is also very important, and all of the pancreatic parenchyma should lie within the focal zone of the transducer (45,56). The highest frequency transducer with the appropriate focal length should be used. The most commonly used are 3.5 or 5 mHz, 13 mm medium internally focused transducers; however, for larger patients a 2.25-mHz, 19 mm long internal focused transducer may be necessary. It should be remembered that the skin-to-pancreas distance usually increases by approximately 1.5 cm with the patient in the erect position (56). One should examine not only the pancreas but also the liver and biliary tract, as pathology is often present in these organs in conjunction with pancreatic disease.

One should start the examination by scanning longitudinally in the midline. The celiac and superior mesenteric branches of the aorta should be identified with the pancreatic parenchyma between them. This is a good area in which to identify the pancreatic duct. The splenic vein should be seen along the posterior superior aspect of the gland, with the splenic artery slightly superior and anterior to it (Fig. 2).

Moving to the right of the midline at 0.5-cm intervals, one will identify the splenic vein joining the superior mesenteric vein behind the neck of the pancreas to form the portal vein. Where the portal vein crosses in front of the inferior vena cava, the proper hepatic artery can be seen on its anterior aspect and the common bile duct can be identified coursing inferiorly to enter the head of the pancreas (Figs. 3 and 5). Once the head of the pancreas is out of view, continue laterally until all of the gallbadder has been evaluated. The remaining portion of the body and tail can be examined longitudinally by using the left lobe of the liver or the water-filled stomach as an acoustic window (Fig. 21B).

Transverse scanning should then be performed. Starting high in the epigastrium, one should proceed inferiorly at 1-cm intervals through the liver. These scans should be performed in deep suspended inspiration. To see the upper most portion of the liver, cranial angulation of the transducer may be necessary. Upon reaching the porta hepatis, the common bile duct and proper hepatic artery should be identified along with the gallbladder. The portal vein should be followed to its bifurcation into the superior mesenteric vein and splenic vein. At this point we should be able to identify pancreatic parenchyma (Fig. 3). More closely spaced (0.5 cm) scans should now be performed. The left lobe of the liver should be used as an acoustic window, and to visualize the lowermost portion of the pancreatic parenchyma it may be necessary to angle the transducer caudally.

In the area of the head of the pancreas, the common bile duct and branches of the gastroduodenal artery should be identified. The sweep of the duodenum marks the outer margin of the head. The uncinate process will be seen posterior to the superior mesenteric vein (Fig. 1). The pancreatic duct is best identified in the portion of the body anterior to the aorta.

If the tail cannot be seen, then a variety of techniques can be used. The patient can be placed in the right decubitus or prone position and the tail identified by using either the spleen or the kidney, respectively, as a sonic window (32). I prefer, however, to have the patient ingest 4 to 6 cups of tap water and then to rescan the pancreas with the patient in the supine position (Fig. 7). In patients who are vomiting or who have acute pancreatitis, the stomach should be filled by using a nasogastric tube, and only the smallest amount of water necessary to visualize the gland should be used. The gain settings and output will have to be adjusted to compensate for the enhanced through transmission of the water-filled stomach (Fig. 7).

If the tail of the pancreas cannot be seen in the supine position, transverse scanning should be performed with the patient in the erect or semi-erect position along the true transverse axis (oblique, horizontal) of this portion of the gland (7,18,65). This is best done by using a multiplanar table that forms a multitude of sitting and semi-reclining positions with the feet in the dependent position. This allows one to alter the position of the water-filled stomach in relationship to the pancreas as ascertained by longitudinal scanning (7). If such a table is not available, the patient should be scanned while sitting upright in a chair. Irrespective of the scanning technique, all tight clothing that might increase intraabdominal pressure should be removed. If the head or body of the pancreas have not been well seen, the position of the water in the stomach can be manipulated by changing the patient's degree of erectness or by having him or her assume the necessary decubitus position in either the semi-erect or supine position. These maneuvers are especially easy with real-time equipment. Using the above method, MacMahon reduced the rate of nonvisualization of the pancreas from 19 to 1% in 100 patients (65). In his series he found the semi-erect position most rewarding and administered Lipomul-Oral or 0.5 to 1.0 unit of glucagon subcutaneously to delay emptying of the stomach.

ACUTE PANCREATITIS

Acute pancreatitis runs the clinical gamut ranging from a benign self-limiting disease to one with a fulminant course associated with rapid shock and death. Because of the absence of a capsule, the inflammatory process takes the path of least resistance and spreads anteriorly into the area of the anterior pararenal spaces and lesser sac. Posterior extension is usually prevented by the anterior renal fascia, which blends with the connective tissue around the great vessels anterior to the spine (109). The disease is usually attributable to alcohol abuse, trauma, biliary tract stones, or hyperlipidemia; however, in a significant proportion of patients (15%) no cause can be found (94). It can be divided into two somewhat overlapping clinical categories. In acute edematous pancreatitis the patient experiences nausea, vomiting, and epigastric pain associated with elevation of the serum and or urinary amylase (1). Acute necrotizing (hemorrhagic, suppurative) pancreatitis represents the severe form of the disease as signalled by a drop in hematocrit and serum calcium, hypotension despite volume replacement, metabolic acidosis, and adult respiratory distress syndrome.

In patients with acute edematous pancreatitis, the pancreas will be sonographically normal in about 29% (17,94). The remaining patients will show either diffuse (52%) or focal (28%) enlargement of the head or adjoining body and tail of the pancreas (20%), and the pancreatic duct may be dilated (Figs. 8 and 9). The portion of the pancreas involved becomes diffusely hypoechoic, presumably as a result of the edema of the gland (88). In some cases of diffuse enlargement, the echo pattern, however, may appear normal (2). In comparing the echo texture of the pancreas with that of the liver, one has to remember the influence of age on pancreatic echogenicity (Fig. 6A), and that associated diffuse liver disease may have altered the hepatic parenchymal echo pattern (66,102).

The mass lesion produced by focal enlargement has reduced echoes in comparison with the unaffected portion of the pancreas, and the through transmission can be so good that it can mimic a pseudocyst. If focal enlargement involves only the body portion of the pancreas directly anterior to the aorta, this is unlikely to be due to focal acute pancreatitis and, therefore, a neoplasm should be considered (94).

The main complications of acute pancreatitis include a phlegmon, pseudocyst, hemorrhage, or abscess (103).

Phlegmonous Pancreatitis

A phlegmon is described as a spreading diffuse inflammatory edema of soft tissues that may proceed to necrosis and even suppuration (82). Phlegmonous extension outside of the gland is only present in 18 to 20% of patients with

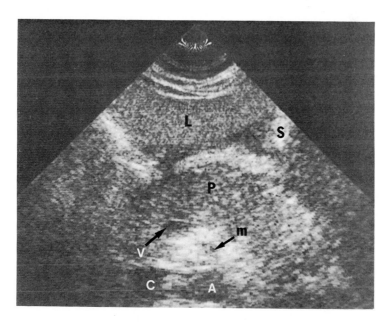

FIG. 8. Transverse sonogram of the pancreas (P) through the left lobe of the liver (L) and stomach (S). The pancreas is diffusely enlarged and hypoechoic due to acute pancreatitis. A, Aorta; C, inferior vena cava; M, superior mesenteric artery; V, superior mesenteric vein.

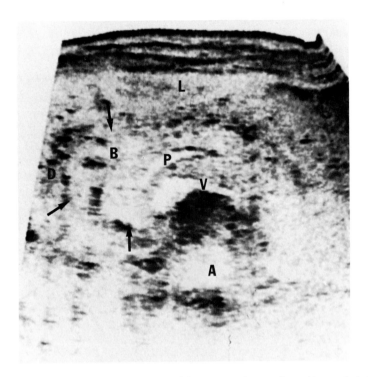

FIG. 9. A: Transverse sonogram of the upper abdomen reveals an enlarged hypoechoic head of the pancreas (*arrows*) caused by acute pancreatitis and a dilated pancreatic duct (P). V, Splenic vein; A, aorta; L, liver; D, duodenum; B, common bile duct.

acute pancreatitis (70,94). The phlegmonous mass is hypoechoic and has good through transmission; however, it does not represent an extrapancreatic fluid collection. The absence of a fluid collection can be proved by percutaneous needle aspiration that yields no fluid and also by examination of the phlegmon at the time of surgery.

The phlegmon usually involves the lesser sac, left anterior pararenal space, and transverse mesocolon (Fig 10). Less commonly it will extend down to involve the small bowel mesentery, lower retroperitoneum, and pelvis (70). The phlegmon in the lesser sac and left anterior pararenal space is often best identified with the patient in the right decubitus position by scanning along the left flank, using the spleen and left kidney as a sonic window. Extension down the right anterior pararenal space can be similarly demonstrated by scanning through the right flank with the patient in the left lateral decubitus position.

Total liquefactive necrosis of the pancreas may occur in which the pancreatic contour is unremarkable, yet the entire central portion of the gland is shelled out and filled with necrotic debris (11). This necrotic center communicates with the ductal system. Sonographically, the gland may appear enlarged and hypoechoic, thus resembling a diffusely edematous gland. The other

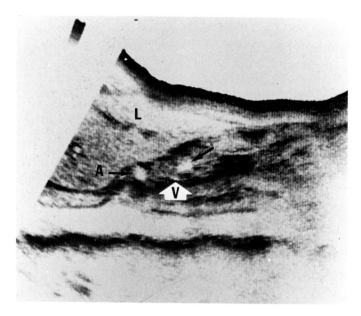

FIG. 9. B: Longitudinal midline sonogram in the same patient. The dilated pancreatic duct (*arrow*) can be seen in the body of the pancreas. V, Splenic vein; A, splenic artery; L, liver.

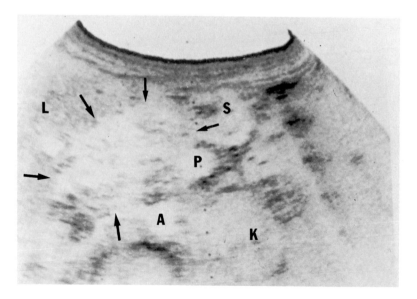

FIG. 10. Transverse sonogram of the upper abdomen in a patient with acute pancreatitis and phlegmonous extension into the area of the lesser sac (*arrows*) with a small pseudocyst (P). L, Liver; S, stomach; A, aorta; K, left kidney.

appearance of this entity consists of a cystic structure within the pancreas with a fluid debris level.

Hemorrhage may occur in the pancreas or associated phlegmon in 2 to 3% of patients with acute pancreatitis (70,94). Such patients usually have a clinical diagnosis of acute necrotizing pancreatitis with a drop in hematocrit and serum calcium. Milder forms of hemorrhage do occur, however, as shown by the presence of hemorrhage in patients with acute edematous pancreatitis and no change in their hematocrit or serum calcium (44). During the acute phase, the area of hemorrhage may appear entirely cystic or may contain varying amounts of echogenic material, and a fluid debris level may be present (Fig. 11). The surrounding wall may be either smooth or irregular. With time these areas of hemorrhage, if they persist, have a cystic appearance that resembles a pseudocyst (44).

Pancreatic Abscess

This complication occurs in about 3% of patients with acute pancreatitis (94). The clinical diagnosis of a pancreatic abscess is extremely difficult, as patients with acute necrotizing pancreatitis can have all the signs and symp-

FIG. 11. A: Longitudinal sonogram to the right of the midline reveals a well-defined complex mass (*arrows*) with focal hyperechoic areas caused by hemorrhagic pancreatitis. L, Liver; P, portal vein; H, hepatic artery; C, inferior vena cava.

toms of an acute suppurative process as part of their disease (103). If a patient has acute necrotizing pancreatitis and persistent fever and leucocytosis, an abscess has to be considered. If there is sonographic evidence of a fluid collection, then percutaneous aspiration should be performed. We performed 27 percutaneous needle aspirations on 19 such patients, and 7 (26%) proved to have an abscess. The sonographic appearance of the wall of the fluid collection varied from being thick and irregular to being entirely smooth and well-defined. The number of internal echoes range from totally echo-free to being entirely filled with echoes (Figs. 13 and 14). Unclotted blood was aspirated in three patients, and these fluid collections were felt to represent focal areas of hemorrhagic pancreatitis (44) (Fig. 11). In the remaining patients, the fluid collections sonographically resembled pseudocysts. Thus, sonography cannot always distinguish an abscess from a pseudocyst or a focal area of hemorrhage.

A gas-containing pancreatic abscess can be sonographically missed because it is hidden behind dilated gas-filled loops of bowel. It may appear as a densely echogenic mass that could be mistaken for gas-filled bowel; however, when doubt exists comparison with plain radiographs of the abdomen and CT will be necessary (25,53,111). The presence of gas in the area of the pancreas in a patient with acute pancreatitis does not necessarily indicate the presence of an abscess. Gas can be present in a pseudocyst that has spon-

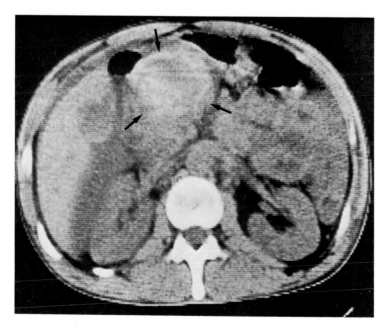

FIG. 11. B: CT scan of the same patient shows evidence of acute hemorrhage (33 EMI units) in the head of the pancreas (*arrows*).

taneously drained into the gastrointestinal tract (69). It can also be found in patients who have necrosis of their pancreatic phlegmon in the absence of suppuration (100). Percutaneous needle aspiration may be necessary to confirm the presence of an abscess in patients with these findings who clinically have acute necrotizing pancreatitis.

Clinical Considerations

Although acute pancreatitis comprises two overlapping clinical categories (acute edematous and acute necrotizing pancreatitis), it is clinically useful to compare the clinical diagnosis with the sonographic findings. Patients with acute edematous pancreatitis may have a normal-appearing pancreas (28%); however, there is usually evidence of focal or diffuse swelling of the gland (61%) or phlegmonous changes (11%) (40). A patient with the diagnosis of acute necrotizing pancreatitis should never have a normal-appearing pancreas and most (89%) will have evidence of phlegmonous pancreatitis (Fig. 10). One should not be alarmed to discover, in a patient who is making an uneventful recovery, continued sonographic findings of phlegmonous pancreatitis, as they may persist for weeks to months long after the patient has made a full clinical recovery (40).

In patients with acute pancreatitis, diagnostic visualization of the pancreas by sonography is 62% compared to 98% for CT (94). Even when the pancreas is seen, the tail is often obscured, and on a follow-up CT pathology may be demonstrated in this area. Using the fluid-filled stomach as an acoustic window in these patients may be contraindicated depending on the severity of their disease. Sonography is nevertheless useful in patients with acute pancreatitis, as it can evaluate the biliary tract to determine if there is any associated biliary tract stones or dilatation while at the same time adequate visualization of the pancreas may be obtained. CT is performed only in those cases where sonography has failed to visualize the pancreas adequately and a complication of acute pancreatitis (phlegmon, pseudocyst, hemorrhage, abscess) is suspected (94).

PANCREATIC PSEUDOCYSTS

Pseudocysts are fluid collections that arise within the pancreas or its adjacent tissues in patients with acute pancreatitis. They are said to occur because of obstruction followed by rupture of a pancreatic duct, which allows the pancreatic juices to escape into the interstitium of the gland. Because of the absence of a pancreatic capsule, the fluid may rupture into and accumulate in the lesser sac; however, it can extend along the retroperitoneal soft tissue planes in any direction (82,93,103). The pseudocyst may contain not only pancreatic juice but also necrotic debris and blood. The formation of an extrapancreatic pseudocyst may be a helpful phenomenon in that it decompresses the pancreas and prevents further damage to the gland (93). Pseudocysts take at least 4 to 6 weeks to become mature. When mature, they have a fibrous tissue lining that differentiates them from true cysts of the pancreas, which are lined by epithelium (82).

Of all patients with acute pancreatitis, 10% will develop pseudocysts (94). They are usually single, oval to round in shape, and vary greatly in size (Figs. 12 and 15). They may arise from any portion of the pancreas and can cause dilatation of the pancreatic duct. The sonographic success rate in detecting them varies from 50 to 92% (33,54,94). Pseudocysts of the head and body are usually seen; however, it is in the area of the tail that trouble occurs, as they may be hidden by gas in the stomach and may even dissect lateral to the left kidney. They can be mistaken for a fluid-filled stomach; however, this dilemma is easily overcome by having the patient drink some water or by injecting some water down the nasogastric tube and by observing the "cystic lesion" with real-time sonography (35).

The wall of the cyst is often smooth; however, it can be somewhat irregular (42) (Fig. 12). The fluid contained within the cyst can have a varying echo pattern but it is usually echo-free. When the pseudocyst is filled with internal echoes or has a fluid debris level superimposed hemorrhage or abscess should be considered (42,57) (Figs. 13 and 14). The amount of debris may be such that through transmission can be reduced so that the pseudocyst has the appearance of a solid mass. Calcification within the wall of the pseudocyst can also be a cause of poor through transmission. A pseudocyst can mimic a cystadenoma or cystadenocarcinoma because of the presence of septations; however, this is uncommon (57) (Fig. 21).

Pseudocysts can dissect anywhere within the abdomen and can mimic cystic lesions of the liver, spleen, or kidney (5,15,20,57,93,101) (Fig. 15). They can extend up into the mediastinum, into the small bowel mesentery, or down the retroperitoneum into the pelvis and may even present as a groin mass (34,93). They can appear within as short a period of time as 6 days and can spontaneously regress by decompressing into either the pancreatic duct or adjoining portion of gastrointestinal tract, usually the stomach (8,86). This occurs in 13 to 20% of cases (8,33). It was believed prior to the development of sonography and CT that 50% of pseudocysts drained spontaneously into the peritoneal space, leading to a high mortality rate from peritonitis and shock. Although this can occur, its incidence is now believed to be exceedingly low.

In the past, persistence of a pseudocyst, especially in symptomatic patients, was an indication for surgery. In recent years, however, there have been reports of successful percutaneous aspiration of pancreatic pseudocysts with the use of sonographic guidance (96). The success rate with no recurrence of the cyst varies from 16 to 80%; however, more than one aspiration

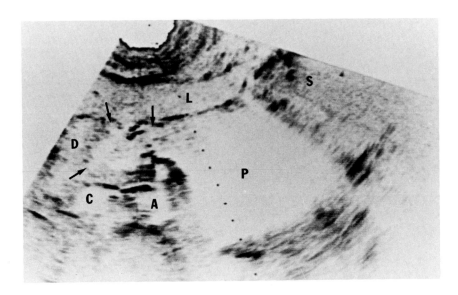

FIG. 12. A: Transverse sonogram of the upper abdomen. A 7.0-cm pseudocyst (P) with well-defined walls and no internal echoes is seen arising from the tail of the pancreas. The remaining pancreas appears hypoechoic (*arrows*). L, Liver; S, stomach; D, duodenum; A, aorta; C, inferior vena cava.

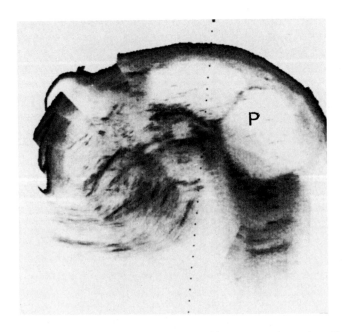

FIG. 12. B: Transverse sonogram of the upper abdomen. A large septated pseudocyst is present in the lesser sac area. It contains a few internal echoes.

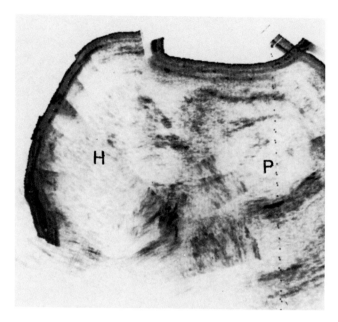

FIG. 13. Transverse sonogram of the upper abdomen. A 7-cm pseudocyst containing internal echoes (P) is visible in the lesser sac area. H, Liver.

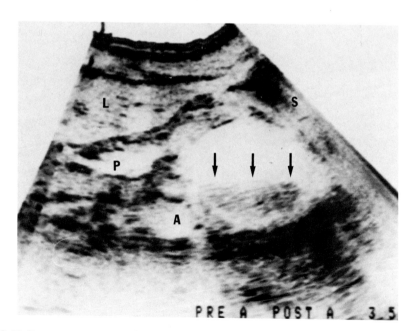

FIG. 14. Transverse sonogram of the upper abdomen. A 6-cm pseudocyst with a fluid debris level (*arrows*) is present. Unclotted blood was percutaneously aspirated. L, Liver; S, stomach; P, portal vein; A, aorta.

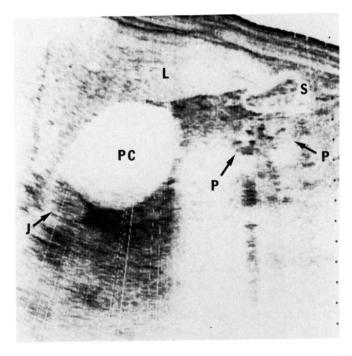

FIG. 15. Longitudinal sonogram just to the left of the midline. A 5-cm pseudocyst (PC) with well-defined walls and minimal echoes in its dependent portion is present along the under surface of the liver (L) anterior to the esophagogastric junction (J). Patient has chronic calcific pancreatitis (P). S, Stomach.

may have to be performed (3,63). Pseudocysts of the lesser sac tend to reaccumulate, and aspiration is usually more successful in pseudocysts that are not contiguous to the pancreas. The amylase level of a pseudocyst successfully treated with percutaneous aspiration is usually lower than in those where the method is unsuccessful (3). Although successful aspiration is more likely in the case of mature pseudocysts, it can be used to relieve symptoms in patients with an acute pseudocyst or where the pseudocyst is causing obstruction of the common bile duct or duodenum. In such cases, it may be necessary to perform a number of percutaneous aspirations over time.

If surgery is contemplated for a pseudocyst then sonography should be performed immediately prior to the operative procedure to confirm that the pseudocyst is still present.

CHRONIC PANCREATITIS

Chronic pancreatitis is caused by progressive destruction of the pancreas by repeated flare-ups of acute pancreatitis (82). The most common cause is alcohol abuse.

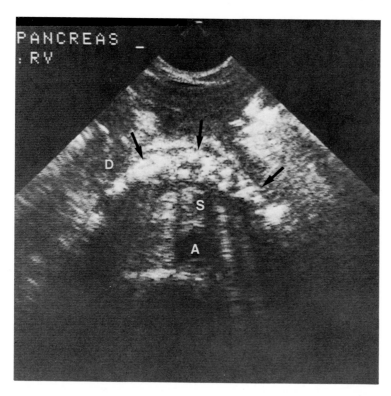

FIG. 16. Transverse sonogram of the pancreas. Diffuse pancreatic calcifications (*arrows*) with acoustical shadowing due to chronic pancreatitis. A, Aorta; S, superior mesenteric artery; D, duodenum.

In chronic pancreatitis there are increased parenchymal echoes resulting from fibrosis and focal calcifications in the gland. These calcifications are due to calculi in the ductal system and may have a focal or diffuse distribution. Sonographically these can be identified as calcifications when acoustical shadowing is present (Fig. 16). The coarse echo texture of the abnormal gland should be distinguished from the homogenous echogenic pattern of the normal pancreatic parenchyma and the surrounding pancreatic fat. Lack of attention to the parenchymal echo pattern may account for the high sonographic false-negative rate (52 to 60%) in detecting chronic pancreatitis as reported by some (28,58). This error is most likely to occur when the gland is not enlarged or atrophied and replaced by fat (75). In one study, Lees was able to detect 91% of patients with chronic pancreatitis as a result of the coarse echo texture of the gland (61).

There may be either focal or diffuse enlargement (27%) of the gland associated with irregularity of outline (45%) (61). The pancreatic duct may be dilated (41%), and this is most likely to be present when there are pancreatic calcifications (92%) (61,106) (Fig. 17). The duct may be dilated secondary to a stricture or as a result of the extrusion of stones from the smaller pancreatic ducts into the main duct. The demonstration that stones are in the main duct is accomplished more easily with sonography than with CT (43). Ductal dilatation is best identified in the area of the body; however, in cases where the ductal dilatation is marked it can be traced into the head and tail. In some instances ductal dilatation can be so marked that it mimics pseudocysts (55). Pseudocysts occur in approximately 20% of patients with chronic pancreatitis (Fig. 15). Thrombosis of the splenic and or portal vein can also be seen (Fig. 18).

When a focal mass is found in chronic pancreatitis, the diagnosis of pancreatic carcinoma has to be considered, since it occurs in 4 to 25% of such patients (76,83). The symptoms of chronic pancreatitis and carcinoma may unfortunately be similar (upper abdominal pain radiating to the back associated with weight loss), which further complicates the issue. Sonography may not be able to distinguish a benign from a malignant mass unless secondary signs of carcinoma such as adjacent lymphadenopathy or liver metastasis are present. The reported accuracy rate in the absence of metastasis varies and is as high as 100% (26,28,61). The presence of calcifications within a mass makes it unlikely to be a tumor (106).

CT has the same difficulty in diagnosing pancreatic carcinoma in the presence of chronic pancreatitis and in such patients endoscopic retrograde cholangiopancreatography (ERCP) may be necessary (28,59,80). If this does not resolve the issue, percutaneous needle aspiration using ultrasonic guid-

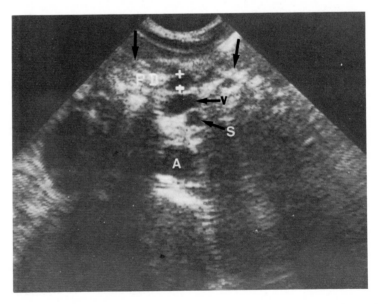

FIG. 17. Transverse sonogram of the upper abdomen reveals a calcified pancreas (*arrows*) with a dilated pancreatic duct (+). A, Aorta; S, superior mesenteric artery; V, splenic vein.

ance may be performed. It should be understood that although biopsies may be performed on all portions of the mass, an underlying carcinoma could be missed.

CYSTIC FIBROSIS

Cystic fibrosis is a genetic disorder involving the exocrine glands, which in this case produce abnormally viscid mucus. This mucus produces obstruction of the draining ducts, leading to cystic dilatation and fibrous atrophy of the involved gland (82). The liver, biliary tract, and especially the pancreas (80%) can all be involved.

As a result of fibrosis and fatty infiltration, the normal echo pattern of the pancreas is increased (60%) (77,108) (Fig. 19). This increased echo pattern is best detected by amplitude analysis as described by Shawker (91). The liver cannot be used as a satisfactory comparison standard because of its involvement in this disease. As in the case of chronic pancreatitis due to other causes,

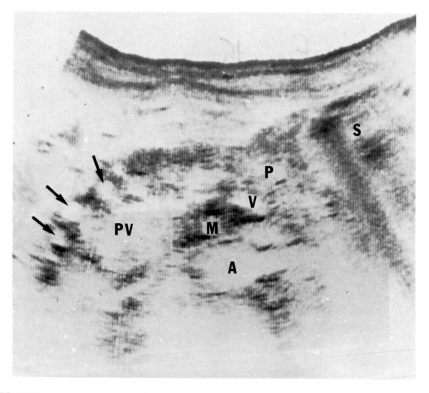

FIG. 18. Transverse sonogram of the upper abdomen reveals evidence of echogenic material in the portal vein (PV) with periportal collaterals (*arrows*). Patient has portal vein thrombosis resulting from chronic pancreatitis (P). V, Splenic vein; M, superior mesenteric artery; A, aorta; S, stomach.

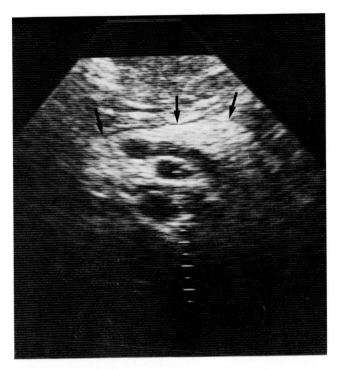

FIG. 19. Transverse sonogram through the pancreas (*arrows*), which is diffusely echogenic in this 24-year-old female with cystic fibrosis. (Courtesy of Dr. Thomas Shawker, National Institutes of Health, Bethesda, Maryland.)

when acute pancreatitis develops, the diffuse hypoechoic echo pattern normally associated with this disease is less likely to be seen (77). Focal hypoechoic areas of enlargement have been described together with pseudocysts and focal calcifications (77,108).

Hepatomegaly with increased parenchymal echoes should be looked for. The gallbladder may not be seen or may be very small and may contain sludge and/or stones. Portal hypertension and splenomegaly are uncommon (108).

PANCREATIC NEOPLASMS

Pancreatic Cysts

Pancreatic cysts are classified as follows (38):

True cysts: These are lined by mucous epithelium and may be
  (a) Congenital. These are believed to result from anomalous development of the pancreatic ducts. Although they may occur singly, they are usually multiple and without septa. They range in size from microscopic to 5 cm in diameter (82).

(b) Acquired. These are comprised of retention cysts (cystic dilatation of a pancreatic duct due to any cause), parasitic cysts, and neoplastic cysts, e.g. cystadenoma, cystadenocarcinoma.

Pseudocysts: They are not lined by epithelium. When mature they have a fibrous wall. They may be postinflammatory or posttraumatic.

Sonography cannot distinguish the different types of single cysts that arise from the pancreas unless other findings are present, e.g., with a pseudocyst there may be focal or diffuse enlargement of the pancreas compatible with acute pancreatitis while multiple septations within the cyst would suggest the presence of a cystadenoma or cystadenocarcinoma (Fig. 21). When multiple cysts are present, both kidneys should be evaluated to look for evidence of adult polycystic kidney disease or Von Hippel Lindau's disease (92).

Pancreatic Carcinoma

The incidence of pancreatic carcinoma has increased in recent years. It arises from the ductal epithelium and presents clinically with weight loss (70%) and epigastric pain (50%) often radiating through into the back (25%) (82). It comprises 95% of all malignant pancreatic neoplasms. As a result of its pattern of silent growth, the cure rate is abysmally low, with tumor spread frequently being present at the time of diagnosis. The 5-year survival rate is only 1 to 2% and the average life expectancy following diagnosis is only 8 months. The tumor may arise from the head (61%), body (13%), or tail (5%), or it may be diffuse (21%) (4). A tumor of the head is more likely to be discovered earlier than one arising from the body or tail, as it may impinge upon the common bile duct and produce jaundice.

The sonographic detection rate of pancreatic carcinoma depends upon the adequacy of visualization of the gland and varies from 81 to 94% (16,48,59,64,78,99). In these studies, the pancreatic nonvisualization rate ranged from 10 to 38% (64,78,99). These technical failures should not be counted as false-negatives, since the failure to detect a pancreatic lesion is obvious and indicates that other diagnostic examinations such as CT and ERCP are necessary. The true false-negative rate (lesions missed in areas of the pancreas that can be seen) is usually attributable to small lesions less than 3 cm in diameter that are often located in the tail (48). When there is adequate visualization of the gland, the likelihood of missing a pancreatic neoplasm is extremely low; Pollock reports a negative predictive value of 99% (78).

Cancer of the pancreas usually has an irregular contour; the mass generally contains low-level echoes with a varying amount of superimposed coarse echoes (97%); on occasion, a mildly echogenic tumor can occur (3%) (107,112) (Fig. 20). Whatever the echo pattern of the tumor, it is usually quite distinct from the homogenous echo pattern of the normal pancreatic parenchyma. Following treatment, the sonographic appearance of the tumor may change,

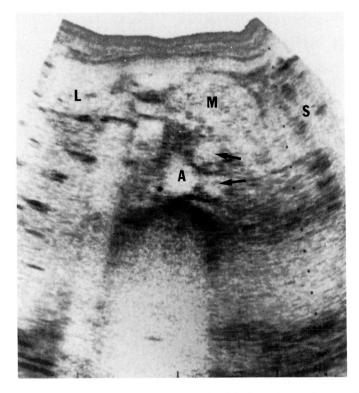

FIG. 20. Transverse sonogram of the upper abdomen. A solid echogenic mass is seen arising from the body and tail of the pancreas (M). Paraaortic adenopathy is present (*arrows*). Diagnosis: Pancreatic carcinoma. S, Stomach; L, liver; A, aorta.

with the mass becoming more sonolucent with a varying degree of central echogenicity (112).

The sizes of pancreatic tumors vary greatly. Those in the head tend to be smaller than those arising in the tail. This is presumably due to compression of the common bile duct with jaundice, which causes tumors in the head to present earlier (82). At surgery, the carcinoma may prove to be smaller than is indicated by sonography. This is due to an adjacent zone of pancreatitis that surrounds the tumor and adds to its sonographic dimensions (112). The shape of the tumor can vary; small tumors tend to be round to oval whereas larger tumors are lobulated (107).

Constant compression of the inferior venae cava will occur (50%) when the tumor arises in the head. Displacement of the superior mesenteric vessels depends upon the site of the tumor. Anterior displacement occurs when the carcinoma arises in the uncinate process and posterior displacement to the left and less often to the right occurs with tumors arising in the head and/or body (112). Soft tissue thickening caused by neoplastic infiltration of perivascular

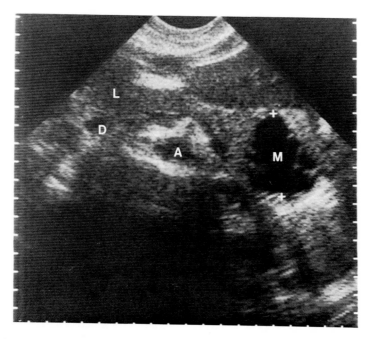

FIG. 21. A: Transverse sonogram of the upper abdomen demonstrates a fairly well-defined 4.5-cm cystic mass (M) with septations arising from the tail of the pancreas. Diagnosis: Pancreatic cystadenoma. L, Liver; D, duodenum; A, aorta.

lymphatics may be seen surrounding the celiac axis or the superior mesenteric artery (68). This occurs more commonly with carcinoma of the body and tail (52%) than with carcinoma of the head (25%). In some patients (11%) it may be the only sign of pancreatic carcinoma, as the primary tumor itself may not be visible. Tumor displacement or invasion of the splenic or portal vein producing thrombosis may occur (12 to 48%) (107,112). Collateral venous channels may develop and be visualized in the periportal area and along the wall of the stomach (19) (Fig. 18).

Obstruction of the extrahepatic common duct can not only be produced by the direct effect of the mass in the head but may also occur as a result of adenopathy in the porta hepatis and along the portal vein (107). If a mass is strategically located near the junction of the common bile duct and the pancreatic duct there may be dilatation of both. Tumors of the head and body can produce dilatation of the pancreatic duct alone. In some instances, ductal dilatation can be seen although the offending mass is invisible because of its small size (36). This can occur not only with carcinoma of the pancreas but also with carcinoma of the ampulla of Vater, cholangiocarcinoma arising near the ampulla, Crohn's disease of the duodenum, and stenosis of the ampulla of Vater (36,50,105). The degree of ductal dilatation does not correlate with the

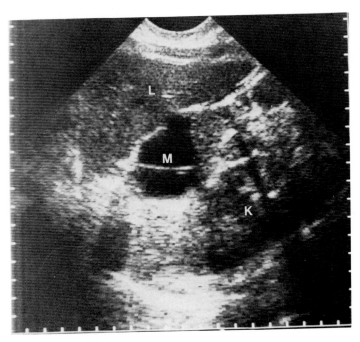

FIG. 21. B: Longitudinal sonogram to the left of the midline in the same patient. The septated cystic mass (M) is seen anterior to the upper left kidney (K). L, Liver.

causative pathology (105). If the cause of pancreatic ductal dilatation cannot be ascertained and if ERCP fails, then percutaneous pancreatic ductography using sonographic guidance should be considered (73). Adenopathy may be found in the paraaortic, pericaval, or portal areas and in the mesentery (Fig. 20). It is usually not as marked as that found with lymphoma. This can be an important differentiating point; if a very hypoechoic pancreatic mass with extensive adenopathy is seen, lymphoma should be considered. Other findings in pancreatic carcinoma include liver metastasis (18%) and ascites.

A focal mass lesion can be found with either acute or chronic pancreatitis. In acute pancreatitis, the mass is usually uniformly hypoechoic with good through transmission unless there is focal hemorrhage or bubbles of gas caused by necrosis and/or suppuration (Figs. 9 and 11). It was previously mentioned that a focal mass confined to the pancreas directly anterior to the aorta is not likely to be inflammatory in nature (94). Other points that help differentiate an inflammatory from a neoplastic mass include the presence of separable adenopathy, which is not found with inflammatory disease, and silhouetting of the wall of the aorta by a contiguous pancreatic mass, which indicates the presence of a neoplasm (41,94).

Differentiating a mass due to acute pancreatitis from carcinoma is usually not a clinical problem, as the clinical history is so different. Although carcinoma can present with acute pancreatitis as a result of obstruction of the pancreatic duct, this is an uncommon occurrence with an incidence of 1% or less (94).

Differentiating a mass due to chronic pancreatitis from one due to carcinoma is difficult, as the clinical history of the two may be similar and the two diseases may coexist in the same patient (26,76,83). More invasive procedures including ERCP, percutaneous aspiration biopsies, and even surgery may be necessary to distinguish these two entities.

Although unlikely, on occasion a pancreatic carcinoma can have the appearance of a pseudocyst with an echo-free center with good through transmission and a well-defined wall (49). CT can also make this mistake. Another error that may occur is where a mass arising from the pancreas insinuates itself posterior to the splenic vein and simulates a left adrenal mass (12).

Nonpancreatic Tumors Simulating Pancreatic Carcinoma

The pancreatic parenchyma is rarely involved by metastasis from other primary malignancies (82). When it is secondarily involved it is most often caused by direct extension of tumor from the stomach, duodenum, colon, or biliary tract (95). The mass thus produced may sonographically be indistinguishable from a primary pancreatic carcinoma (107). The peripancreatic lymph nodes can be involved by local or distant primary carcinomas and by lymphoma.

CYSTIC NEOPLASMS OF THE PANCREAS

These are divided into two groups (110):

(a) *Microcystic adenoma and adenocarcinoma (cystadenoma, cystadenocarcinoma)*: These uncommon slow-growing tumors of the pancreas arise from the ductal epithelium. They are round, lobulated cystic masses that are usually multiloculated and are lined by a single layer of cuboidal or columnar epithelium that is thrown up into papillary projections (82). They contain serous or mucinous material, and calcification (10 to 18%) may be present in their walls. The size of the tumor varies greatly from a few centimeters to 20 cm. The body or the tail is the most frequent site of origin; however, they may arise from the pancreatic head. They usually occur in middle-aged females who present with a palpable abdominal mass and pain presumably resulting from the pressure effects of the mass. They have been associated with diabetes mellitus and an increased incidence (40%) of other benign and malignant tumors (97). The sonographic appearance of this neoplasm is fairly characteristic. A well-defined multiloculated cystic mass with good through transmission is identified arising from the pancreas (13,29,

62,110) (Fig. 21). The internal septations are usually thin, and as the gain is increased the cystic areas tend to fill in with echoes (13,51). Septations when present may not be seen on CT (110). Highly echogenic areas owing to a variable amount of soft tissue component can be identified along the cyst walls (13) (Fig. 22). Focal areas of echogenicity with acoustical shadowing due to calcifications may be present (29). It should be remembered that sonography cannot determine whether the lesion is benign or malignant unless, of course, metastases, such as in the liver, are present (29). One should have no difficulty differentiating focal pancreatitis or pancreatic carcinoma from these cystic neoplasms. The main differential for this neoplasm is a pancreatic pseudocyst that uncommonly can have septa. If the clinical history is not helpful or associated signs of acute or chronic pancreatitis are not present then angiography may help. Most cystadenomas or cystadenocarcinomas have enlarged feeding arteries with neovascularity and a tumor stain. The degree of vascularity is related to the relative proportion of cystic to solid tumor (29). Hypovascular areas are seen as a result of the cystic component of the neoplasm. Some tumors can be entirely hypovascular and as such may be indistinguishable from a pseudocyst.

(b) *Microcystic adenoma (glycogen-rich cystadenoma)*: This is a benign

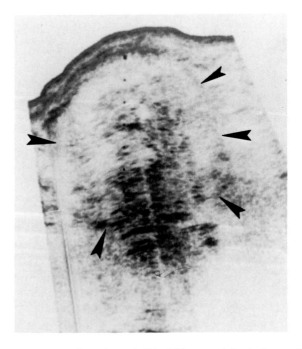

FIG. 22. Transverse sonogram through a palpable right upper abdominal mass. This 7-cm fairly well-defined solid mass (*arrows*) has a more echogenic center with good through transmission compatible with a necrotic center. Diagnosis: Pancreatic cystadenocarcinoma.

tumor composed of many small cysts (less than 2 cm) that contains little or no mucin (110). It often contains a central stellate scar that may be calcified from which radiating bands of connective tissue arise. The tumor is usually well-circumscribed and oval in configuration. Like the macrocystic variety, it usually involves the body and tail of the pancreas (60%) but may arise from the head (30%) or be diffuse (10%). Sonographically the tumor may be identified as being composed of multiple small cysts. When the cysts are very small, however, the tumor appears as a solid echogenic mass and as such may be difficult to distinguish from a pancreatic carcinoma (29,110).

ISLET CELL TUMORS

These rare tumors are divided into functioning (85%) and nonfunctioning (15%) islet cell tumors. They may be either benign or malignant, with the majority of nonfunctioning tumors being malignant (92%) (51). The presence of malignancy is difficult to determine at the cellular level and can be diagnosed only when there is invasion of surrounding structures or liver metasta-

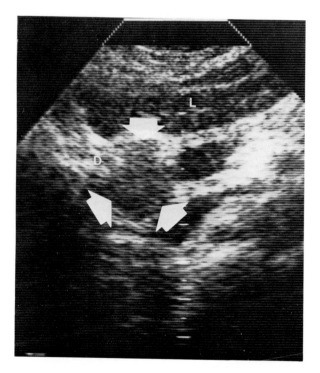

FIG. 23. Transverse sonogram through the pancreas. A 1.5 cm by 2 cm well-defined hypoechoic mass (*arrows*) is present in the head of the pancreas. Patient had a history of hypoglycemia. Diagnosis: Insulinoma. L, Liver; D, duodenum. (Courtesy of Dr. Thomas Shawker, National Institutes of Health, Bethesda, Maryland.)

sis. Unlike pancreatic carcinoma, malignant islet cell tumors are slow-growing and have a 5-year survival rate of 44% (51).

The most common functioning islet cell tumor is the insulinoma (60%) followed by the gastrinoma (18%); the remaining 7% are composed of tumors that produce vasoactive intestinal peptide (WDHA syndrome), somatostatin, etc. (51). Functioning tumors are more difficult to detect than nonfunctioning tumors, as they present while still very small as a consequence of their hormonal activity (Fig. 23). The rate of malignancy is higher for gastrinomas (25 to 60%) than for insulinomas (10%) (89). Each of the functioning tumors, with the exception of somatostatinoma, has been described in association with the multiple endocrine neoplasia Type I syndrome. Nonfunctioning tumors are easier to detect, as they often reach a large size before causing symptoms (31) (Fig. 24). These tumors are usually solitary and present with pain (36%), jaundice (28%), or a palpable mass (8%) (51).

Shawker et al. have reported on the sonographic findings in 45 patients with unresected islet cell tumors (90). In general they tend to be spherical, well marginated, homogenous solid masses frequently with low-amplitude echoes; however, some of the larger tumors have high-amplitude echoes (Fig. 23). Tumor calcifications and cystic areas can occasionally be present (31,79,90). The overall detection rate was 51% and was lowest for insulinomas (30%)

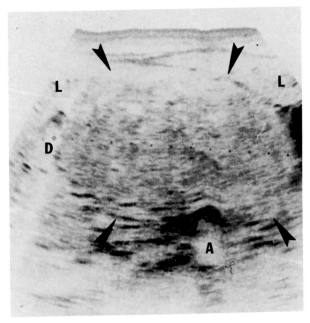

FIG. 24. Transverse sonogram of the upper abdomen. An 8 cm by 12 cm echogenic solid mass (*arrows*) is present anterior to the aorta (A). Diagnosis: Malignant nonfunctioning islet cell tumor. L, Liver; D, duodenum.

because of their small size and patient obesity produced by overeating in response to hypoglycemia (90). These difficulties are compounded by the fact that 10% of insulinomas may be multiple (89).

Thirty-one percent of patients had liver metastasis (90). Metastases from functioning tumors were echogenic whereas those from nonfunctioning tumors ran the gamut from high-level to low-level echoes, with target lesions being present in some patients (Fig. 25). Metastatic calcifications with acoustical shadowing and central cystic areas of necrosis were occasionally encountered. One patient had portal vein invasion.

Computed tomography locates only 32 to 43% of functioning islet cell tumors. Dunnick ascribes this low rate of detection to the ectopic location of some adenomas and to cellular hyperplasia rather than a discrete tumor in some patients (24). The highest detection rate has been reported with angiography; however, it appears not to be as successful with gastrinomas (Zollinger-Ellison syndrome) as with other endocrine tumors of the pancreas (23). Venous sampling has also been reported as being helpful; however, it may be misleading in patients with multiple adenomas (22).

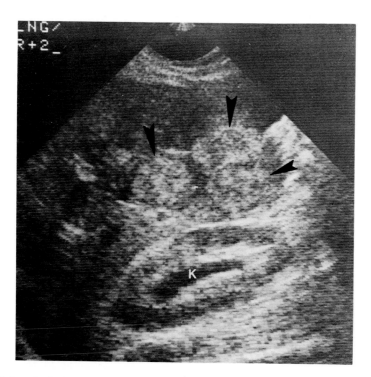

FIG. 25. Longitudinal sonogram through the right lobe of the liver. Focal echogenic masses (arrows) due to metastatic insulinoma are present. K, Kidney.

REFERENCES

1. Ammann, R. *Acute Pancreatitis in Gastroenterology,* edited by H. L. Bockus, pp. 1020–1039, 1976. W. B. Saunders, Philadelphia.
2. Arger, P. H., Mulhern, C. B., Bonavita, J. A., Stauffer, D. M., and Hale, J. An analysis of pancreatic sonography in suspected pancreatic disease. *J. Clin. Ultrasound,* 7:91–97, 1979.
3. Barkin, J. S., Smith, F. R., Pereiras, R., Isikoff, M., Levi, J., Livingstone, A., Hill, M. C., and Rogers, A. I. Therapeutic percutaneous aspiration of pancreatic pseudocysts. *Dig. Dis. Sci.,* 26:585–586, 1981.
4. Beazley, R. M., and Cohn, I. Pancreatic carcinoma. *Cancer J. Clinicians,* 31:346–358, 1981.
5. Bellon, E. M., George, C. R., Schreiber, H., and Marshall, J. B. Pancreatic pseudocysts of the duodenum. *AJR,* 133:827–831, 1979.
6. Berger, M., Smith, E. H., Bartrum, R. J., Holm, H. H., and Mascatello, V. False-positive diagnoses of pancreatic tail lesions caused by colon. *J. Clin. Ultrasound,* 5:343–345, 1977.
7. Bowie, J. D., and MacMahon, H. Improved techniques in pancreatic sonography. *Semin. Ultrasound,* 1:170–177, 1980.
8. Bradley, E. L., and Clements, L. J. Spontaneous resolution of pancreatic pseudocysts: Implications for timing of operative intervention. *Am. J. Surg.,* 129:23–28, 1975.
9. Bryan, P. J. Appearance of normal pancreatic duct: A study using real-time ultrasound. *J. Clin. Ultrasound,* 10:63–66, 1982.
10. Burrell, M. I., Avella, J., Spiro, H. M., and Taylor, K. J. W. Diagnostic imaging procedures in acute pancreatitis. *JAMA,* 242:342–343, 1979.
11. Burrell, M., Gold, J. A., Simeone, J., Taylor, K., and Dobbins, J. Liquefactive necrosis of the pancreas. *Radiology,* 135:157–160, 1980.
12. Callen, P. W., Breiman, R. S., Korobkin, M., Martini, W. J., and Mani, J. R. Carcinoma of the tail of the pancreas: An unusual CT appearance. *AJR,* 133:135–137, 1979.
13. Carroll, B., Sample, W. F. Pancreatic cystadenocarcinoma: CT body scan and gray scale ultrasound appearance. *Am. J. Roentgenol.,* 131:339–341, 1978.
14. Churchill, R. J., Reynes, C.J., and Love, L. Pancreatic pseudotumors: Computed tomography. *Gastrointest. Radiol.,* 3:251–256, 1978.
15. Conrad, M. R., Landay, M. J., and Khoury, M. Pancreatic pseudocysts: Unusual ultrasound features. *AJR,* 130:265–268, 1978.
16. Cotton, P. B., Lees, W. R., Vallon, A. G., Cottone, M., Croker, J. R., and Chapman, M. Gray-scale ultrasonography and endoscopic pancreatography in pancreatic diagnosis. *Radiology,* 134:453–459, 1980.
17. Cox, K. L., Ament, M. E., Sample, W. F., Sarti, D. A., O'Donnell, M., and Byrne, W. J. The ultrasonic and biochemical diagnosis of pancreatitis in children. *J. Pediatr.,* 96:407–411, 1980.
18. Crade, M., Taylor, K. J. W., and Rosenfield, A. T. Water distention of the gut in the evaluation of the pancreas by ultrasound. *Am. J. Roentgenol.,* 131:348–349, 1978.
19. Dach, J. L., Hill, M. C., Pelaez, J. C., LePage, J. R., and Russell, E. Sonography of hypertensive portal venous system: Correlation with arterial portography. *AJR,* 137:511–517, 1981.
20. De Graaff, C. S., Taylor, K. J. W., Rosenfield, A. T., and Kinder, B. Gray scale ultrasonography in the diagnosis of pseudocysts of the pancreas simulating renal pathology. *J. Urol.,* 120:751–753, 1978.
21. De Graaff, C. S., Taylor, K. J. W., Simonds, B. D., and Rosenfield, A. J. Gray-scale echography of the pancreas. *Radiology,* 129:157–161, 1978.
22. Doppman, J. L., Brennan, M. F., Dunnick, N. R., Kahn, C. R., and Gorden, P. The role of pancreatic venous sampling in the localization of occult insulinomas. *Radiology,* 138:557–562, 1981.
23. Dunnick, N. R., Doppman, J. L., Mills, S. R., and McCarthy, D. M. Computed tomographic detection of nonbeta pancreatic islet cell tumors. *Radiology,* 135:117–120, 1980.
24. Dunnick, N. R., Long, J. A., Krudy, A. Shawker, T. H., and Doppman, J. L. Localizing insulinomas with combined radiographic methods. *AJR,* 135:747–752, 1980.
25. Federle, M. P., Jeffrey, R. B., Crass, R. A., and Dalsem, V. V. Computed tomography of pancreatic abscesses. *AJR,* 136:879–882, 1981.

26. Ferrucci, J. T., Wittenberg, J., Black, E., Kirkpatrick, R. H., and Hall, D. A. Computed body tomography in chronic pancreatitis. *Radiology*, 130:175–182, 1979.
27. Filly, R. A., and London, S. S. The normal pancreas: Acoustic characteristics and frequency of imaging. *J. Clin. Ultrasound*, 7:121–124, 1979.
28. Foley, W. D., Stewart, E. T., Lawson, T. L., Geenan, J., Loguidice, J., Mather, L., and Unger, G. F. Computed tomography, ultrasonography, and endoscopic retrograde cholangiopancreatography in the diagnosis of pancreatic disease: A comparative study. *Gastrointest. Radiol.*, 5:29–35, 1980.
29. Freeny, P. C., Weinstein, C. J., Taft, D. A., Allen, F. H. Cystic neoplasms of the pancreas: New angiographic and ultrasonographic findings. *AJR*, 131:795–802, 1978.
30. Ganong, W. F. Exocrine portion of the pancreas; Endocrine function of the pancreas. In: *Review of Medical Physiology*, 5th ed., pp. 245–260; 366–368, 1971. Lange Medical Publication, Los Altos, California.
31. Gold, R., Rosenfield, A. T., Sostman, D., Burrell, M., and Taylor, K. J. W. Nonfunctioning islet cell tumors of the pancreas: Radiographic and ultrasonographic appearances in two cases. *Am. J. Roentgenol.*, 131:715–717, 1978.
32. Goldstein, H. M., and Katragadda, C. S. Prone view ultrasonography for pancreatic tail neoplasms. *AJR*, 131:231–234, 1978.
33. Gonzalez, A. C., Bradley, E. L., and Clements, J. L. Pseudocyst formation in acute pancreatitis: Ultrasonographic evaluation of 99 cases. *Am. J. Roentgenol.*, 127:315–317, 1976.
34. Gooding, G. A. W. Pseudocyst of the pancreas with mediastinal extension: an ultrasonographic demonstration. *J. Clin. Ultrasound*, 5:121–123, 1977.
35. Gooding, G. A. W., and Laing, F. C. Rapid water infusion: A technique in the ultrasonic discrimination of the gas-free stomach from a mass in the pancreatic tail. *Gastrointest. Radiol.*, 4:139–141, 1979.
36. Gosink, B. B., and Leopold, G. R. The dilated pancreatic duct: Ultrasonic evaluation. *Radiology*, 126:475–478, 1978.
37. Haber, K., Freimanis, A. K., and Asher, W. M. Demonstration and dimensional analysis of the normal pancreas with gray-scale echography. *AJR*, 126:624–628, 1976.
38. Haubrich, W. S., and Berk, J. E. Cysts of the pancreas. In: *Textbook of Gastroenterology*, edited by H. L. Bockus, pp. 1154–1166, 1976. W. B. Saunders, Philadelphia.
39. Higashi, Y., Sakazaki, T., Hirata, T., Murakami, K., and Matsuura, K. Pancreatic thickness during ultrasonography—Variations with respiration and scanning planes. *Presented (Paper 304) at The American Institute of Ultrasound in Medicine Meeting, San Francisco, Aug. 1981.*
40. Hill, M. C., Barkin, J., Isikoff, M. B., Silverstein, W., and Kalser, M. H. Acute pancreatitis: Clinical versus CT findings. AJR (*in press*) 1982.
41. Hill, M. C., Isikoff, M. B., and Weiner, C. Sonic silhouetting of the aorta. *Int. Am. J. Radiol.*, 5:109–111, 1980.
42. Hill, M. C., and Sanders, R. C. Gray scale B scan characteristics of intra-abdominal cystic masses. *J. Clin. Ultrasound*, 6:217–222, 1978.
43. Isikoff, M. B., and Hill, M. C. Ultrasonic demonstration of intraductal pancreatic calculi: A report of two cases. *J. Clin. Ultrasound*, 8:449–452, 1980.
44. Isikoff, M. B., Hill, M. C., and Barkin, J. The clinical significance of acute pancreatic hemorrhage. *AJR*, 136:679–684, 1981.
45. Jaffe, C. C., and Harris, D. J. Sonographic tissue texture: Influence of transducer focusing pattern. *AJR*, 135:343–347, 1980.
46. Jeanty, P., Van Gansbeke, D., Kuhn, G., and Struyven, J. The replaced right hepatic artery: Ultrasonic demonstration. *Presented at the Radiological Society of North America Meeting, Chicago, Nov. 1981.*
47. Johnson, M. L., and Mack, L. A. Ultrasonic evaluation of the pancreas. *Gastrointest. Radiol.*, 3:257–266, 1978.
48. Kamin, P. D., Bernardino, M. E., Wallace, S., and Jing, B. S. Comparison of ultrasound and computed tomography in the detection of pancreatic malignancy. *Cancer*, 46:2410–2412, 1980.
49. Kaplan, J. O., Isikoff, M. B., Barkin, J., and Livingstone, A. S. Necrotic carcinoma of the pancreas: The pseudo-pseudocyst. *J. Comput. Assist. Tomogr.*, 4:166–167, 1980.

50. Kaude, J. V., Wood, M. B., Cerda, J. J., and Nelson, E. W. Ultrasonographic demonstration of the pancreatic duct. *Gastrointest. Radiol.*, 4:239–244, 1979.

51. Kent, R. B., van Heerden, J. A., and Weiland, L. H. Nonfunctioning islet cell tumors. *Ann. Surg.*, 193:185–190, 1981.

52. Kreel, L., Sandin, B., and Slavin, G. Pancreatic morphology: A combined radiological and pathological study. *Clin. Radiol.*, 24:154–161, 1973.

53. Kressel, H. Y., and Filly, R. A. Ultrasonographic appearance of gas containing abscesses in the abdomen. *AJR*, 130:71–73, 1978.

54. Kressel, H. Y., Margulis, A. R., Gooding, G. W., Filly, R. A., Moss, A. A., and Korobkin, M. CT scanning and ultrasound in the evaluation of pancreatic pseudocysts: A preliminary comparison. *Radiology*, 126:153–157, 1978.

55. Kuligowska, E., Miller, K., Birkett, D., and Burakoff, R. Cystic dilatation of the pancreatic duct simulating pseudocysts on sonography. *AJR*, 136:409–410, 1981.

56. Kwa, A., and Bowie, J. D. Transducer selection for pancreatic ultrasound based on skin to pancreas distance in the supine and upright position. *Radiology*, 134:541–542, 1980.

57. Laing, F. C., Gooding, G. A. W., Brown, T., and Leopold, G. R. Atypical pseudocysts of the pancreas: An ultrasonographic evaluation. *J. Clin. Ultrasound*, 7:27–33, 1979.

58. Lawson, T. L. Sensitivity of pancreatic ultrasonography in the detection of pancreatic diseases. *Radiology*, 128:733–736, 1978.

59. Lawson, T. L., Irani, S. K., and Stock, M. Detection of pancreatic pathology by ultrasonography and endoscopic retrograde cholangiopancreatography. *Gastrointest. Radiol.*, 3:335–341, 1978.

60. Lee, J. K. T., Stanley, R. J., Melson, G. L., and Sagel, S. S. Pancreatic imaging by ultrasound and computed tomography: A general review. *Radiol. Clin. North Am.*, 16:105–117, 1979.

61. Lees, W. R., Vallon, A. G., Denyer, M. E., Vahl, S. P., and Cotton, P. B. Prospective study of ultrasonography in chronic pancreatic disease. *Br. Med. J.* 1:162–164, 1979.

62. Lloyd, T. V., Antonmattei, S., and Freimanis, A. K. Gray scale sonography of cystadenoma of the pancreas: Report of two cases. *J. Clin. Ultrasound*, 7:149–151, 1979.

63. MacErlean, D. P., Bryan, P. J., and Murphy, J. J. Pancreatic pseudocyst: Management by ultrasonically guided aspiration. *Gastrointest. Radiol.*, 5:255–257, 1980.

64. Mackie, C. R., Bowie, J., Cooper, M. J., Kunzmann, A., and Moossa, A. R. Prospective evaluation of gray scale ultrasonography in the diagnosis of pancreas cancer. *Am. J. Surg.*, 136:575–581, 1978.

65. MacMahon, H., Bowie, J., and Beezhold, C. Erect scanning of pancreas using a gastric window. *AJR*, 132:587–591, 1979.

66. Marks, W. M., and Filly, R. A. Ultrasonic evaluation of normal pancreatic echogenicity and its relationship to fat deposition. *Radiology*, 137:475–479, 1980.

67. McCarten, K. M., and Littlewood Teele, R. Preduodenal portal vein: Venography, ultrasonography, and review of the literature. *Ann. Radiol.*, 21:155–160, 1978.

68. Megibow, A. J., Bosniak, M., Ambos, M. A., and Beranbaum, E. R. Thickening of the celiac axis and/or superior mesenteric artery: A sign of pancreatic carcinoma on computed tomography. *Radiology*, 141:449–453, 1981.

69. Mendez, G., and Isikoff, M. B. Significance of intrapancreatic gas demonstrated by CT: A review of nine cases. *AJR*, 132:59–62, 1979.

70. Mendez, G., Isikoff, M. B., and Hill, M. C. CT of acute pancreatitis: Interim assessment. *AJR*, 135:463–469, 1980.

71. Moss, A. A., Federle, M., Shapiro, H. A., Ohto, M., Goldberg, H., Korobkin, M., and Clemett, A. The combined use of computed tomography and endoscopic retrograde cholangiopancreatography in the assessment of suspected pancreatic neoplasm: A blind clinical evaluation. *Radiology*, 134:159–163, 1980.

72. Newman, C. H., and Hessel, S. J. CT of the pancreatic tail. *AJR*, 135:741–745, 1980.

73. Ohto, M., Saotome, N., Saisho, H., Tsuchiya, Y., Ono, T., Okuda, K., and Karasawa, E. Real-time sonography of the pancreatic duct: Application to percutaneous pancreatic ductography. *AJR*, 134:647–652, 1980.

74. Parulekar, S. G. Ultrasonic evaluation of the pancreatic duct. *J. Clin. Ultrasound*, 8:457–463, 1980.

75. Patel, S., Bellon, E. M., Haaga, J., and Park, C. H. Fat replacement of the exocrine pancreas. *AJR*, 135:843–845, 1980.

76. Paulino-Netto, A., Dreiling, D. A., and Baronofsky, I. D. The relationship between pancreatic calcification and cancer of the pancreas. *Ann. Surg.*, 151:530–537, 1960.
77. Phillips, H. E., Cox, K. L., Reid, M. H., and McGahan, J. P. Pancreatic sonography in cystic fibrosis. *AJR*, 137:69–72, 1981.
78. Pollock, D., and Taylor, K. J. W. Ultrasound scanning in patients with clinical suspicion of pancreatic cancer: A retrospective study. *Cancer*, 47:1662–1665, 1981.
79. Raghavendra, B. N., and Glickstein, M. L. Sonography of islet cell tumor of the pancreas: Report of two cases. *J. Clin. Ultrasound*, 9:331–333, 1981.
80. Ralls, P. W., Halls, J., Renner, I., and Juttner, H. Endoscopic retrograde cholangio-pancreatography (ERCP) in pancreatic disease. *Radiology*, 134:347–352, 1980.
81. Ralls, P. W., Quinn, M. F., Rogers, W., and Halls, J. Sonographic anatomy of the hepatic artery. *AJR*, 136:1059–1063, 1981.
82. Robbins, S. L., and Cotran, R. S. The pancreas. In: *Pathological Basis of Disease*. 2nd ed., pp. 1092–1114, 1979. W. B. Saunders, Philadelphia.
83. Robinson, A., Scott, J., and Rosenfeld, D. D. The occurrence of carcinoma of the pancreas in chronic pancreatitis. *Radiology*, 94:289–290, 1970.
84. Sample, W. F. Techniques for improved delineation of normal anatomy of the upper abdomen and high retroperitoneum with gray-scale ultrasound. *Radiology*, 124:197–202, 1977.
85. Sanders, R. C., and Chang, R. Splenic artery mimicking the pancreatic duct. *J. Clin. Ultrasound (In press)*.
86. Sarti, D. A. Rapid development and spontaneous regression of pancreatic pseudocysts documented by ultrasound. *Radiology*, 125:789–793, 1977.
87. Sarti, D. A. *Ultrasonography of the Pancreas in Diagnostic Ultrasound: Text and Cases*, edited by D. A. Sarti and W. F. Sample, pp. 168–225, 1980. G. K. Hall, Boston.
88. Sarti, D. A., and King, W. The ultrasonic findings in inflammatory pancreatic disease. *Semin. Ultrasound*, 1:178–190, 1980.
89. Service, F. J., Dale, A. J., Elveback, L. R., and Jiang, N. Insulinoma: Clinical and diagnostic features of 60 consecutive cases. *Mayo Clin. Proc.*, 51:417–429, 1976.
90. Shawker, T. H., Doppman, J. L., Dunnick, N. R., and McCarthy, D. M. Ultrasound investigation of pancreatic islet cell tumors. *Submitted for publication*.
91. Shawker, T. H., Parks, S. I., Linzer, M., Jones, B., Lester, L. A., and Hubbard, V. S. Amplitude analysis of pancreatic B-scans: A clinical evaluation of cystic fibrosis. *Ultrasonic Imaging*, 2:55–66, 1980.
92. Shirkhoda, A., and Mittelstaedt, C. A. Demonstration of pancreatic cysts in adult polycystic diseases by computed tomography and ultrasound. *AJR*, 131:1074–1076, 1978.
93. Siegelman, S. S., Copeland, B. E., Saba, G. P., Cameron, J. L., Sanders, R. C., and Zerhouni, E. CT of fluid collections associated with pancreatitis. *AJR*, 134:1121–1132, 1980.
94. Silverstein, W., Isikoff, M. B., Hill, M. C., and Barkin, J. Diagnostic imaging of acute pancreatitis: Prospective study using CT and sonography. *AJR*, 137:497–502, 1981.
95. Simeone, J. F., Dembner, A. G., and Mueller, P. R. Invasion of the pancreas by gastric carcinoma: Ultrasonic appearance. *J. Clin. Ultrasound*, 8:501–503, 1980.
96. Slovis, T. L., VonBerg, V. J., and Mikelic, V. Sonography in the diagnosis and management of pancreatic pseudocysts and effusions in childhood. *Radiology*, 135:153–155, 1980.
97. Soloway, H. B. Constitutional abnormalities associated with pancreatic cystadenoma. *Cancer*, 18:1297–1300, 1965.
98. Sommer, G., and Filly, R. A. Patient preparations to decrease bowel gas: Evaluation by an ultrasonographic measurement. *J. Clin. Ultrasound*, 5:87–88, 1977.
99. Taylor, K. J. W., Buchin, P. J., Viscomi, G. N., and Rosenfield, A. T. Ultrasonographic scanning of the pancreas: Prospective study of clinical results. *Radiology*, 138:211–213, 1981.
100. Torres, W. E., Clements, J. L., Sones, P. J., and Knopf, D. R. Gas in the pancreatic bed without abscess. *AJR*, 137:1131–1133, 1981.
101. Vick, C. W., Simeone, J. F., Ferrucci, J. T., Wittenberg, J., and Mueller, P. R. Pancreatitis-associated fluid collections involving the spleen: Sonographic and computed tomographic appearance. *Gastrointest. Radiol.*, 6:247–250, 1981.
102. Walters, M. N. Adipose atrophy of the exocrine pancreas. *J. Pathol. Bacteriol.*, 92:547–557, 1966.

103. Warshaw, A. L. Inflammatory masses following acute pancreatitis: Phlegmon, pseudocyst and abscess. *Surg. Clin. North Am.*, 54:621–635, 1974.
104. Weill, F., Schraub, A., Eisenscher, A., and Bourgoin, A. Ultrasonography of the normal pancreas. *Radiology*, 123:417–423, 1977.
105. Weinstein, D. P., and Weinstein, B. J. Ultrasonic demonstration of the pancreatic duct: An analysis of 41 cases. *Radiology*, 130:729–734, 1979.
106. Weinstein, B. J., Weinstein, D. P., and Brodmerkel, G. J. Ultrasonography of pancreatic lithiasis. *Radiology*, 134:185–189, 1980.
107. Weinstein, D. P., Wolfman, N. T., and Weinstein, B. J. Ultrasonic characteristics of pancreatic tumors. *Gastrointest. Radiol.*, 4:245–251, 1979.
108. Willi, U. V., Reddish, J. M., and Teele, R. L. Cystic fibrosis: Its characteristic appearance on abdominal sonography. *AJR*, 134:1005–1010, 1980.
109. Williams, P., and Warwick, R. The pancreas. In: *Gray's Anatomy*, 36th ed., pp. 1368–1374, 1980. W. B. Saunders, Philadelphia.
110. Wolfman, T., Ramquist, A., Karstaedt, N., and Hopkins, M. B. Cystic neoplasms of the pancreas: CT and sonography. *AJR*, 138:37–41, 1982.
111. Woodard, S., Kelvin, F. M., Rice, R. P., and Thompson, W. M. Pancreatic abscess: Importance of conventional radiology. *AJR*, 136:871–878, 1981.
112. Wright, C. H., Maklad, F., and Rosenthal, S. J. Grey-scale ultrasonic characteristics of carcinoma of the pancreas. *Br. J. Radiol.*, 52:281–288, 1979.

Ultrasound Annual 1982,
edited by Roger C. Sanders.
Raven Press, New York © 1982.

Renal Medical Disorders:
The Role of Sonography

Hedvig Hricak

*Department of Diagnostic Radiology, Henry Ford Hospital,
Detroit, Michigan 48202*

Renal failure, defined as the degree of renal insufficiency causing substantial alteration in plasma biochemistry, is considered acute if it develops over days or weeks, and chronic if it spans months or years. The third category, "acute on chronic renal failure," refers to a rapid reduction in glomerular filtration rate (GFR) in a patient with previously stable chronic renal disease (31).

In the azotemic patient, the value of sonography in assessing the renal size and contour and in diagnosing hydronephrosis is well accepted (14,51). Recently, attempts have been made to use ultrasonography for tissue characterization (46). The role of diagnostic ultrasound has been extended to include the evaluation of the renal parenchyma (21,38,44,46). With current commercially available gray-scale units, normal renal anatomy can be well displayed, and the dense central sinus echoes and renal parenchyma, divided into cortex and medulla, can be recognized (45). Although diagnostic ultrasound yields gross anatomic information, by imaging the interface between the tissue structures, it reflects the histopathologic characteristics of an organ. When the sonographic findings are combined with the clinical data, this noninvasive modality can play a significant role in the management of the azotemic patient.

ANATOMIC AND PATHOLOGIC CHARACTERISTICS
OF RENAL SONOGRAMS

On sonography, the kidney can be divided into two distinct zones, the peripheral renal parenchyma and the central renal sinus.

Renal Parenchyma

The renal parenchyma contains the cortex and the medulla. In the adult kidney, the cortex generates low-level echoes that are less intense than those of the spleen, liver, or renal sinus. The medulla is located central to the cortex and is displayed as a triangular, hypoechoic area. Intense, punctate echoes at the corticomedullary junction represent arcuate arteries and veins (45). In neonates, the echogenicity of the renal cortex approaches that of the liver (3,16,24). The medulla in the neonate is prominent and more hypoechoic than the cortex, and the corticomedullary boundary is accentuated (3,24). Renal cortical echogenicity can be visually analyzed on standard longitudinal supine B-scans by comparing the image echo strength of the renal cortex with the image echo strength of the adjacent liver. Comparison of the renal cortex and the liver is valid only if hepatic disease is not present (44). Normal liver echogenicity can be evaluated on A-scans because the normal liver has an echo amplitude of one-third to one-half the height of the diaphragmatic echoes (52). Comparison of hepatic and renal echogenicity should always be performed at the same depth to avoid difficulties caused by the transducer beam pattern. Scans in which the gallbladder is anterior to the kidney should be avoided to eliminate the distal sound enhancement. Renal cortical echogenicity can be classified into four groups (Fig. 1):

Grade 0:   Normal—the echo intensity of the cortex of the right kidney is less than that of the liver.

Grade I: The echo intensity of the cortex of the right kidney is equal to that of the liver.

Grade II: The echo intensity of the cortex of the right kidney is greater than that of the liver, but less than that of the renal sinus.

Grade III: The echo intensity of the renal cortex is equal to that of the renal sinus.

A statistically significant correlation is seen between the cortical echogenicity and the severity of histopathologic changes (33), especially with prevalence of global sclerosis, focal tubular atrophy, hyaline casts, and leukocytic infiltration (21). A correlation between cortical echogenicity and interstitial changes has been reported previously (38,44,46). However, it has been stated that no correlation is seen between the glomerular changes and the cortical echogenicity (44,46). Early in the course of the renal medical disease, the histologic changes may be isolated within the nephron, vessels, or interstitium. As glomeruli occupy only approximately 8% of the cortex in the adult kidney, in the early stage when the histopathologic changes are minimal, changes restricted to glomeruli will not influence the cortical echogenicity whereas abundant cellular infiltration in the interstitium can cause increased echogenicity. However, as the disease progresses, changes in all three compartments are interrelated. Tubular and interstitial changes usually follow glomerular changes, so that the degree of involvement in all three compartments is similar and the increase in cortical echogenicity in the advanced stage of the disease is the result of combined changes within the glomeruli, tubules, and interstitium (21). There is a negative correlation between cortical echo-

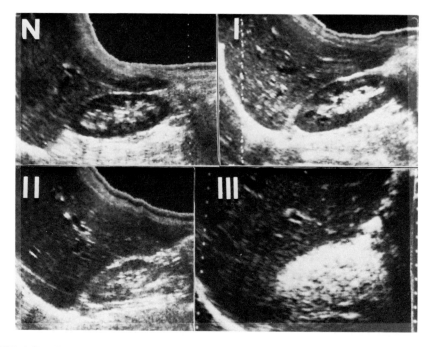

FIG. 1. Renal cortical echogenicity classified into four groups: N, normal; the echo intensity of the cortex of the right kidney is less than that of the liver; I, the echo intensity of the cortex of the right kidney is equal to that of the liver; II, the echo intensity of the cortex of the right kidney is greater than that of the liver, but less than that of the renal sinus; III, the echo intensity of the renal cortex is equal to that of the renal sinus.

genicity and edema, indicating that edema in any given organ will cause increased transonicity.

The renal cortex and medulla can be clearly differentiated in approximately one-half of adults (45). The detectability and sonographic characteristics of the medulla change in response to the diuretic status of the kidney (20). With increased diuresis, the medulla becomes prominent, anechoic, and readily visible (20). With increased edema surrounding the peritubular capillaries in the medullary region, with medullary congestion (25), and with increased blood flow through the kidney, the medullary pyramids are also readily visible, prominent, and anechoic. When the detectability of the medullary pyramids is correlated with the histopathologic changes in the cortex, there appears to be a correlation between the number of hyaline casts per glomerulus and nonvisualization of the medullary pyramids (21). Medullary pyramids may be echogenic, with the amplitude of the medullary echoes being higher than that of the adjacent cortex. The finding is described in nephrocalcinosis (3) (Fig. 2). It can also be seen in chronic rejection, where it is probably a result of deposition of microcalcifications and collagen fibrous tissue (Fig. 3).

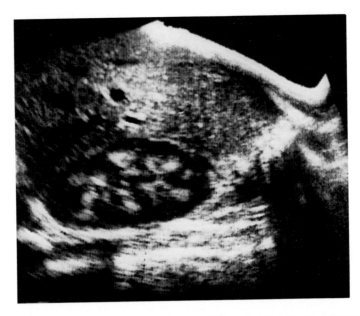

FIG. 2. Milk-alkali syndrome. **A:** Longitudinal parasagittal section through the right upper quadrant. Normal echo intensity of the renal cortex; the amplitude of the cortical echo is less than that of the adjacent liver. Echogenic medullary pyramids with an amplitude of the medullary echoes being higher than that of the cortex.

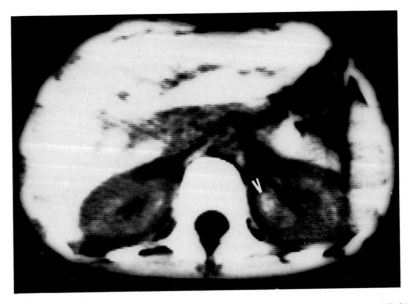

B: Computed tomography demonstrating nephrocalcinosis (*arrowhead*). (Courtesy of T. Slovis, Childrens Hospital, Detroit, Michigan.)

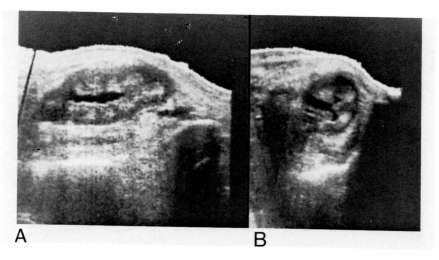

FIG. 3. Chronic rejection. **A:** Longitudinal scan. **B:** Transverse scan. Increased echogenicity of the renal pyramids. On pathologic specimen, microcalcification and collagen fibrous tissue was present in the medulla.

Renal Sinus

Intense central sinus echoes are primarily caused by hilar adipose tissue, whereas blood vessels and collecting structures are secondary contributors (5). The renal sinus echoes are compact and homogeneous, and the amplitude of the echoes is equal to the amplitude of the renal capsule echoes. The high-amplitude echoes are due to the inherent scattering properties of fat cells and not attributable to coexisting fibrous tissue septa (22). Regardless of etiology, any infiltrative process, such as edema, mononuclear cell infiltration, or fibrosis (as seen during allograft rejection), leukocytic infiltration and fibrosis (present in pyelonephritis), or tumor cell infiltration (as seen in neoplasms) will produce a change in the sonographic appearance of the renal sinus. The earliest architectural changes that result from infiltration cause uneven widening of the interlobar septa. Minor cellular infiltration causing a separation of the fat cells is seen as well. With increased septal thickness, the spatial pattern of the renal sinus changes. The sinus echoes become inhomogeneous, patchy, and coarse in distribution. As the infiltrative process progresses, fibrosis follows and atrophy and loss of adipose tissue cells are concomitantly seen. This results in accentuation of the septation of the sinus fat compartment. In the more advanced stages, fibrous tissue predominates and a single cell or small group of residual adipose tissue cells become widely separated. As the area of the renal sinus is replaced by fibrous tissue and the recognizable fat cells become rare, the echogenicity of the central sinus decreases, with loss of distinction between the renal sinus and the renal parenchyma (Fig. 22).

Distention of the pyelocaliceal system (hydronephrosis) will result in separation of the central sinus echoes.

Renal Size

On sonography, the kidneys can be visualized in the prone, coronal, or supine position. The longitudinal axis of the kidney is inclined and the kidney is rotated with its hilus facing forward and medial. The upper pole of the kidney lies posteriorly and the lower pole more anteriorly. Therefore, sagittal supine or prone sections of the kidney are inaccurate in measuring true renal length. To obtain the renal length, either the posterior oblique longitudinal scan, as advocated by Bazzocchi (4); or the same frontal plane of the kidney, the coronal view (6) or "bi-valve" frontal sections (10) can be obtained with the patient in the decubitus position. The coronal view provides a longitudinal dimension of the renal hilus, which best displays the pyelocaliceal system and assesses the contour abnormalities of the kidney and also provides optimal assessment of the renal length. As compared with radiography, excellent results are achieved. One must take into consideration 12% magnification on radiographic measurements. It was reported that sonographic measurements foreshorten the renal length (34). However, the renal length was measured on prone scans and the discrepancy was probably caused by inclination of the kidney, which may vary from 5 to 64% (34). The renal length may be assessed in comparison with the distance between the transverse process of the first four lumbar vertebrae. As measured by prone paraspinal scans, in 95% of normal people (mean $\pm$ 2 SD), the renal length will fall within that range (34). In absolute numbers, in adults a sonographic renal length of less than 10 cm is considered small (38). In assessment of the renal growth measurement, the renal volume is a more sensitive parameter than renal length. The integration method and a calculation based on the assumption of an ellipsoid renal shape using two dimensions have been employed (39,41). When assessing the renal size, observation of renal symmetry and uniformity of the parenchyma should always be done as well.

ACUTE RENAL FAILURE

Acute renal failure (ARF) may be prerenal (renal hypoperfusion due to a systemic cause), renal (as the result of renal medical disease, also referred to as intrinsic renal failure), or postrenal (as a consequence of outflow obstruction) (37). Prerenal failure can be distinguished from the other two entities by clinical and laboratory data alone. A problem arises in distinguishing renal from postrenal causes of acute renal failure. The postrenal causes (5% incidence) (31) are surgically correctable, and the diagnosis should be made rapidly and with a high degree of accuracy. Once the diagnosis of acute renal failure has been established and the prerenal causes excluded, an ultrasound examination should be performed as a primary screening modality (19). Obstruction causes azotemia only if there is simultaneous obstruction of both ureters, or unilateral ureteric obstruction in the absence of a contralateral kidney or with a severely diseased contralateral kidney. The sensitivity of sonography in detecting hydronephrosis is 93 to 98% (14,51).

Renal medical disease is the most common cause of acute renal failure. Most commonly ARF is caused by acute tubular necrosis (48). Other entities include injury to the glomerular capillaries, interstitium, small renal vessels, or rarely major blood vessel disease. ATN may be caused by renal ischemia or by the presence of nephrotoxins (Table 1). The most common cause of ATN is renal ischemia, for which a variety of clinical conditions can be responsible (48). Acute renal failure has been associated with a spectrum of surgical, obstetrical, and medical insults. Rhabdomyolysis as a cause of ATN is known to occur in patients with crush injuries, extensive burns, and muscle inflammation. Also known is nontraumatic rhabdomyolysis, occurring in a variety of conditions in which muscle blood flow and metabolism are disturbed or muscle energy production is increased. Aminoglycoside nephrotoxicity is the

TABLE 1. *Medical causes of acute renal failure*

1. Acute tubular necrosis
   Ischemic disorders
       Major trauma
       Massive hemorrhage
       Compartmental syndrome
       Septic shock
       Transfusion reactions
       Myoglobinuria
       Postpartum hemorrhage
       Cardiac, aortic, and biliary surgery
       Pancreatitis, gastroenteritis
   Nephrotoxicities, including hypersensitivity reactions to
       Heavy metals: mercury, arsenic, lead, bismuth, uranium, cadmium
       Organic solvents: carbon tetrachloride, ethylene glycol
       X-ray contrast media
       Pesticides
       Fungicides
       Antibiotics: aminoglycosides, penicillins, tetracyclines, amphotericin
       Other agents: phenytoin, phenylbutazone, uric acid, calcium
2. Cortical necrosis
3. Acute interstitial nephritis
4. Diseases of glomeruli and small blood vessels
       Acute poststreptococcal glomerulonephritis
       Systemic lupus erythematosus
       Polyarteritis nodosa
       Schönlein-Henoch purpura
       Subacute bacterial endocarditis
       Serum sickness
       Goodpasture's syndrome
       Malignant hypertension
       Hemolytic-uremic syndrome
       Drug-related vasculitis
       Abruptio placentae, abortion with or without Gram-negative sepsis, postpartum renal failure
       Rapidly progressive glomerulonephritis of unknown etiology

leading nephrotoxic cause of ATN. Toxic ARF associated with the use of radiographic contrast media is well recognized.

Acute renal failure may develop as a result of acute interstitial nephritis, which is often drug-induced. While the reactions may be caused by any member of the penicillin group, methicillin sensitivity has been described most often (31). Other agents known to cause acute interstitial nephritis include sulfonamides, diuretics, allopurinol, and phenylbutazone.

Acute renal failure can be caused by glomeruli and small blood vessel disease (Table 1). β-Hemolytic streptococcal infections are responsible for nearly all cases of acute glomerulonephritis; the disease also may occur in association with subacute bacterial endocarditis and possibly with certain virile infections (48). Acute glomerulonephritis may be seen with rapidly progressive nephritis, Goodpasture's syndrome, diffuse lupus nephritis, polyarteritis, and Wegener's granulomatosis (31).

In our experience, in the vast majority of cases of acute renal failure, the renal sonogram is unremarkable. Specifically, in a patient with uncomplicated acute tubular necrosis, the echo characteristic of the renal parenchyma is normal. The kidney may be globular in appearance. The renal size can be normal or enlarged (Fig. 4). On pathologic examination, the kidney is enlarged and swollen, with a tense capsule and parenchyma that bulges through the cut surface of the capsule suggesting interstitial edema. The cortex is pale and the medullary area is congested. The predominant histologic findings are interstitial edema (some authors consider this the critical point in diagnosing the etiology of oliguria during acute renal failure) and tubular dilatation with flattening and degenerative changes of the tubular epithelium. Necrotic tubular debris and precipitated protein (cast) can be seen (Fig. 4B). The histologic findings of tubular necrosis are absent in many cases (48). While we have found that there is no change in cortical echogenicity in cases of uncomplicated ATN, it is reported in the literature that ATN may demonstrate increased cortical echogenicity and preservation of the corticomedullary boundary (3,46). In our experience, increased cortical echogenicity in acute renal failure was seen in cases of ARF caused by acute interstitial nephritis (Fig. 5). In contrast with classic acute tubular necrosis, acute interstitial nephritis histologically shows diffuse infiltration of the interstitium with lymphocytes, macrophages, plasma cells, and polymorphonuclear leukocytes (48). Such increased cellular infiltration of the interstitium is probably responsible for the increased echogenicity seen sonographically. The ultrasonic appearance of acute poststreptococcal glomerulonephritis as reported by Rochester resembles multiple solid renal masses within an enlarged kidney (42). LeQuesne reported a diffuse increase in the cortical echogenicity during acute glomerulonephritis. After the renal function had returned to normal, the cortical echogenicity reversed to normal as well (33). When acute glomerulonephritis is found as a part of streptococcal septicemia, in addition to the changes in the glomeruli, consisting of increased cellularity caused by proliferation of mesangial and endothelial cells and the infiltration of the tuft

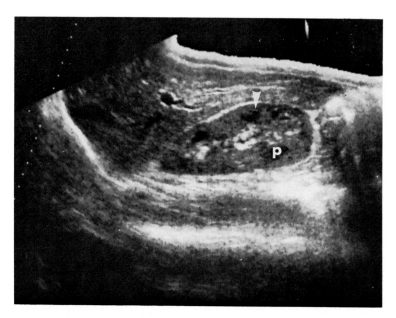

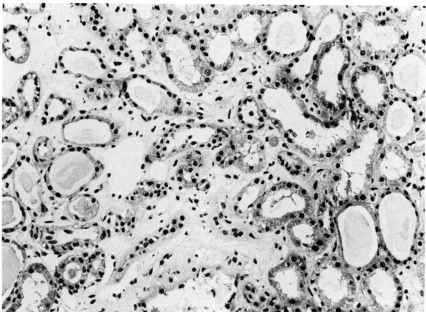

FIG. 4. Acute tubular necrosis. **Top:** Longitudinal parasagittal section through the right upper quadrant. The right kidney is enlarged and globular in configuration. The renal cortical echogenicity is normal. The medullary pyramids (p) are visible and hypoechoic. The arcuate arteries are seen at the corticomedullary boundary (*arrowhead*). **Bottom:** Photomicrograph of the renal cortex (H & E, 110×). Mild interstitial edema with minor leukocytic infiltration of the interstitium. There are tubular epithelial changes, uneven tubular dilatation, presence of cellular debris within the lumen of tubules, and also seen are occasional pigmented casts within the tubular lumen.

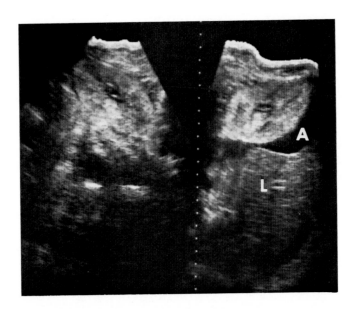

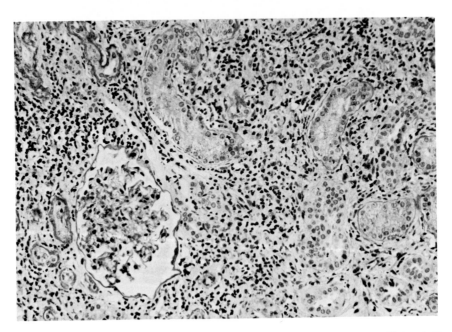

FIG. 5. Interstitial nephritis. Acute renal failure developed as a result of acute interstitial nephritis caused by methicillin sensitivity. **Top:** Transverse scan obtained in a prone position. Marked bilateral increase in renal cortical echogenicity. The amplitude of the renal cortical echoes is considerably higher than that of the liver (L). Ascites (A) was due to congestive heart failure. **Bottom:** Photomicrograph (PAS, 110×). Marked diffuse infiltration of the interstitium with poly-morphonuclear leukocytes, macrophages, and lymphocytes. Minor concomitant tubular and glomerular changes are also present.

by neutrophils and monocytes, also present are a large number of polymorphonuclear cells in the interstitium (18). Such increased echogenicity seen during the acute stage of glomerulonephritis supports the premise that increased interstitial cellularity will influence renal parenchymal echogenicity.

The corticomedullary boundary in acute renal failure is usually well preserved, and the medullary pyramids are readily visible and hypoechoic. In addition to the medullary congestion reported in human and animal studies, acute renal failure is associated with alteration of the renal blood flow (8). The cortical flow fraction decreases from 80% to about 10% of the total renal blood flow whereas the corticomedullary fraction increases from 14% to 80% (37). Coexisting congestion, edema, and increased medullary blood flow are probably responsible for ready visualization of the medullary pyramids during acute renal failure.

In acute renal failure, the kidney is often globular in configuration. The renal size may be either normal or enlarged.

Major blood vessel disease such as renal artery thrombosis, embolism, or stenosis, or bilateral renal vein thrombosis, can cause acute renal failure. In experimental work, we found that a uniform decrease in echogenicity throughout the renal cortex developed 48 hr after complete ligation of the renal arteries (26). Similar findings were not seen in humans. In two patients with angiographically proven complete thrombosis of the renal artery, the kidney remained sonographically normal 3 and 5 days following angiography.

In the case of renal vein thrombosis in both experimental and clinical cases, there is a wide spectrum of sonographic findings that depends on the duration of the venous occlusion (23,47). In the acute stage, the findings consist of immediate renal enlargement (23,47), increased cortical thickness, sparsely distributed cortical echoes with increased transonicity (23), an indistinct corticomedullary boundary, and parenchymal anechoic areas (23,47) caused by hemorrhage and hemorrhagic infarct (23,47). Also reported were dilated renal veins and, in animal work only, renal rupture (23). The findings are not specific to renal vein thrombosis and should always be closely correlated with the clinical presentation.

Acute renal failure may be precipitated by acute pyelonephritis. A urinary infection may cause acute renal failure in patients with obstruction, analgesic nephropathy, calculi, and papillary necrosis. Outside of these conditions, pyelonephritis is rarely the cause of acute renal failure (31). Sonographically, the kidney is swollen and may show an increased anechoic corticomedullary area with multiple scattered low-level echoes (12). Also, an increased through transmission can be seen as a result of increased fluid content in the renal parenchyma (12). The renal sinus may be compressed because of marked parenchymal swelling, or slight dilatation of the pyelocaliceal system secondary to atony can be seen. In the appropriate clinical setting, such a combination of sonographic findings may be suggestive, but not diagnostic, of acute pyelonephritis.

In summary, the main role of sonography in acute renal failure is to exclude the presence of hydronephrosis. In the vast majority of cases, when the ARF is renal in origin the sonographic features of the renal anatomy are unaltered (Fig. 6). The cortical echogenicity most commonly will be normal, with the echo amplitude of the renal cortex being less than that of the adjacent liver. Increased cortical echogenicity, resulting from abundant cellular infiltration of the interstitium, can be present. Isoechoic or hypoechoic renal parenchyma with increased through transmission secondary to edema or hemorrhage can be seen as well. However, the specific diagnosis of a disease cannot be made.

The medulla in acute renal failure is readily visible. This is probably a result of congestion and edema as seen in acute tubular necrosis and, in addition, of redistribution of the blood flow characteristically seen during acute renal failure. Because of marked swelling of the parenchyma, the renal sinus may be compressed, or as a result of atony, a slight distention of the pyelocaliceal system can be seen. Outside of these conditions, the sonographic appearance of the renal sinus during acute renal failure is usually unremarkable. The kidney is commonly enlarged and globular in configuration, but normal renal size can be seen as well.

It should be stressed that the specific diagnosis of the disease by sonography is not possible. If, after clinical evaluation, the etiology of acute azotemia remains unknown, a systemic disease or acute interstitial nephritis is suspected but not proven, or if the oliguric phase of ARF extends beyond 4 or 5 weeks, a percutaneous or open renal biopsy for the purpose of diagnosing the disease is indispensable.

RENAL BIOPSY

Renal biopsy is the single most important tool in the diagnosis of renal medical disease. Adequate analysis of the biopsy specimen requires light microscopy, electron microscopy, and immunofluorescent studies. Either upon biopsy, as part of the surgical procedure, or percutaneous biopsy can be performed. The latter is currently used more often. The reported success rate of percutaneous renal biopsy ranges between 89.8 and 94.9% (11,54). The most important factor increasing the success of the biopsy is accurate localization of the biopsy site. The lower pole of the kidney is usually chosen for biopsy because latex perfusion studies indicate that this area contains the smallest number of large vessels and is the least vascular portion of the renal parenchyma. Either the right or left kidney may be chosen for biopsy. With the patient in the prone position, it is generally more comfortable for a right-handed operator to perform a biopsy on the left kidney. However, some physicians regularly choose the right kidney to avoid proximity to the great vessels. Also, in the case of splenomegaly, the right side is preferable. The lower pole of the kidney was formerly localized on a plain radiograph of the

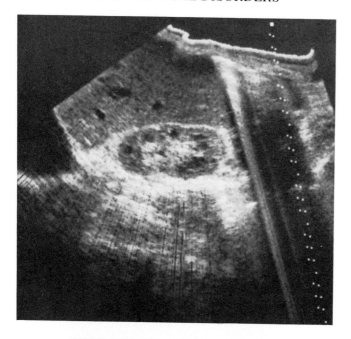

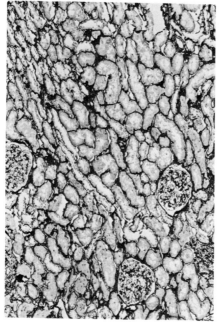

FIG. 6. Proliferative lupus nephritis. **Top:** Longitudinal scan through the right kidney. Normal appearing kidney with good delineation between the cortex and medulla. Unremarkable renal sinus echoes. **Bottom:** Photomicrograph (Jones stain, 70×). Minor tubular and glomerular changes. A minor diffuse mesangial hypercellularity characteristic of mesangial proliferative lupus nephritis.

abdomen. Greater precision has been achieved with high-dose urography and direct fluoroscopic visualization. With the development of ultrasound, further progress in accurate localization has been made. In scanning the patient for biopsy purposes, both the right and the left kidney have to be identified, and the renal and perirenal areas have to be well displayed bilaterally. The renal size has to be evaluated. Possible renal or perirenal masses, as well as any congenital abnormalities, should be excluded. This is important, since the absolute contraindications for renal biopsy include unilateral kidney or marked congenital abnormalities. Relative contraindications are small kidneys, atypical position, or any renal or perirenal masses. Supine or decubitus scans of the right kidney, in conjunction with the liver, and of the left kidney and spleen for purposes of assessing renal echogenicity should also be obtained. While the diagnosis of the renal disease cannot be made by ultrasound, such information is extremely useful in follow-up of the patient. With the patient in a prone position over sandbags or a firm pillow with a soft pillow on the top, care must be taken in respect to the position of the body. The body has to be bent forward at the level of the diaphragm, the shoulders have to be down, and it is imperative that the horizontal level of the shoulder girdle be absolutely parallel to the bed or stretcher and the spine be as straight as possible. The patient should be instructed to take the same size breath he or she will be taking during the biopsy. We prefer to instruct the patient "to take a breath halfway into a deep breath." Scanning should be done in a single linear pass in both sagittal and transverse projections. No sector scanning or angulation of the transducer is allowed. The lower pole of the kidney along its lateral, caudal, and medial margins is marked on the skin surface. The distance from the skin surface to the posterior aspect of the kidney is measured. Also given is the thickness of the renal parenchyma along the lower pole. To minimize the complication rate and to maximize the accuracy of the biopsy specimen, the biopsy should be performed in the ultrasound department immediately after localization. A real-time biopsy transducer that is currently being developed will aid in further improvement of the quality and accuracy of percutaneous renal biopsies.

Percutaneous renal biopsy has a definite attendant risk, with the complication rate ranging from 0.7 to 8.1% (11,50). The complications include oliguria, decrease in the hematocrit reading in 44% of cases (7), hematuria in 10 to 40% of cases (54), AV fistula, renal abscess, Gram-negative septicemia, and perforation of hollow viscera (29). Postbiopsy hematomas are seen in over 50% of the patients, as documented by computed tomography (CT) (2,43). By serial scanning of postbiopsy patients, we have been able to identify hematomas in 75% of patients. Most commonly the hematomas are small and not clinically apparent. Intrarenal hematomas are identified as hypoechoic areas within the renal parenchyma associated with localized renal enlargement (Fig. 7). The features of the normal renal anatomical distinction between the cortex and medulla become distorted (Fig. 7). Subcapsular hematomas char-

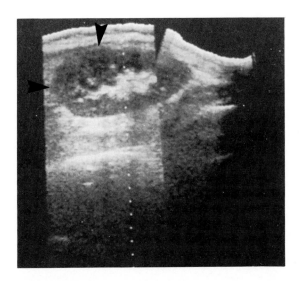

FIG. 7. Intrarenal postbiopsy hematoma in a transplanted kidney. Biopsy was performed because of prolonged posttransplant renal failure. Biopsy proven acute tubular necrosis. A small intrarenal hematoma is identified as a hypoechoic area within the superior pole of the kidney causing localized enlargement of the kidney (*arrowheads*). The allograft is otherwise sonographically unremarkable consistent with the diagnosis of acute tubular necrosis.

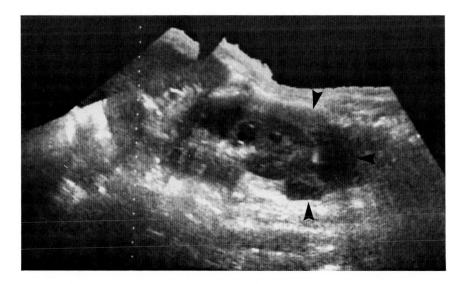

FIG. 8. Coronal view, left kidney. Postbiopsy perirenal hematoma (*arrowheads*).

acteristically compress the adjacent renal parenchyma, and at the same time cause outward bulging of the renal capsule (40). Perirenal hematomas are seen surrounding the lower pole of the kidney (as the biopsy site is always restricted to the lower pole) (Fig. 8). Posterior pararenal hematomas typically displace the kidney anteriorly (Fig. 9). On the 24-hr postbiopsy scan, the echo

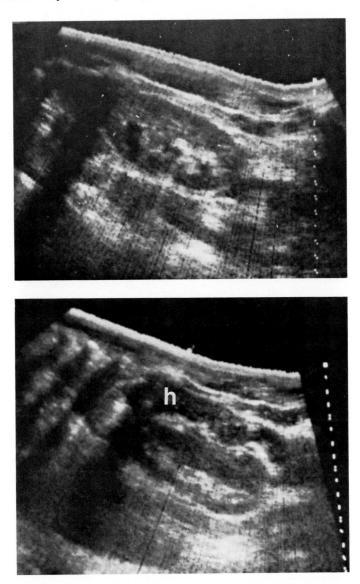

FIG. 9. Postbiopsy posterior pararenal hematoma. **Top:** Prone parasagittal scan, left kidney. The scanning was done for biopsy localization. **Bottom:** One day postbiopsy. Anterior displacement of the kidney by a posterior pararenal hematoma (h). The echo characteristic of the hematoma is similar to the surrounding tissue.

characteristics of hematoma is similar to those of the surrounding soft tissue and the diagnosis is often suggested solely on the basis of anterior displacement of the kidney (Fig. 9B). Approximately 3 days postbiopsy a hematoma is seen as a hypoechoic area and high-amplitude, punctate echoes within it can be identified. At 5 to 7 days postbiopsy, the hematoma presents as a hypoechoic area, and occasionally septations within the hematoma can be seen

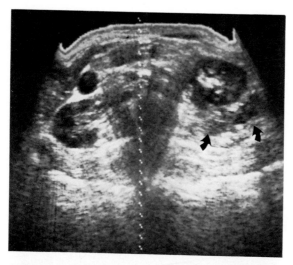

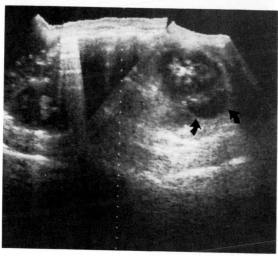

FIG. 10. Transverse supine scan through the upper abdomen. **Top:** Scan obtained 1 day postbiopsy. There is anterior displacement of the left kidney by a perirenal hematoma (*arrows*). The sonographic characteristic of the hematoma is similar to the surrounding soft tissue. **Bottom:** Scan obtained 4 days postbiopsy. The hematoma now appears as hypoechoic, probably secondary to liquification.

(Fig. 10). Clinically significant hematomata occur in approximately 0.4% of cases (Fig. 11). Massive bleeding following renal biopsy necessitated the performance of nephrectomy in 5 out of 8,000 cases and splenectomy in 2 out of 8,000 cases (54). Reported cases of mortality range from 0.07 to 0.17% (29,54).

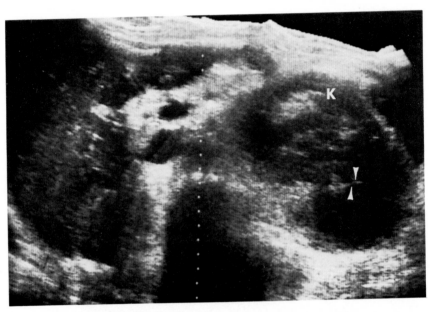

FIG. 11. Large perirenal and posterior pararenal hematoma. The hematoma was clinically apparent. The patient experienced severe flank pain, increasing over the next 48 hr. The vital signs remained stable, but persistent falling hematocrit was present. The patient required transfusion with seven units of packed red cells before the hematocrit stablized. **A:** Supine transverse scan through the upper abdomen. Anterior displaced left kidney (K) by a large hematoma. The hematoma is both perirenal and pararenal in location. In correlation with computed tomography, the echogenic line separating the perirenal from pararenal hematoma represents Gerota's fascia (*arrowhead*).

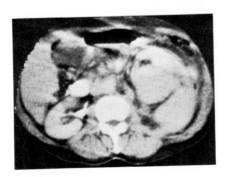

FIG. 11. B: Computed tomography of the right upper quadrant, post injection study showing a normally excreting right kidney. No excretion in the left kidney is seen. The left kidney is displaced anteriorly by large perirenal and pararenal hematomas.

Because progression of renal disease has to be followed closely, particularly when there is a discrepancy between the apparent histologic features and the course of the disease, or when the effects of therapy need to be closely monitored, a noninvasive method that reduces or obviates the need for serial biopsies would assume great importance. Therefore, the use of diagnostic ultrasound in monitoring the progression of the disease in these patients is an important advance (Fig. 12).

CHRONIC RENAL FAILURE

In the wide spectrum of renal medical diseases that cause chronic renal failure (Table 2), three main types can be histologically recognized: nephron, vascular, and interstitial disorders (28). In the chronic stage of the disease, after the kidney has been injured for a long period of time, pathologic changes in all three compartments are interrelated. Nephron changes may result from vascular disease. In turn, vascular disease follows nephron and interstitial tissue changes. In a similar fashion, interstitial tissue changes invariably follow and accompany vascular and nephron disease. The kidney that has been seriously injured for a long period of time tends to lose the distinguishing features of the specific disease regardless of whether the original disease was glomerular, tubular, interstitial, or vascular. Eventually, all of the component

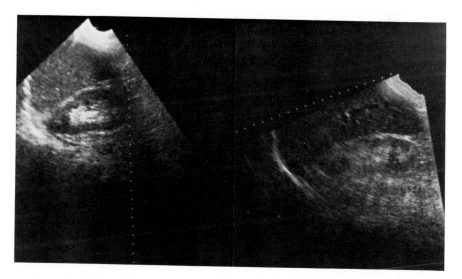

FIG. 12. Membranous lupus nephritis, follow-up. **Left:** Longitudinal section through the right upper quadrant. Normal appearing kidney with an echo amplitude of the renal cortex less than that of the adjacent liver. **Right:** Longitudinal scan of the right kidney obtained 7 months later. Increased echo intensity of the renal cortex with the amplitude of the renal cortical echoes being greater than the adjacent liver, but less than the renal sinus (cortical echogenicity Grade II).

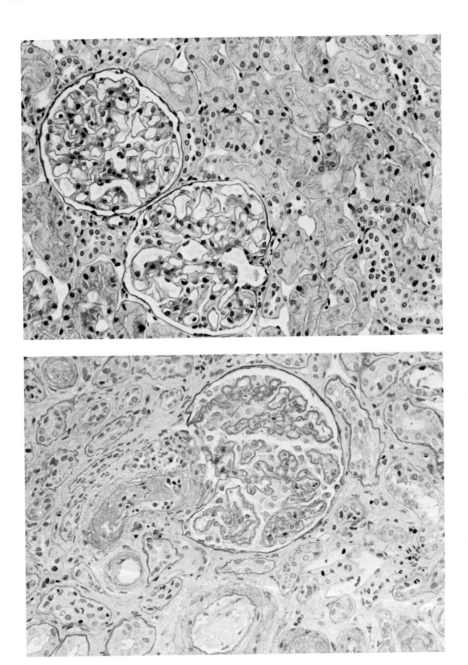

FIG. 12. continued. **Top:** Photomicrograph (PAS, 140 ×). Biopsy obtained at the time of the initial sonogram. Membranous lupus nephritis showing minimal atrophy and parenchymal scarring. **Bottom:** Photomicrograph from a biopsy specimen obtained 7 months later at the time of the second sonographic examination (PAS, 140 ×). Marked interval progression of the disease, and changes of the membranous lupus nephritis are now more advanced and are accompanied by lupus vasculitis. Tubular atrophy and interstitial fibrosis are apparent.

TABLE 2. *Common medical causes of chronic renal failure*

1. Glomerulonephritis
2. Chronic pyelonephritis
3. Renal vascular disease
4. Metabolic causes
   Diabetes mellitus
   Gout
   Hypercalcaemia
   Hyperoxaluria
   Cystinosis
   Angiokeratoma corporis diffusum (Fabry's disease)
5. Nephrotoxins
6. Renal tuberculosis
7. Sarcoidosis
8. Dysproteinaemia
   Myeloma
   Amyloidosis
   Mixed IgA-IgM cryoglobulinaemia
   Waldenstrom's macroglobulinaemia
9. Hereditary or congenital
   Polycystic disease
   Nephronophthisis (medullary cystic disease)
   Alport's syndrome
   Cystinosis
   Hyperoxaluria
   Chronic tubular acidosis
   Infantile nephrotic syndrome
   Dysplastic kidneys
10. Miscellaneous
    Radiation
11. Major blood vessel diseases
    Renal artery thrombosis, embolism, or stenosis
    Bilateral renal vein thrombosis
12. Hepatorenal syndrome
13. Acute pyelonephritis

structures—vessels, nephrons, and interstitial tissue—become distorted in a complicated picture. Distinction between different patterns of disease becomes a problem even for the pathologist faced with end-stage kidney disease (28).

Sonographically, with the exception of cystic disease, a specific diagnosis of renal medical disease cannot be made. In polycystic kidney disease (PKD), sonography has been shown to be more sensitive and more specific than excretory urography (49) (Fig. 13). The kidney in polycystic kidney disease may be normal in size, but in the stage that renal failure is apparent, the kidney is usually enlarged (53). The increase in renal volume actually starts before the renal function is impaired. The renal growth continues as renal function deteriorates. In the initial stage of adult polycystic disease, the volume of the polycystic kidneys on two sides is often different. However, with progressive renal failure the difference becomes insignificant. The growth rate of polycystic kidneys seems to follow a linear course when determined in patients with

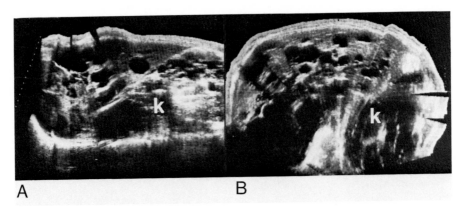

FIG. 13. Polycystic kidney and liver disease. **A:** Longitudinal and **B:** transverse sections through the upper abdomen in a patient with severe chronic renal failure. Both kidneys are markedly enlarged (k). The renal parenchyma is extensively replaced by numerous varying sized cysts. Multiple cysts within the liver are seen as well.

variable degrees of renal function impairment (53). In patients with chronic renal failure caused by PKD, sonographically bilaterally enlarged kidneys with numerous cysts of varying sizes within the cortex and medulla are observed. The remaining renal parenchyma appears echogenic. The pathologic characterizations of polycystic disease are decrease in the number of nephrons and considerable increase in interstitial connective tissue. Cysts within the liver (50%) and pancreas (9%) can be seen as well (49).

Aside from cystic diseases, a specific diagnosis of renal medical disease cannot be made by sonography. There are reports of the characteristic appearance of chronic atrophic pyelonephritis (30). Even on pathology the contracted kidney of chronic pyelonephritis is not always distinguishable from a shrunken kidney caused by glomerulonephritis, arterial and arterionephrosclerosis, periarteritis nodosa, and multiple healed infarcts (1). Further, there is general agreement that the histopathologic changes of chronic interstitial nephritis resulting from bacterial infection are not pathognomonic of response to infection, but are common to a great variety of renal diseases resulting in interstitial inflammation. The sonographic appearance of the renal parenchyma responds to the degree of histopathologic changes and advanced changes in the glomeruli, interstitium, and tubules will cause an increase in the echo intensity of the renal cortex. However, a specific diagnosis of a disease cannot be made (Fig. 14).

In chronic renal failure, the renal medulla can be seen when the cortical echogenicity reads Grade I or Grade II, but very rarely with a marked increase in cortical echogenicity is the medulla identified. Usually in the end stage of the disease there is loss of distinction between the cortex and medulla, and in addition the area of the central sinus echoes often blends with

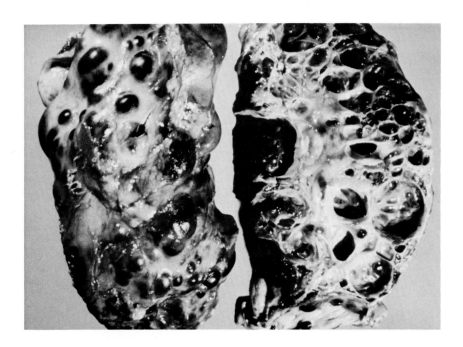

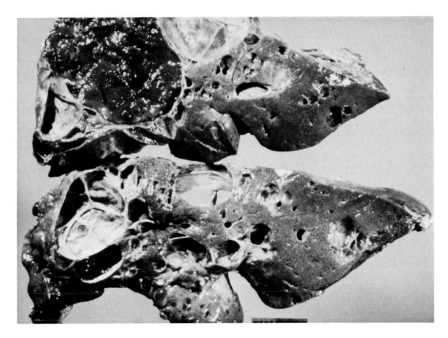

FIG. 13. continued. **Top:** Gross specimen. Markedly enlarged kidneys with numerous cysts throughout. Hemorrhage within some of the cysts is apparent. **Bottom:** Gross specimen of the liver. There are multiple cysts within the liver.

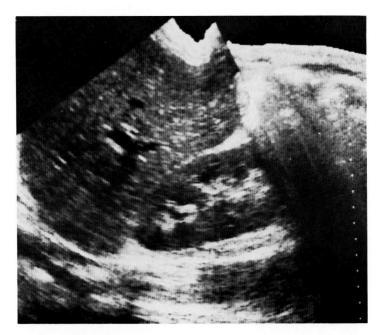

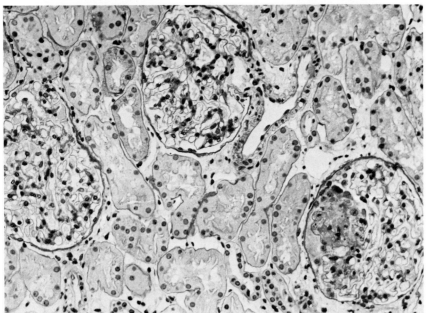

FIG. 14. Wegener's granulomatosis. **Top:** Longitudinal sections through the right upper quadrant. Normal appearing kidney. The echo amplitude of the renal cortex is less than that of the adjacent liver. **Bottom:** Photomicrograph (PAS, 110 ×). Focal segmental glomerular hypercellularity, leukocytic infiltration, and necrosis. The tubules and interstitium are unremarkable. As the involvement of the kidney on histopathologic changes was minimal, the renal sonogram is normal.

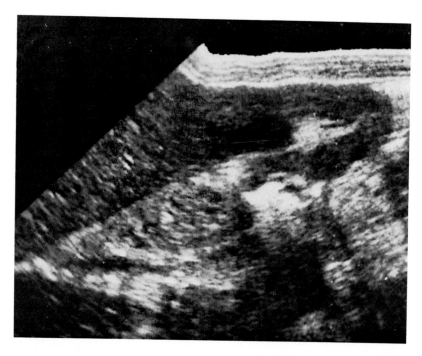

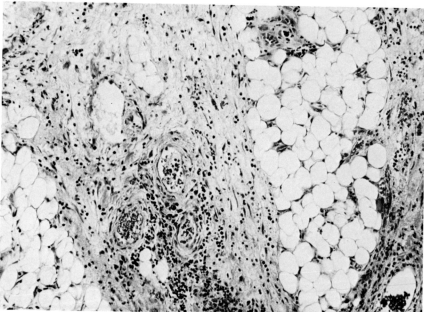

FIG. 15. Chronic atrophic pyelonephritis. **Top:** Longitudinal scan of the right kidney. Small kidney with increased cortical echogenicity (Grade III). Intense echoes of the renal sinus are seen in an irregular pattern. **Bottom:** Photomicrograph of the hilar adipose tissue (H & E, 70 ×). Residual fat cells are seen separated by widened interlobar septa. Widening of the interlobar septa is caused by fibrous inflammatory tissue.

the adjacent parenchyma. Also, in diseases causing a deposition of calcium (nephrocalcinosis) or collagen fibrous tissue the medulla will appear echogenic. As recently noted in chronic pyelonephritis, inflammatory changes present in the cortex can be seen within the hilar adipose tissue as well (Fig. 15). Further studies regarding the appearance of the renal hilar adipose tissue in the remaining renal medical diseases are needed.

Correlating the renal length with the severity of the pathologic changes shows a relationship between the renal length and prevalence of global sclerosis, focal tubular atrophy, and the number of hyaline casts per glomerulus. The latter suggests that renal fibrosis is associated with this finding.

DIALYSIS PATIENT

Patients on hemodialysis have an incidence of cystic disease that reaches 43.5% when they are on dialysis for less than 3 years and 79.3% when the dialysis is maintained for more than 3 years (27) (Fig. 16). In addition, these patients have a significantly higher incidence of infection, papillary necrosis, and renal adenocarcinoma (Fig. 17). Sonography is the ideal imaging modality for evaluating the kidney status of dialysis patients. An excretory urogram, save for information regarding renal size, usually provides no further data.

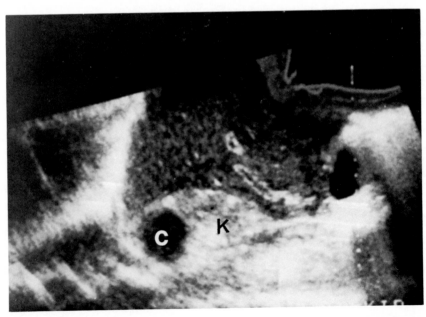

FIG. 16. Renal cyst (c) in the upper pole of the right kidney (k). A small, highly echogenic kidney (echogenicity, Grade III). The patient was on dialysis for 2 years. Chronic renal failure was caused by diabetic nephropathy.

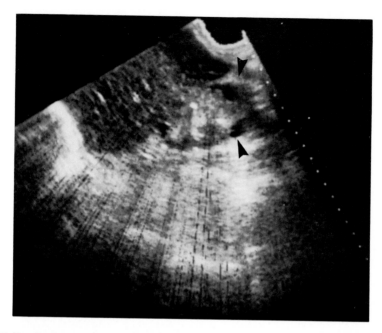

FIG. 17. Renal adenocarcinoma. **A:** Longitudinal sections through the right kidney. The echo amplitude of the renal cortex is greater than the adjacent liver, but less than the renal sinus (Grade II). There is enlargement of the lower pole of the right kidney. Hypoechoic areas are present within the mass.

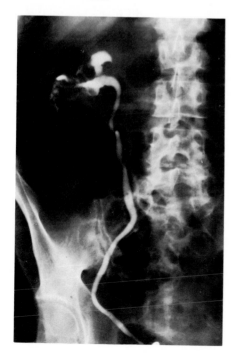

FIG. 17. B: Retrograde pyelogram. The calices to the lower pole of the right kidney are displaced and destroyed with extravasation of the contrast media into the tumor. On gross specimen, necrotic renal adenocarcinoma in the lower pole of the kidney was found. The patient was on chronic dialysis for 5 years. Chronic renal failure was caused by systemic lupus erythematosus.

Retrograde pyelography will delineate the pyelocaliceal system, but the renal parenchyma cannot be assessed. CT can be utilized, but without the benefit of injection of contrast media information is limited. Nuclear magnetic resonance (NMR) is an emerging modality that promises to overcome the drawbacks of other radiologic procedures. By imaging a hydrogen, NMR combines the advantages of CT (by offering tomography with an excellent resolution) and nuclear medicine (by providing information on metabolic processes). It matches ultrasound by avoiding the danger of ionizing radiation. In the future, NMR will benefit from the use of contrast media that will follow metabolic pathways. Currently, ultrasound is the ideal modality for evaluating patients on long-term dialysis. In addition to the information regarding renal size and status of the collecting system, the renal parenchyma can be well displayed, and the sonic texture of the renal parenchyma can be assessed. Further, evaluation for possible renal masses can be achieved. Having serial studies available for comparison increases the value of the examination. Obtaining periodic renal sonograms in patients on dialysis is advised. This is particularly true in the case of a change in urine sediment, especially with the appearance of hematuria.

RENAL TRANSPLANTATION

Over the past 2 decades, renal transplantation has evolved into a standard treatment modality for irreversible renal failure. To date, more than 27,000 renal transplants have been performed in the United States alone. Although the technical aspects of organ perfusion in transplantation has been nearly perfected, the search for effective methods to control the immunologic mechanism responsible for rejection continues. From the moment of implantation, the renal allograft exists and functions in a precarious balance between the rejection process and survival on immunosuppressive treatment. A prompt and accurate diagnosis of the cause of acute posttransplantation renal failure is therefore mandatory.

Urologic complications causing acute posttransplantation renal failure are reported as high as 12% (13). The diagnosis of urologic complications is often difficult, since concomitant rejection may mimic or mask clinical signs and symptoms. Urologic complications consist of either obstruction or urinary leak and formation of urinoma. In any case, sonography is a noninvasive, accurate, and sensitive modality in diagnosis of both hydronephrosis and perirenal fluid collections, and in acute posttransplantation renal failure diagnostic ultrasound should be a screening modality. When dilatation of the collecting system is sonographically present and obstruction is clinically suspected, antegrade pyelography using both a sonographic and a radiologic technique is advocated (17). Although ultrasound is a sensitive modality in diagnosing perirenal collections (9,17), the sonographic appearance of fluid collections is nonspecific and if the exact diagnosis is mandatory, under ultrasonic guidance a percutaneous puncture and aspiration of the fluid can be obtained. Sonography has proven to be more sensitive than either excretory urography or a radionuclide study in detecting perirenal fluid collections. The accuracy of sonography in diagnosing perirenal fluid collections approaches 100% whereas a radionuclide study shows a sensitivity of 86% and excretory urography, 68% (17). In the early postoperative period, vascular complications occur at a lower rate. An anastomotic leak and secondary massive hemorrhage show an incidence of 1.9 to 4.4% (32). Arterial thrombosis occurs in 1 to 2% of renal allografts, and the incidence of venous thrombosis has been reported as high as 3.6% (13,32). Sonographically, a spectrum of findings can be present, but the findings are not specific, and a definite diagnosis requires angiography. However, sonography, combined with clinical and laboratory correlation, is most helpful in suggesting the route of further radiologic evaluation.

In the vast majority of cases, acute posttransplant renal failure is caused by acute tubular necrosis or rejection.

Acute Tubular Necrosis

Acute tubular necrosis is a common cause of acute posttransplant renal failure, with a reported incidence in as many as 50% of cadaver kidney recipients. Clinically, ATN presents with elevated serum creatinine levels. Urine output may be low from the time of transplantation, or urine volume may be initially good and then be followed by oliguria. Uncomplicated ATN is usually reversible; the only therapy needed is maintenance of hydration and immunosuppression. It is important to recognize uncomplicated ATN and distinguish it from acute rejection, which requires an increase in immunosuppressive therapy. Clinically, the diagnosis of ATN is most commonly a diagnosis of exclusion. Although sonographically there have been reports of sonographic findings in acute tubular necrosis, mainly increase in cortical echogenicity (44), as reported by Maklad (35) and in our experience, during the course of acute tubular necrosis the appearance and anatomic characteristics of the kidney remain unchanged from the baseline study. The kidney may be globular in appearance but the increase in renal volume remains within the range of normal renal hypertrophy (20).

After an intact vascular supply is verified by a radionuclide study, the normal appearing sonogram, in a clinical setting of acute renal failure, suggests the diagnosis of acute tubular necrosis (Fig. 7).

Allograft Rejection

After the first posttransplant week, acute rejection is the most common cause of renal failure. On the basis of clinical and laboratory data alone, the diagnosis of acute rejection often remains in question. Radiologic studies that utilize contrast media can contribute to further renal damage and should therefore be avoided. Radionuclide studies are noninvasive and provide functional and anatomical information about the renal allograft. Although the diagnosis of acute rejection can be suggested, often the results are inconclusive and differentiation between acute rejection and acute tubular necrosis is not possible.

Sonographically there is a spectrum of findings accompanying rejection (20,25,35,36). Rejection is a dynamic process, and different sonographic findings correspond to the spectrum of histopathologic changes. An early finding in acute rejection is increase in renal size caused by edema, congestion, and mononuclear cell infiltration. Sonographically, increase in renal volume is a common early finding (88% of acute rejection). Enlargement of the medullary pyramids as a result of a combination of edema surrounding the peritubular capillaries and marked congestion at the corticomedullary junction is another early feature of acute rejection. Enlargement of the medullary pyramids, in our experience, was usually visualized in acute rejection when the conventional vascular changes of rejection were a predominant feature (Fig. 18). When mononuclear cell infiltration was the major feature of acute rejection, a sonographically indistinct corticomedullary boundary was a common finding (Fig. 19). In addition, the cortical echoes may become sparsely distributed or increased during acute rejection. The latter finding is probably a result of abundant mononuclear cell infiltration of the interstitium. A uniform or localized decrease in parenchymal echogenicity corresponding to the area of edema, hemorrhage, and hemorrhagic infarction has been reported. While the rejection starts and is most pronounced at the corticomedullary boundary, it will progress and involve the entire renal unit. Within the hilar adipose tissue, the earliest change is due to edema, causing uneven widening of the perilobular and intralobar adipose tissue septa. When the involvement of the hilar adipose tissue was only minimal, no sonographic abnormalities could be detected (Fig. 20). However, when the changes within the hilar fat were considered moderate and histologically mononuclear cell infiltrates were seen causing further widening of the intralobular septa, with separation and decrease in size of the fat cells, sonographically there was a change in the spatial pattern distribution. The dense echoes of the renal hilum were not homogeneously distributed, but rather the renal sinus echoes appeared in a blotchy and coarse distribution (Figs. 21 and 22). Assuming that each fat cell functions as an independent scattering unit, with characteristic reflectivity, separation of the group of cells by widened perilobular and intralobar adipose tissue septa could be responsible for change in the spatial pattern of the renal sinus. With further progression of rejection, subsequent fibrosis, fat cell atrophy, and fat cell loss accentuate adipose tissue septations. In the severely involved specimen, an area of edema and/or fibrosis is predominant, and adipose tissue cells form a minor portion of the volume of the hilar fat

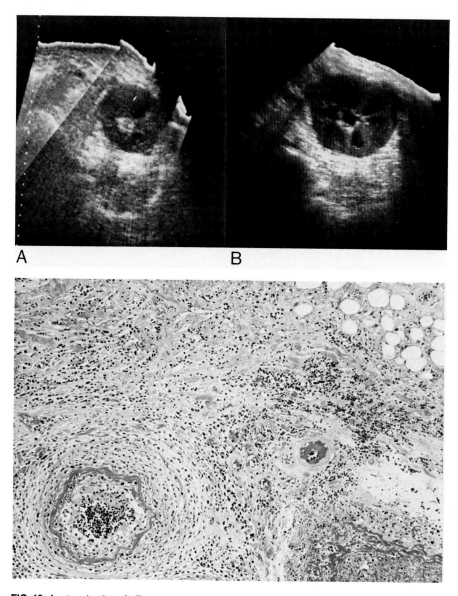

FIG. 18. Acute rejection. **A:** Transverse scan. A normal baseline study. **B:** Two weeks later during acute renal failure. The kidney shows marked interval enlargement. The medullary pyramids are enlarged. There is mild dilatation of the central pyelocaliceal system. At the time of nephrectomy, no obstruction was seen. Few remaining bright central sinus echoes can be seen. **C:** Photomicrograph of the sinus fat compartment (H & E, 70 ×). Residual adipose tissue cells are decreased in size. Fat cells form a minor component of the sinus fat compartment, which now consists mainly of fibrous tissue with mononuclear cell infiltrate. Within the cortex as well as within the hilar adipose tissue, the marked conventional artery rejection changes were apparent.

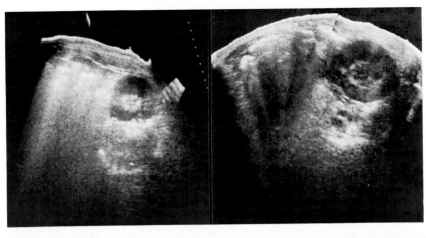

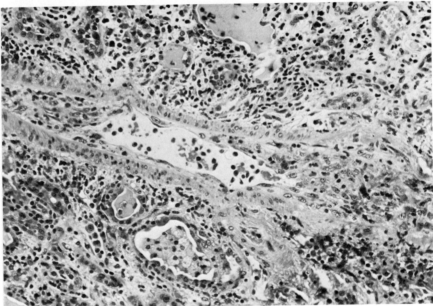

FIG. 19. Acute rejection. **Top left:** Transverse sonogram. Normal baseline study. **Top right:** Ten days later at the time of biopsy proven acute rejection. Renal enlargement with indistinctness of the corticomedullary boundary. There is also a change in the appearance of the central sinus echoes which are now inhomogeneous and patchy in distribution. **Bottom:** Photomicrograph of the renal cortex (H & E, 140 ×). There is a marked mononuclear cell infiltration of the interstitium. Also seen is edema and separation of the tubules. The conventional artery changes of rejection are minor to moderate.

compartment. Sonographically, the area of the renal sinus blends with the adjacent parenchyma and no distinction between the two compartments can be made. A few remaining fat cells reflecting strong echoes can be found in a very irregular pattern (22).

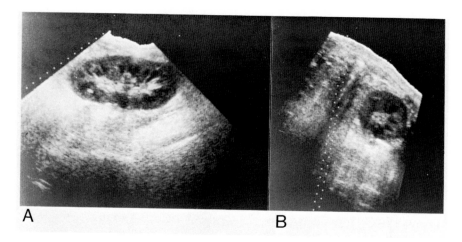

FIG. 20. A: Longitudinal and **B:** transverse scans obtained 40 days posttransplantation. The patient had no clinical evidence of renal failure. Sonographically, this was a normal transplant kidney. The central sinus echoes are of high intensity and are homogeneously distributed. Two days later, the patient died from pulmonary embolus.

Perirenal fluid collections representing either hematoma or lymphocele can be seen during acute rejection. Renal rupture secondary to acute rejection is a serious complication occurring in 3.6 to 6% of cases. Most ruptures occur during the early posttransplant period—80% within the first 2 weeks (32,39). Many authors believe rejection itself plays a major role in initiating the stress and rupture. Most commonly rupture occurs longitudinally at the convex border of the kidney at a depth of several millimeters. When scanning a patient in acute rejection, evaluation of the renal as well as perirenal area is of extreme importance.

As the rejection process involves the entire renal unit, it will be seen within the pelvic wall and ureter. Minimal to mild dilatation of the collecting system as a part of the rejection process can be seen as well.

While there is a spectrum of findings present during acute rejection, not a single finding per se is specific to acute rejection. Sonographically, to suggest a diagnosis of rejection, at least two of the above described features should be present. Further, the findings should always be correlated with clinical and laboratory data as well as results from radionuclide studies, if available. The correct diagnosis of rejection based on sonographic criteria was achieved in 84% of the patients as reported by Meier et al. (36). In a series reported by Frick, false-negative results of rejection were obtained in 8% (15).

Recent attempts have been made to differentiate acute from chronic rejection. In Meier's experience (36), a failure to differentiate between acute and chronic rejection occurred only 7% of the time. In their experience, the initial findings of acute rejection were increased renal size with at least 20%, enlargement of the medullary pyramids, and indistinctness of the cortico-

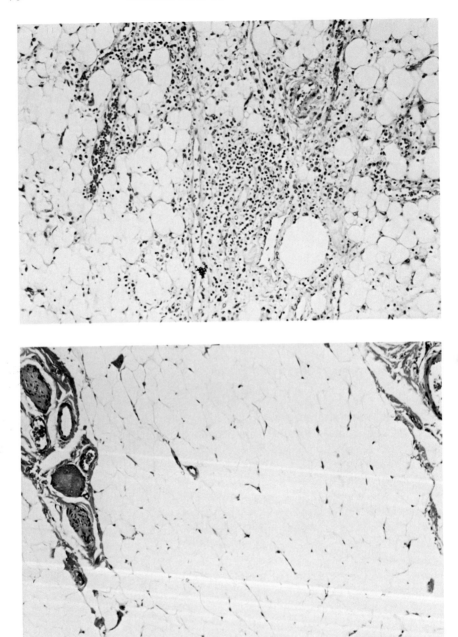

FIG. 20. continued. **Top:** Photomicrograph of the hilar adipose tissue (H & E, 70 ×). The lobular and intralobular septa are unevenly expanded by edema and mononuclear cell infiltrates. Minor mononuclear cell infiltration causing separation of the fat cells is also seen. Changes of acute rejection are considered minimal. **Bottom:** Microscopic section of the normal hilar adipose tissue (H & E, 70 ×). Closely packed fat cells are enclosed by interlobar septa. The septa contain arteries, veins, and nerves.

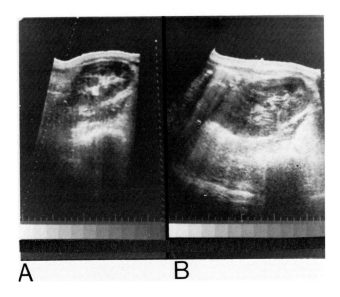

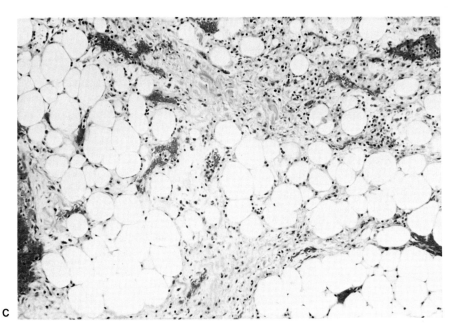

FIG. 21. A: Longitudinal scan. Normal baseline study. **B:** Scan obtained 28 days later during acute renal failure. The kidney is enlarged and globular in appearance. There is indistinctness of the corticomedullary boundary. There is change in the spatial pattern of the central sinus echoes which are now inhomogeneous and patchy in distribution. **C:** Photomicrograph, hilar adipose tissue (H & E, 70 ×). Widening of the intralobular and interlobular septa by mononuclear cells and collagen fibrous tissue. Some of the fat cells show a decrease in size. The degree of rejection is histologically considered moderate.

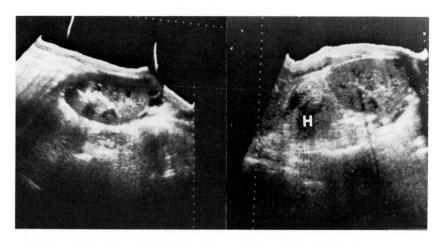

FIG. 22. Acute rejection with rupture. **Left:** Longitudinal scan obtained during acute renal failure. The kidney is globular. There is uneven enlargement of the medullary pyramids. The renal sinus echoes are patchy in distribution. **Right:** Oblique scan obtained at the day of nephrectomy. There is a mixed echo pattern mass-hematoma (H) inferior to the upper pole of the transplanted kidney. The kidney shows severe changes of rejection with indistinctness of the corticomedullary boundary and blending of the renal sinus area with remaining parenchyma.

medullary boundary. In the advanced stage of acute rejection, there were localized hypoechoic areas within the cortex and medulla. They describe the renal contour in the end stage of acute rejection as large, confluent, hypoechoic areas in the medulla and cortex. By contrast, during chronic rejection the initial finding consists of renal enlargement with blending of the cortex and medulla and nondifferentiation of the central sinus echoes. There was a diffuse increase in echodensity throughout the kidney with a coarse spatial distribution of the echoes and smooth kidney contour. In the end stage of chronic rejection, the kidney was small, very echogenic, and had an irregular contour because of scarring; differentiation between the parenchyma and central sinus echoes was not possible.

REFERENCES

1. Allen, A. C. *The Kidney, Medical and Surgical Diseases*, 1951. Grune & Stratton, New York.
2. Alter, A. J., Zimmerman, S., and Kirachaiwanich, C. Computerized tomographic assessment of retroperitoneal hemorrhage after percutaneous renal biopsy. *Arch. Intern. Med.*, 140:1323–1326, 1980.
3. Babcock, D. S. Medical diseases of the urinary tract and adrenal glands. *Clin. Diagn. Ultrasound*, 8:113–134, 1981.
4. Bazzocchi, M., and Giorgio, R. The value of the posterior oblique longitudinal scan in renal ultrasonography. *Urol. Radiol.*, 1:221–225, 1980.
5. Behan, M., and Kazam, E. The echographic characteristics of fatty tissues and tumors. *Radiology*, 129:143–151, 1978.

6. Behan, M., Wixson, D., and Kayam, E. Sonographic evaluation of the non-functioning kidney. *J. Clin. Ultrasound*, 7:449–458, 1979.
7. Bolton, W. K. Nonhemorrhagic decrements in hematocrit values after percutaneous renal biopsy. *JAMA*, 238:1266–1268, 1977.
8. Cattell, W. R. Acute renal failure. In: *Recent Advances in Renal Disease*, edited by N. F. Jones, pp. 1–47, 1975. Churchill Livingstone, London.
9. Coyne, S. S., Walsh, J. W., Tisnado, J., et al. Surgically correctable renal transplant complications: An integrated clinical and radiologic approach. *AJR*, 136(6):1113–1119, 1981.
10. Dana, A., and Moreau, J. F. Interet de la coupe echographique frontale vraie du rein, dite "bivalve." *J. Radiol.*, 62(1):59–62, 1981.
11. Diaz-Buxo, J. A., Donadio, J. V., Jr. Complications of percutaneous renal Biopsy: An analysis of 1,000 consecutive biopsies. *Clin. Nephrol.*, 4:227, 1975.
12. Edell, S. L., and Bonavita, J. A. The sonographic appearance of acute pyelonephritis. *Radiology*, 132:683–685, 1979.
13. Ehrlich, R. M., and Smith, R. B. Surgical complications of renal transplantation. *Urology (Suppl.)*, 10(1):43–55, 1977.
14. Ellenbogen, P. H., Scheible, F. W., Talner, L. B. et al. Sensitivity of gray-scale ultrasound in detecting urinary tract obstruction. *Am. J. Roentgenol.*, 130:731–733, 1978.
15. Frick, M., Feinberg, S., Sibley, R., et al. Ultrasound in acute renal transplant rejection. *Radiology*, 138:657–660, 1981.
16. Haller, J. O., Berdon, W. E., and Friedman, A. P. Increased renal cortical echogenicity: A normal finding in neonates and infants. *Radiology*, 142:173–174, 1982.
17. Heckemann, R., Hartmann, H. G., and Eickenberg, H-U. Combined ultrasound—radiographic detection of ureteral obstruction in renal transplants. *Urol. Radiol.*, 1:233–235, 1980.
18. Heptinstall, R. H. Pathology of acute glomerulonephritis. In: *Diseases of the Kidney*, edited by Maurice B. Strauss and Louis G. Welt, pp. 405–462, 1971.
19. Hricak, H. Ultrasound and the azotemic patient. In: *Diagnostic Radiology*, edited by A. Margulis and C. A. Gooding, pp. 385–392, 1982.
20. Hricak, H., Cruz, C., Eyler, W. R., et al. Acute post-transplantation renal failure: Differential diagnosis by ultrasound. *Radiology*, 139:441–449, 1981.
21. Hricak, H., Cruz, C., Romanski, R., et al. Renal parenchymal disease: Sonographic-histological correlation. *Radiology*, 144:141–147, 1982.
22. Hricak, H., Romanski, R., and Eyler, W. R. Renal sinus during allograft rejection: Sonographic histopathologic correlation. *Radiology*, 142:693–699, 1982.
23. Hricak, H., Sandler, M. A., Madrazo, B. L., et al. Sonographic manifestations of acute renal vein thrombosis: An experimental study. *Invest. Radiol.*, 16(1):30–35, 1981.
24. Hricak, H., Slovis, T. L., and Romanski, R. N. Neonatal kidneys-sonographic anatomic correlation. (*In preparation*).
25. Hricak, H., Toledo-Pereyra, L. H., Eyler, W. R., et al. The role of ultrasound in the diagnosis of kidney allograft rejection. *Radiology*, 132:667–672.
26. Hricak, H., Zayat, P., Romanski, R. N., et al. Sequence of sonographic changes following renal arterial occlusion-Experimental study. *Presented at 81st ARRS, San Francisco, 1981*.
27. Ishikawa, I., Saito, Y., Onouchi, Z., et al. Development of acquired cystic disease and adenocarcinoma of the kidney in glomerulonephritic chronic hemodialysis patients. *Clin. Nephrol.*, 14(1):1–6, 1980.
28. Jenis, E. H., and Lowenthol, D. T. *Kidney Biopsy Interpretation*. F. A. Davis, Philadelphia, 1977.
29. Kark, R. M. Renal biopsy. *JAMA*, 205(4):80–86, 1968.
30. Kay, C. J., Rosenfield, A. T., Taylor, K. J. W., et al. Ultrasonic characteristics of chronic atrophic pyelonephritis. *AJR*, 132:47–49, 1979.
31. Kerr, D. N. S. Acute renal failure. In: *Renal Disease*, edited by Sir Douglas Black and N. F. Jones, pp. 437–493, 1979.
32. Lee, H. M., Madge, G. E., Mendez-Picon, G., et al. Surgical complications in renal transplant recipients. *Surg. Clin. North Am.* 58(2):285–304, 1978.
33. LeQuesne, G. W. Assessment of glomerulonephritis in children by ultrasound. In: *Ultrasound in Medicine*, edited by D. White and E. A. Lyons, 1978. Plenum Press, New York.
34. Lewis, E., and Ritchie, W. G. M. A simple ultrasonic method for assessing renal size. *J. Clin. Ultrasound*, 8:417–420, 1980.

35. Maklad, N. F., Wright, C. H., Rosenthal, S. J. Gray scale ultrasonic appearances of renal transplant rejection. *Radiology*, 131:711–717, 1979.
36. Meier, J., Otto, R., and Binswanger, U. Die sonographie bei funktionseinschrankung von nierentransplantaten. *Fortschr. Rontgenstr.*, 134(2):142–147, 1981.
37. Merrill, J. P. Acute renal failure. In: *Diseases of the Kidney*, edited by M. B. Strauss and L. G. Welt, pp. 637–666, 1971. Little Brown and Co., Boston, Mass.
38. Moccia, W. A., Kaude, J. V., Wright, P. G., and Gaffney, E. F. Evaluation of chronic renal failure by digital gray-scale ultrasound. *Urol. Radiol.*, 2:1–7, 1980.
39. Moskowitz, P. S., Carroll, B. A., and McCoy, J. M. Ultrasonic renal volumetry in children. *Radiology*, 134:61–64, 1980.
40. Ralls, P. W., Colletti, P., Boger, D. C., et al. Ultrasonographic diagnosis of post-percutaneous renal biopsy hematoma. *Urol. Radiol.*, 2:23–24, 1980.
41. Rasmussen, S. N., Hasse, L., Kjeldsen, H., and Hancke, S. Determination of renal volume by ultrasound scanning. *J. Clin. Ultrasound*, 6(3):160–164, 1978.
42. Rochester, D., Aronson, A. J., Bowie, J. D., and Kunzmann, A. Ultrasonic appearance of acute poststreptococcal glomerulonephritis. *J. Clin. Ultrasound*, 6:1–72, 1978.
43. Rosenbaum, R., Hoffsten, P. E., Stanley, R. J., et al. Use of computerized tomography to diagnose complications of percutaneous renal biopsy. *Kidney Int.*, 14:87–92, 1978.
44. Rosenfield, A. T., and Siegel, N. J. Renal parenchymal disease: Histopathologic–sonographic correlation. *AJR*, 137:793–798, 1981.
45. Rosenfield, A. T., Taylor, K. J. W., Crade, M., et al. Anatomy and pathology of the kidney by gray-scale ultrasound. *Radiology*, 128:737–744, 1978.
46. Rosenfield, A. T., Taylor, K. J. W., and Jaffe, C. C. Clinical applications of ultrasound tissue characterization. *Radiol. Clin. North Am.*, 18(1):31–50, 1980.
47. Rosenfield, A. T., Zeman, R. K., Cronan, J. J., et al. Ultrasound in experimental and clinical renal vein thrombosis. *Radiology*, 136:735–741, 1980.
48. Schrier, R. W. Acute renal failure: Pathogenesis, diagnosis, and management. *Hosp. Pract.*, 7:93–112, 1981.
49. Shirkhoda, A. Gray-scale patterns and ultrasonic spectrum of adult polycystic kidney disease. *Appl. Radiol. Ultrasound*, 7:99–102, 1981.
50. Slotkin, E. A., and Madsen, P. O. Complication of renal biopsy: Incidence in 5,000 reported cases. *J. Urol.*, 87:13–15, 1962.
51. Talner, L. B., Scheible, W., Ellenbogen, P. H., et al. How accurate is ultrasonography in detecting hydronephrosis in azotemic patients? *Urol. Radiol.*, 3(1):1–6, 1981.
52. Taylor, K. J. W., Carpenter, D. A., Hill, C. R., et al. Gray-scale ultrasound imaging: The anatomy and pathology of the liver. *Radiology*, 119:415–423, 1976.
53. Thomsen, H. S., Madsen, J. K., Thaysen, J. H., et al. Volume of polycystic kidneys during reduction of renal function. *Urol. Radiol.*, 3(2):85–89, 1981.
54. Welt, L. Cited in Kark, R. M. Renal Biopsy. *JAMA*, 205:86, 1968.

Ultrasound Annual 1982,
edited by Roger C. Sanders.
Raven Press, New York © 1982.

Neonatal Intracranial Ultrasound

Thomas L. Slovis

*Departments of Radiology and Pediatrics, Wayne State University School of
Medicine and Children's Hospital of Michigan, Detroit, Michigan 48201*

Ultrasonic evaluation of intracranial contents originated in the late 1950s
with the A-mode (amplitude mode) technique (50,77,78). A-mode had drawbacks in that it was operator-dependent, time-consuming, and did not give
precise anatomical detail. Improved ultrasonic techniques became available
in the late 1960s and early 1970s with B-mode (brightness mode) sonography
(13). B-mode, soon to be followed by gray-scale sonography, gave better
anatomical detail but was still quite time-consuming and demanded that the
patient be brought to the X-ray department (5,35,40,41,43). With the development of the real-time sector scanners, at the end of the 1970s, a new era of
intracranial sonographic evaluation began. It is the advances with this technique that are described in this chapter.

SONOGRAPHY TECHNIQUE AND INSTRUMENT CONSIDERATIONS

The ideal approach for evaluation of intracranial contents involves availability of (a) portable equipment, (b) small transducers that can fit over the
anterior fontanelle, and (c) high-frequency transducers to obtain good resolution without loss of penetration. If these ideal criteria are to be approached,
one of the various real-time sector scanners must be utilized. There are
several types, and either a mechanical sector scanner or an electrical phased
array unit may be used (85). A mechanical sector scanner may be either a
single-element system, in the form of a "wobbler," or a multiple-element unit
on a spinning wheel. These units are portable and are of the type generally
used. The phased array units are more expensive and less mobile, although
equivalent detail can be obtained. It is most important that the transducer fit
comfortably on the fontanelle without bony overlap and that high-frequency
transducers are available with the equipment (10,18,51). Most examinations
are done with 5.0-megahertz (MHz) short-to-medium focus transducers, but
newer 6.0- and 7.5-MHz transducers may, in fact, prove more useful if
penetration is adequate. A 3.5-MHz transducer will not give optimal resolution in most premature babies; it is occasionally of use in a full-term infant or
in infants with small anterior fontanelles. Linear array transducers are usually
too large to fit comfortably over the fontanelle. Conventional, static, articu-

lated arm equipment can give superb views of intracranial contents but is limited by its lack of portability and the length of time it takes to do a complete examination.

Even with optimal equipment, it is important to achieve good contact between the infant's fontanelle and the transducer. If a neonate is on a radiant warmer, the head is generally quite accessible. However, when an infant is in a small isolette, it is difficult to manipulate any of these transducers. Therefore, the infant is moved to a position in which good contact can be obtained (Fig. 1) (73,74). This may require opening the isolette doors or moving the

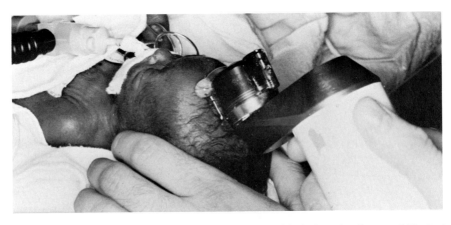

FIG. 1. Sonographic technique. **A:** The infant is covered, and the isolette door is opened. The bed is pulled toward the examiner so that optimal contact can be made between the transducer and the fontanelle.

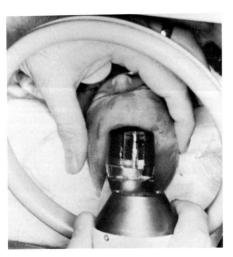

FIG. 1. B: Sonographic technique. This is a smaller premature infant, and adequate contact is made by working through the porthole.

infant toward the portholes for optimal examination. The lights should be dimmed in the nursery for optimal viewing of the TV monitor. The ultrasonic examination of the neonate is a team effort (30). Without the help of the nurses to position the baby and adjust the respirator or other life-support systems, the examination becomes much more difficult, dangerous, and perhaps impossible. The baby should be kept warm during the examination; if the isolette is opened, the infant should be covered. If, as is commonplace, one has to remove the oxygen hood to do a study, supplemental oxygen must be kept adjacent to the patient's nares. Despite these precautions, the various parameters of heart rate, respiratory rate, and oxygenation are altered when the baby is moved. Therefore, the examination should be rapid and accurate. It can be completed within less than 5 min.

Coronal and sagittal sections are obtained through the anterior fontanelle (Fig. 2) (73,74). Others have tried using the sutures and posterior fontanelle as means of seeing intracranial contents, but I do not think they have added any

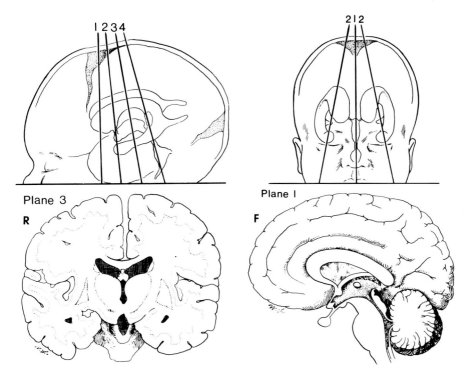

FIG. 2. Schematic of the planes of study. **Left:** The first four coronal sections through the anterior fontanelle are depicted. The patient's right (R) is always on the examiner's left. Whenever possible, a view of the occipital horns is added as well as an angled view of the parenchyma. **Right:** Sagittal study through the anterior fontanelle. The position of the transducer is the midline and parasagittal planes are presented. The frontal horns (F) are always directed to the examiner's left. (From ref. 73, with permission).

additional information (7,8). In the past, the axial view has been performed, but the resolution in axial scanning is considerably less than the anterior fontanelle approach.

The examination is recorded on videotape in addition to either Polaroid or radiographic hard copy. It is important to have a permanent record of some type for review by the neonatalogist, the radiologist, and the neurosurgeon. I feel that with the use of the videotape, a well-trained sonographic technician can perform the examination. The ultrasonic physician then can review the tapes for the "sweep" (caudate, choroid, etc.; *see below*) to obtain the 2-D image. I find this procedure tedious and prefer to be present and frequently do the examination. It is, after all, the "sweep" that gives the most information; static images alone should not be considered optimal for reading the examination. Many institutions use videotape only to record the examination for review at the time of the dictation. I prefer to keep the tape examination for future review as more information becomes available. Many tapes are now 60 to 120 min in length, so storage is less of a problem than with the initial small, 20-min tapes. Videotape recorders may be either ½-inch or ¾-inch, of any of the accepted formats. The ½-inch, however, is smaller and more portable. It is quite useful to have a good freeze-frame apparatus and digital recall.

ANATOMICAL CONSIDERATIONS

The infant is studied in two planes: the coronal and the sagittal (Fig. 2). Although in describing the planes the ventricles are used as the primary landmarks, it is essential to remember that it is the state of the parenchyma that will determine the patient's status and neurological development. The coronal section is studied initially. The patient's right is placed on the screen to the examiner's left as if the patient were facing him. The transducer is placed on the anterior fontanelle and angled from the most anterior aspect of the brain back until the frontal horns are visualized (Fig. 3). The ventricles are sonolucent, the pia-arachnoid echogenic, and the brain tissue has specific echogenic characteristics (architecture). The brain region adjacent to the lateral aspect of the frontal horn is the caudate nucleus. These caudate nuclei are slightly more echogenic than the frontal cortex and clearly separable from the finer textured and less echogenic thalamus. On this first section, the anterior longitudinal cerebral fissure, the cingulate sulcus, and the corpus callosum all can be visualized as echogenic structures (11).

The second coronal plane is obtained by angling the transducer posteriorly until the foramen of Monro is visualized (Fig. 4). At this point, the posterior aspect of the frontal horns and the anterior aspect of the body of the lateral ventricles can be seen. The choroid plexus on the floor of the lateral ventricles is noted to enter the foramen of Monro (28). The caudate nuclei are still visible along the lateral aspects of the frontal horns, and the thalamus is now seen

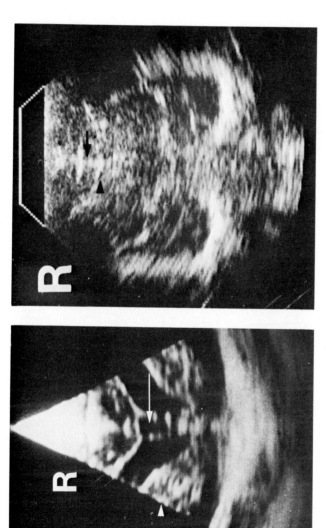

FIG. 3. Coronal plane at the level of the frontal horn of the lateral ventricle. **Left:** A child with dilated ventricles who shows both the dilated frontal horns and a cavum septi pellucidi (*arrow*). The echogenic caudate nucleus (*arrowhead*) is seen along the lateral ventricular wall. **Right:** A normal premature infant in whom the anterior longitudinal fissure (*arrow*) and cingulate sulcus are seen. The *arrowhead* is on the very small, barely perceptible lateral ventricle.

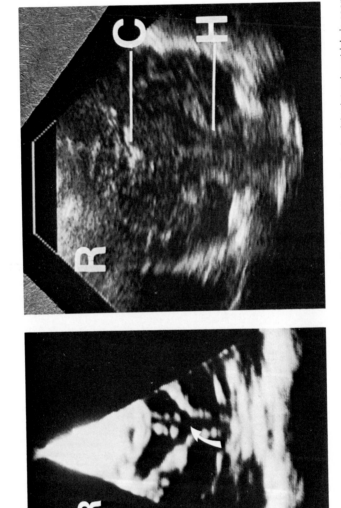

FIG. 4. Coronal plane at the level of the foramen of Monro. **Left:** In a child with dilated ventricles, the body of the lateral ventricle is seen as it enters the foramen of Monro (*arrow*). Again the cavum septi pellucidi is noted. The near-gain is too high, so that the caudate nucleus appears sonolucent. This is a technical artifact. **Right:** In a normal child, it is difficult to see the slit-like ventricles. The density at the base of the ventricle (C) is the choroid plexus as it merges into the foramen of Monro. The curvilinear structure inferiorly is the hippocampal sulcus (H) of the temporal lobe.

NEONATAL INTRACRANIAL ULTRASOUND

inferior to the caudate nuclei. The hippocampal sulcus of the temporal lobe is seen. It is recognized by the curvilinear echogenicity of the pia and arachnoid. The temporal horns of the lateral ventricles are not seen unless enlarged.

Since the ventricles are normally slit-like, the septum pellucidum is not always seen. However, in many premature infants, there is a fluid-filled cavum septi pellucidi (26). In premature infants with normal ventricles, the cavum septi pellucidi is much more prominent than the lateral ventricles.

In this plane particularly, many pulsating blood vessels are easily noted on the TV monitor or videotape (33,74):

(a) The middle cerebral artery is seen twice, once near its origin medial to the hippocampal sulcus and again in the sylvian fissure.

(b) The pericallosal artery (of the anterior cerebral) is noted above the corpus callosum. In the first coronal plane, the anterior cerebral artery is seen below the genu of the corpus callosum. The posterior cerebral arteries are seen just above the tentorium in the third plane.

The third coronal plane is obtained by angling the transducer posteriorly once again and visualizing the body of the lateral ventricles (Fig. 5). The thalamus is clearly seen medially, and *if* the third ventricle is *enlarged*, it can be seen in the midline. The temporal *lobes* and hippocampal sulci are again seen with a bright linear echo (? middle temporal sulcus). The sylvian fissure is visualized laterally, and the tentorium and posterior fossa are seen inferiorly.

The fourth coronal plane is obtained by angling the transducer more posteriorly and visualizing the glomus of the choroid plexus (Fig. 6). On this scan, the parietal region on each side can be seen. Occasionally a fluid-filled structure is seen medial to the glomus of the choroid plexus. It is a normal structure that may communicate with the cavum septi pellucid, and is called the cavum Vergae (8). The quadrigeminal plate cistern may also be visualized in this region (73). If the fontanelle is large enough, an additional plane is obtained even more posteriorly to see the occipital ventricle.

A fifth coronal plane has been added to see the lateral brain parenchyma (Fig. 7). The transducer is angled to each side and gently rocked from front to back so that the parenchyma on each side of the calvarium is visualized. This projection also gives another view of the caudate nuclei and the choroid plexus, which is important when one is trying to diagnose intracranial hemorrhage.

Remember, these are the idealized planes sought in every patient. In those patients with mild ventricular enlargement, all of these structures easily can be seen. In the normal infant, particularly the premature infant, the ventricles are slit-like and difficult to see. The major identifiable structures are the cavum septi pellucidi, the hippocampal sulci, the tentorium, posterior fossa, and the glomus of the choroid plexus.

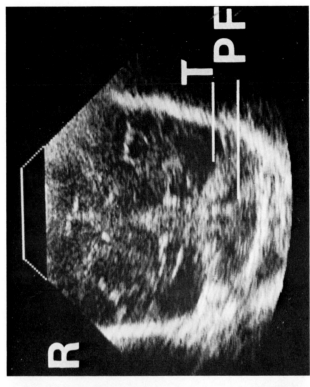

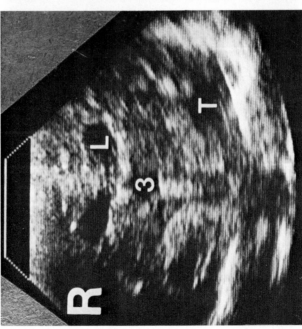

FIG. 5. Coronal plane at the level of the body of the lateral ventricle. **Left:** In a child with dilated ventricles, the body of the lateral ventricle (L), the third ventricle (3), and the mildly dilated temporal horn (T) are easily visible. **Right:** In the normal child, it is difficult to see the lateral ventricle, and the third ventricle is never seen on coronal sections. Rather, at this level, the tentorium (T) and the posterior fossa (PF) are noted.

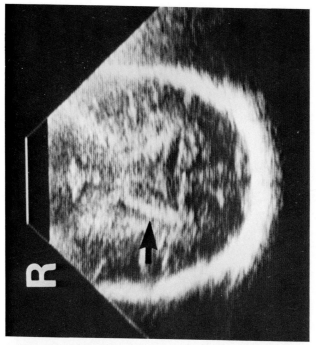

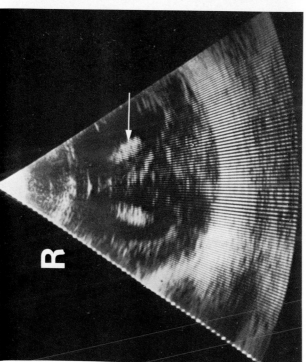

FIG. 6. Coronal planes through the choroid plexus. **Left:** In a child with dilated ventricles, the homogenous glomus of the choroid plexus (*arrow*) and the enlarged anechoic ventricles surrounding it are identified. The glomus of the choroid plexus is attached medially and does not float in the ventricles. **Right:** In a normal child, the major structure seen in this plane is the echogenic choroid plexus (*arrow*).

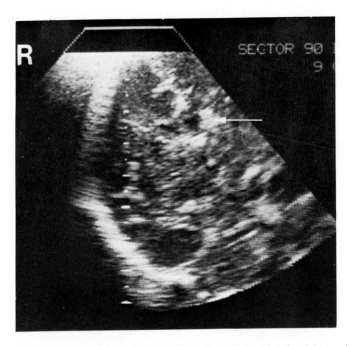

FIG. 7. An angled parenchymal view. The transducer is angled so that the right cerebral parenchyma is seen. In addition, another view is obtained of the caudate nucleus, which in this child is dense on the left (*arrow*) and represents a caudate nucleus bleed.

On sagittal sections, three planes are routinely obtained (Fig. 2). The patient's frontal lobe is always to the viewer's left. The transducer is first placed in the midline, and midline structures are noted (Fig. 8). This means that the cavum septi pellucidi, when present, the foramen of Monro, and the third ventricle with the anterior ophthalmic and pituitary recesses and posterior pineal and habenular recesses are evaluated. The aqueduct is seen, and the echogenic cerebellum is noted. The fastigium of the fourth ventricle is seen as a triangular sonolucent structure immediately below the echogenic cerebellum. Because of the position of the transducer, the floor of the fourth ventricle is rarely seen as a straight line. The anechoic region, which is the fourth ventricle, is between the echogenic cerebellum posteriorly and brain stem anteriorly. The basilar artery is seen anterior to the pons, and in this particular view, the internal carotid artery and the anterior and middle cerebral arteries are noted pulsating. Behind the cerebellum, the cisterna magna can frequently be seen.

The transducer is then angled toward each side (Fig. 9). The frontal horn, the body of the lateral ventricle, and glomus of the choroid plexus are seen. The temporal horn is frequently seen. Because of the posterior angulation, the large size of the transducer, and the small size of the fontanelle, the occipital horn is frequently *not* seen (26). Anterior and superior to the thalamus is the

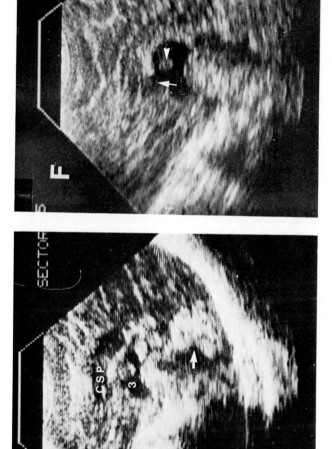

FIG. 8. Midline evaluation in the sagittal plane. **Left:** In this normal child, the cavum septi pellucidi (CSP) is seen above the third ventricle (3). Both the anterior and posterior recesses are suggested. The massa intermedia is seen directly behind the numeral 3. The aqueduct is a small gently curving lucency extending posteriorly and inferiorly from the third ventricle to the fourth ventricle. The festigium of the fourth ventricle (arrow) is seen immediately below the very echogenic cerebellum. **Right:** In another normal child, the foramen of Monro (arrow) and massa intermedia (arrowhead) are readily identified. Again, the aqueduct, fourth ventricle, and cerebellum are seen.

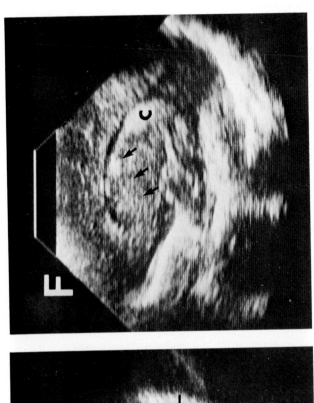

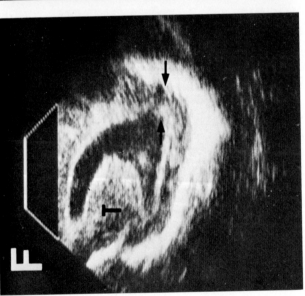

FIG. 9. Sagittal view off the midline. **Left:** In a child with enlarged ventricles, the frontal horn, body, temporal horn, and occipital horn of the lateral ventricle are easily identified. The thalamus (T) and the echogenic glomus of the choroid plexus stand out. There is decreased occipital cortical mantle (*arrow*). **Right:** A normal child in this plane reveals the echogenic glomus of the choroid plexus (C), the thalamus directly anterior to the glomus, and the caudate nucleus superior and anterior to the thalamus (*arrow*). The ventricles are slit-like but are easier to see than on coronal sections.

caudate nucleus. The thalamus is less echogenic than the caudate. The brain parenchyma is then noted on each side lateral to the ventricular system. In addition, the frontal and occipital cortical mantle are measured by observing the distance anterior to the frontal horn and the distance posterior to the occipital horn of the lateral ventricle.

In normal neonates, particularly premature ones, the occipital horns are seen infrequently (less than 15% of the time) (76). While the ventricles are still slit-like, they are easier to see in the sagittal sections than in the coronal sections. There is little difficulty in viewing the midline structures and seeing the posterior fossa in either premature or full-term babies.

There has been little anatomical confirmation of the ultrasonic findings described. For this reason, a rudimentary attempt was made to inject various ventricular and extraventricular structures to confirm their identity. Under ultrasonic guidance in cadavers, large-bore needles were placed either through the anterior fontanelle or through cisternal taps into the appropriate structures, and a combination of esophotrast and silicone was injected. Radiographs were obtained, and casts were made of the ventricles and of the extraventricular structures. These studies provide a firm anatomical basis for the description of the structures presented (Figs. 10–12). The normal ventricle is difficult to see and should not be confused with abnormal ventricular enlargement. It has been well documented that when infants have ventricular enlargement the occipital horn is the first to enlarge (27,39,71). For this reason, although there are multiple references denoting ratios and sizes of ventricles, one can conclude that if the occipital horns are not seen, the ventricles, except under very unusual circumstances, are not enlarged (32,40,41,53,76). As a routine, I do not measure ventricular size. If the occipital horn is seen and the occipital cortical mantle is equal to or greater than two-thirds the size of the frontal cortical mantle, this portion of the ventricle is still considered to be within normal limits (76). Change in the shape of the medial aspect of the body of the lateral ventricles has been described as an early sign of ventricular dilatation (27). However, in our experience, this happens only after occipital enlargement has occurred.

NEONATAL INTRACRANIAL HEMORRHAGE

The real impetus for intracranial evaluation in neonates came with the initial pathological studies followed by computerized tomographic (CT) recognition of the high incidence of intracranial hemorrhage in premature infants (14,15,44,46,49,57,58,61). Since the majority of these infants do not initially have any clinical symptoms referrable to the intracranial hemorrhage, an imaging procedure is mandatory in high-risk groups (47). Knowledge of the predisposing factors, as well as the pathogenesis of the different kinds of

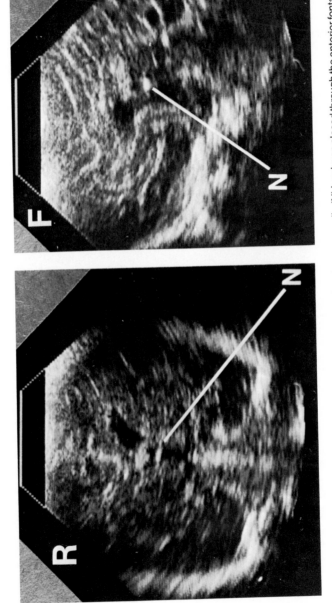

FIG. 10. Postmortem evaluation of the third ventricle. **Left:** Coronal section shows a needle (N) has been placed through the anterior fontanelle and into the third ventricle by ultrasonic guidance. Note the dilated left frontal horn. **Right:** Sagittal section confirms the presence of the needle (N) in the third ventricle.

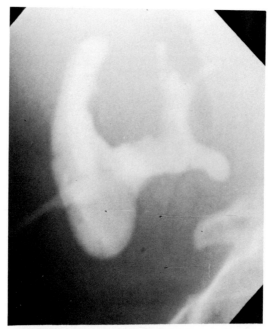

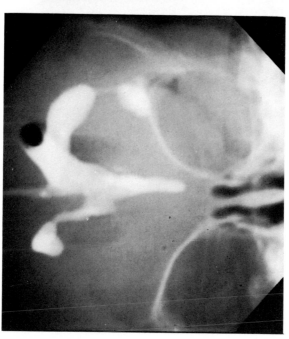

FIG. 10. continued. **Left:** Silicone (TRV-U by General Electric) mixed 3:1 with esophotrast was then injected, and this frontal radiograph was obtained. The tip of the catheter is within the third ventricle, and contrast fills the third ventricle and also spills into the lateral ventricles. The dilated left frontal horn is again seen. **Right:** A lateral view confirms the presence of the catheter and shows the third ventricle as well as the aqueduct.

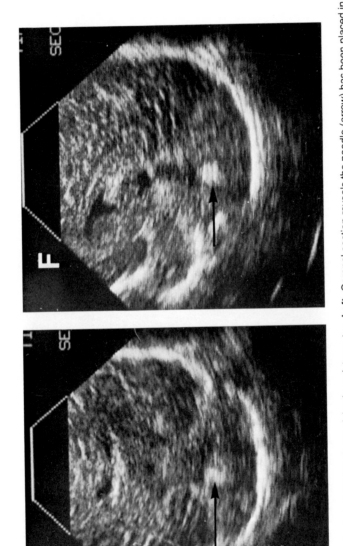

FIG. 11. Postmortem evaluation of the fourth ventricle via a cisterna tap. **Left:** Coronal section reveals the needle (*arrow*) has been placed in the fourth ventricle under ultrasonic control. **Right:** Sagittal midline section confirms this placement (*arrow*).

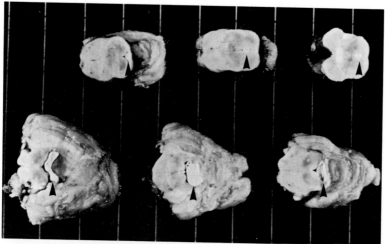

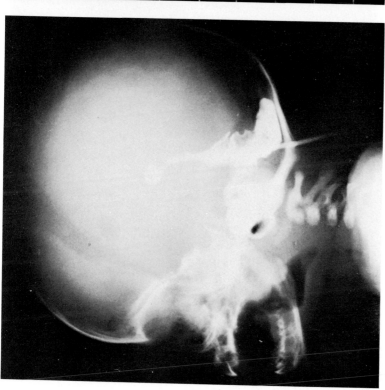

FIG. 11. continued. **Left:** Lateral radiograph taken after injection of the silicone-esophotrast mixture reveals filling of the fourth ventricle, the aqueduct, and the posterior aspect of the third ventricle. There is some spillover into the cisterna magna. **Right:** Examination of the specimen reveals contrast within the ventricular system (*arrowheads*).

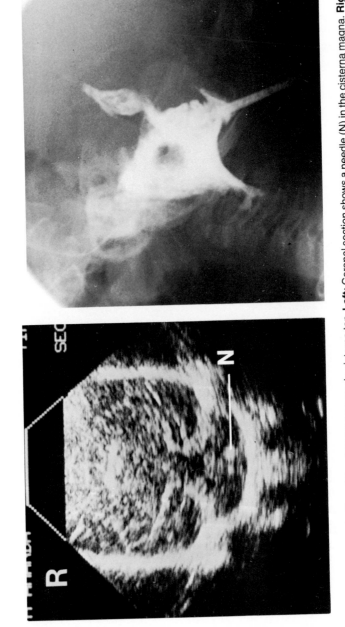

FIG. 12. Postmortem examination of the cisterna magna via cisterna tap. **Left:** Coronal section shows a needle (N) in the cisterna magna. **Right:** Lateral radiograph after injection of contrast reveals filling of the cisterna magna and supracerebellar cisterns. There is some spill into the spinal subarachnoid space.

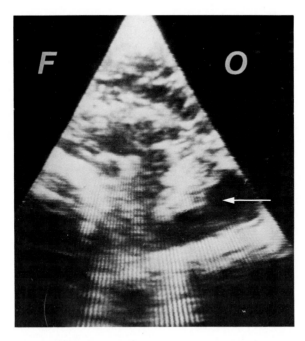

FIG. 12. continued. In another child, a sagittal view demonstrates the echogenic cerebellum and the enlarged cisterna magna posteriorly (*arrow*).

bleed, is crucial to charting a logical approach to treatment of these neonates (22,80–84).

Volpe (84) lists four types of intracranial bleeding: (a) subdural, (b) primary subarachnoid, (c) intracerebellar, and (d) periventricular-intraventricular (PVH-IVH). The first two types, subdural and primary subarachnoid, are more often found in full-term babies with a history of obstetrical trauma. The remaining types of bleeding are more often found in premature infants and relate to "hypoxic-ischemic events." It is fair to say that the first three forms of intracranial hemorrhage are of relatively minor concern compared with the large volume of PVH-IVH in prematures. A subdural hematoma has not been detected in our neonatal unit in 3 years, although the water bath technique is quite capable of diagnosing such lesions (75). Primary subarachnoid bleeding is difficult to detect with current ultrasonic techniques and appears to have little clinical significance. This will be discussed at the end of this section when evaluation of full-term infants is considered. An intracerebellar hemorrhage is rare and generally occurs in conjunction with PVH-IVH and not as an isolated finding (20,60). There have been few reported CT diagnoses of cerebellar lesions and even fewer sonographic diagnoses; most intracerebellar hemorrhages have been detected at autopsy.

The majority of neonatal intracranial hemorrhages that can be detected by the ultrasonic technique are of the PVH-IVH type and occur in premature infants less than 32 weeks gestation (there is an incidence of 40 to 50% in this high-risk group; these infants usually weigh less than 1,500 g) (1,3,36,56,86). These neonates have an anatomical predisposition in that the supporting glial tissue and type of cerebral vascularity differ from those in the older premature and term infant. The basal ganglion develops well before the cortex and the cerebral white matter (86). Glial cells are found above the basal ganglion, specifically above the caudate nucleus, and below the ependymal portion of the lateral ventricle—this is the subependymal germinal matrix. Since the basal ganglion is so large in the preterm infant, its vascular supply is greater than in the rest of the hemisphere. These vessels are supported by this subependymal germinal matrix. The arterial supply is from Huebner's artery, a branch of the anterior cerebral artery, the lateral striate arteries (from the middle cerebral artery), and the anterior choroidal artery (from the internal carotid artery). This abundant arterial supply feeds a heterogeneous capillary bed (the microcirculation) within the germinal matrix that "by light microscopic features cannot be categorized as arterial, capillaries or venules" (84). This microcirculation has been called "a persistent, immature vascular rete in the subependymal germinal matrix" that is remodelled into a definite capillary bed only when the germinal matrix disappears between 32 weeks and term (86). The venous drainage from this region differs from that in the term infant in that "a major flow of blood changes direction sharply at that junction"—a circuitous course of the choroidal, thalamostriate, and the terminal vein form the internal cerebral vein (84).

Between 32 weeks and term, the subependymal germinal matrix disappears, the vascularity of the cerebral cortex increases, and the cerebral cortex predominates over the basal ganglion region; a more mature capillary system is formed. In infants of less than 32 weeks in whom this anatomical predisposition exists, it is not hard to understand why "ischemic-hypoxic events" precipitate bleeding in the microcirculation of the germinal matrix in the region of the caudate nucleus. In addition, stressed infants between 32 and 35 weeks are also at high risk for PVH-IVH, as they are compromised during their transition period from the immature to mature microcirculation.

Ischemia and hypoxia alter the permeability of these susceptible "rete" and cause hemorrhage. In addition, once the vessel has been damaged, it cannot respond appropriately to the stresses of hypo- or hypertension. This loss of *autoregulation* compounds the problem (84).

Several other extrauterine stresses have been shown to correlate with PVH-IVH. Certainly hyaline membrane disease with hypoxia and elevated P_{CO_2} accentuates the risk. An infant with a pneumothorax may have a greater chance for a PVH-IVH, as a pneumothorax causes hypoxia, hypercapnia, and instability of the intracerebral vascular contents (22,52). A second extrauterine stress that is common in these infants, particularly infants with hyaline membrane disease, is the opening of a left-to-right shunt, such as a patent ductus arteriosus. This causes decreased cerebral circulation and may alter the autoregulatory mechanism of the premature vessels (22). Finally, a predisposing factor may be maternal ingestion of aspirin during the last weeks of pregnancy (63).

The initial bleeding, then, is in the microcirculation of the germinal matrix adjacent to the head of the caudate nucleus and the foramen of Monro or in the choroid plexus (19). The bleeding, however, does not necessarily remain contained in these locations; in up to 80% of these infants bleeding will extend into the ventricles. When the insult is severe enough, the entire ventricular system is filled with blood (a cast of the ventricles), and there may be further extension of hemorrhage into the cerebral parenchyma. The ventricles may enlarge, and posthemorrhagic hydrocephalus can occur. This is a dynamic, ongoing process.

The timing of the initial bleed has been well documented; the majority occur within the first 72 hr of life (62). Few babies have intrauterine PVH-IVH, and it is very rare to have the initial PVH-IVH after 2 weeks of age.

The sonographic appearance of blood is that of echogenic material. The caudate nucleus itself appears echogenic because of bleeding into the subependymal germinal matrix (Fig. 13). There is not only increased echogenicity but also increased size of the caudate; the germinal matrix-caudate nucleus complex enlarges and becomes convex toward the ventricle. In some infants, the bleeding is about the choroid plexus on the floor of the lateral ventricle. This can be confusing unless it is noted that the thickest portion of the choroid is at the glomus. It thins out as it approaches the foramen of Monro. Because the echogenic choroid may be confused with hemorrhage, it is essential that the bleeding be identified in both coronal and sagittal sections. When blood extends into the ventricle, it is again echogenic, although nonhomogeneous. It does not adhere to any specific anatomical location (so it should not be confused with a homogeneous, medially attached glomus of the choroid plexus). Blood is inhomogeneous and, in fact, may give a halo appearance as the clot retracts (Fig. 14) (24). The ventricles may initially be of normal size and then enlarge. An adhesive, basilar arachoiditis causing an obstruction to the outlet of the fourth ventricle has been postulated as a cause of posthemorrhagic hydrocephalus (46). In our experience, few cases have enlargement of the third and fourth ventricles. In fact, frank ventricular dilatation may be secondary to the increased ventricular pressure at the time of hemorrhage or stasis of cerebrospinal fluid (CSF) within the ventricular system itself as a result of sludging. Another possibility is the altered production or resorption of CSF.

When the insult is severe enough, blood forms a cast of the ventricular system (Fig. 15A and B). In these cases, it is frequently noted that the brain parenchyma itself is poorly echogenic. This correlates well with the autopsy findings of severe cerebral edema. The most severe form of hemorrhage is

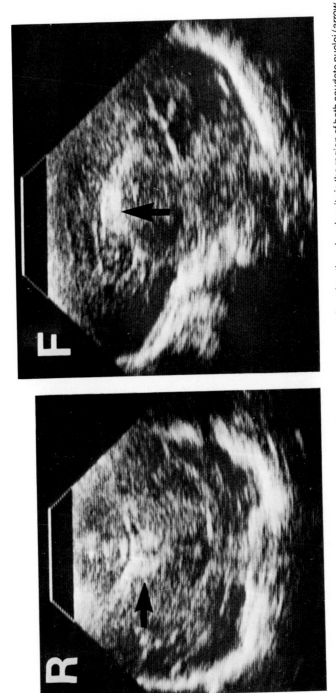

FIG. 13. Bleeding into the microcirculation of the germinal matrix. **Left:** On this coronal section, there is a density in the region of both caudate nuclei (*arrow on right*). The ventricles are normal and slit-like. **Right:** A sagittal section shows the increased echogenicity and increased thickness of "the caudate nucleus." (From ref. 69, with permission.)

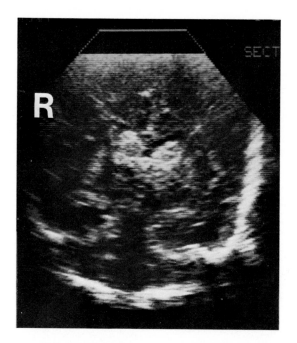

FIG. 13. continued. Another preterm infant with bilateral subependymal germinal matrix bleeding. Note how convex the lesion is toward the ventricles. To make the diagnosis of subependymal germinal matrix bleeding, it is mandatory to see the bleed in two views.

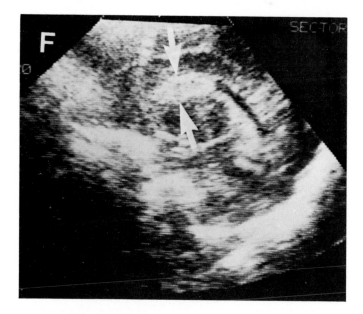

FIG. 13. continued. Sagittal section off the midline reveals the bleed on the caudate nucleus (*arrows*) and normal thalamus and choroid.

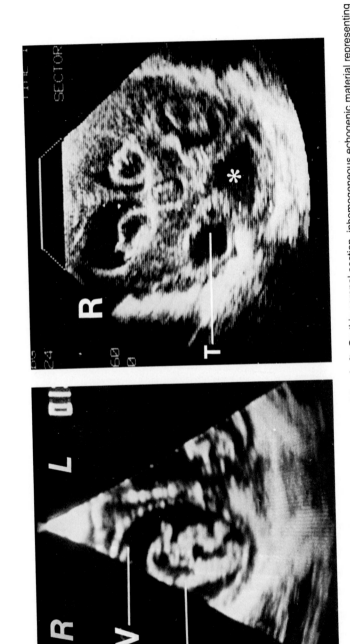

FIG. 14. Bleeding into the ventricular system with enlarged ventricles. **Left:** On this coronal section, inhomogeneous echogenic material representing retracting clot (B) is seen floating within the enlarged ventricular system (V). The centers of these densities are sonolucent—a halo effect. **Right:** Coronal section of another child with blood in enlarged frontal and temporal horns as well as in the third ventricle. The third ventricle is enlarged, so it is visualized on coronal section. The fourth ventricle (star) is also enlarged. T, Temporal horn.

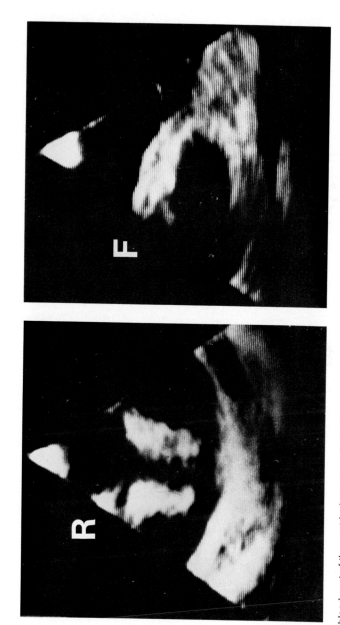

FIG. 15. A blood cast of the ventricular system. **Left:** Coronal section through the frontal horn shows echogenic material filling the entire aspect of the ventricles. There is cerebral edema (proven at autopsy), so the rest of the brain is poorly echogenic. **Right:** Sagittal section shows blood filling the lateral ventricle.

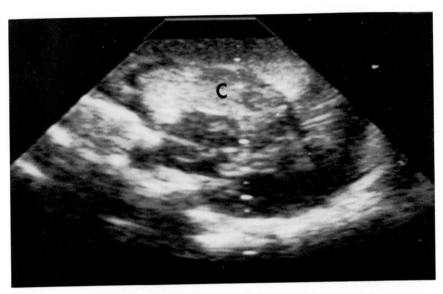

FIG. 15. continued. Sagittal section of another child with cast of blood (C) filling entire ventricle.

that of parenchymal extension of the bleeding (Fig. 16). In this case, the parenchyma of the brain becomes very echogenic in a nonuniform manner with irregular borders (34,69). This echogenicity is most often seen in the parietal region, in contrast to those cases where there appears to be increased echogenicity behind the junction of the body of the lateral ventricle and the occipital horn of the enlarged ventricles. This pattern is homogeneous—like through transmission—and may be merely a technical accentuation of echoes.

The bleeding has been classified in various ways. The pioneer work of Papile et al. (58) is based on initial CT scans done during the first week of life. Subsequently, CT and sonographic classifications have been presented, based on either location of the bleed, size of the ventricles, or amount of blood within the ventricles (Table 1) (54,58,71). However, these classifications have two shortcomings when one tries to adapt them to the sonographic evaluation. First, they are based on a single study, and, as described above, this is a dynamic process occurring over a 2-week period. In the second instance, most are based on the axial CT study and not on the ultrasonic examination in the coronal and sagittal planes. After 7 days, blood may become isodense and then CT is not as accurate as a sonogram. There has been only one classification based on sonographic criteria; this is not a prognostic classification (9).

In contrast, Shankaran et al. (69) have proposed a sonographic classification based on the most severe form of hemorrhage reached by 2 weeks of age (since almost all initial bleeding and initial ventricular enlargement occurs by this time). These infants were then followed sonographically and neu-

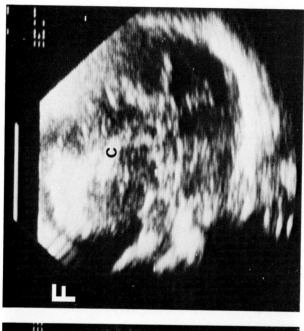

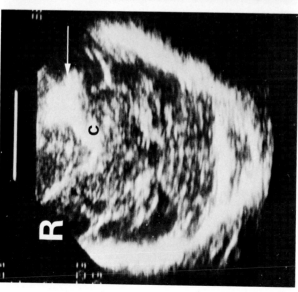

FIG. 16. Cerebral extension of intracranial bleeding and subsequent porencephaly. **Left:** Coronal section at the level of the body of the lateral ventricle reveals dense echogenic material in the region of the caudate nucleus (C), involving the lateral ventricle and extending in a nonuniform, irregular manner into the parenchyma (*arrow*). **Right:** Sagittal section at the same time reveals the dense parenchymal extension as well as the dense caudate bleed (C). The ventricle is full of blood and is not seen as an isolated structure.

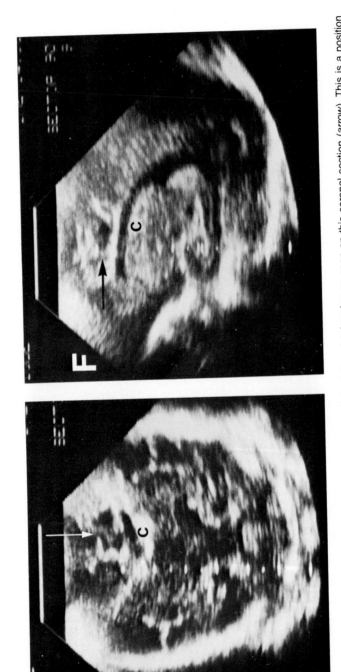

FIG. 16. continued. **Left:** Four weeks later, there is liquefaction of the left parietal region as seen on this coronal section (*arrow*). This is a position identical to the parenchymal extension. The ventricular enlargement is now visualized, and there remains density of the left caudate nucleus (C). **Right:** Sagittal section shows the liquefaction of the brain parenchyma (*arrow*) and also the caudate nucleus bleed (C).

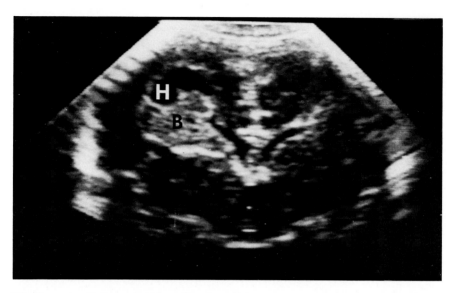

FIG. 16. continued. Coronal section of another child with parenchymal bleed (B), with a porencephalic cyst developing at its peripheral edge (H).

rologically up to 18 months of age. The classification correlates morbidity, mortality, and short-term neurologic outlook (Table 2). What is fascinating is that of the infants in the moderate and severe group, persistent ventricular dilatation did not correlate statistically with neurological deficits. Once again this points out the fact that although we look at ventricles as landmarks of various planes, it is the parenchyma that is the most important determinant of neurological outcome. When the most severe form of bleeding with parenchymal extension of the hemorrhage occurs, it is invariably associated with necrosis of the cerebral parenchyma and the formation of porencephalic dilatation of the ventricles developing between 2 and 6 weeks postbleed (Figs. 16 and 17) (66). Porencephaly has been well shown to have the highest incidence of neurological sequelae (45).

Several severely affected infants have had a different parenchymal sequela: liquefaction or at least destruction of the parenchyma in areas where bleeding is never detected. These "holes" are in the vascular distribution of the middle cerebral artery (Fig. 18). From the early ultrasonic studies, it is difficult to anticipate this rare phenomenon if there is no bleeding or change of the parenchyma. In the two instances where we have seen this kind of parenchymal destruction, the first clue was seen on the echogram at 4 to 6 weeks following birth.

The efficacy of the ultrasonic diagnosis of parenchymal lesions, except for the extension of bleeding and subsequent porencephaly, has not been investigated. On the other hand, CT investigation of "periventricular leukomalacia"

TABLE 1. *Classification of PVH-IVH*

Division	Imaging procedure	Prognostic correlation	Reference
I Subependymal germinal matrix	CT	Survival, persistent ventricular dilatation	Papile et al. (58)
II Intraventricular without enlargement			
III Intraventricular with enlargement			
IV Intraventricular with parenchymal extension			
I Subependymal and/or 10% ventricle with blood	CT	Survival, hydrocephalus	Montavani et al. (54)
II 10–50% Ventricular area with blood			
III >50% Ventricular area with blood			
SEH (subependymal)	CT	Survival, persistent ventricular dilatation	Silverboard et al. (71)
Mild IVH: blood in ¼ or less of ventricles, 3rd and 4th normal			
Moderate IVH: blood in ¼–½ ventricles			
Severe IVH: blood in >½ ventricles			
Mild: SEH or IVH without ventricular dilation	Ultrasound	—	Bejar et al. (9)
Major: IVH with dense echoes and/or ventricular dilation			

TABLE 2. Dynamic prognostic sonographic classification[a]

Divisions	Mortality as related to intracranial hemorrhage (%)	Posthemorrhagic hydrocephalus in survivors (%)	Persistent ventricular dilatation (%)	Short-term[c] neurologic abnormality (%)
Mild	0	0	0	0
Moderate	27	79	47	53
Severe				
Cast	75	100[d]	67	33
Parenchymal extension	43	100[b]	100	71

[a]Modified from ref. 69.
[b]100% Porencephaly
[c]Neurological defects include spastic diplegis, quadriplegia, hemiplegia, some increase in muscle tone.
[d]Only three patients in the group.

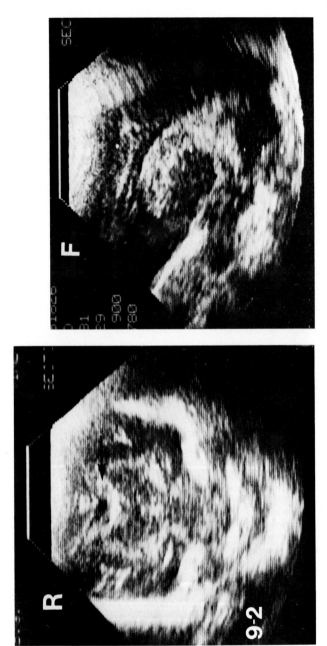

FIG. 17. Sequence of events in a premature with PVH-IVH with progression and subsequent ventricular shunting. **Left:** Coronal section on day of birth reveals bilateral caudate nucleus bleeding (*arrow*) with no ventricular enlargement. The cavum septi pellucidi is seen. **Right:** At 6 days of age, the bleeding has extended into the ventricles and forms a cast of the ventricles, as seen on this sagittal section.

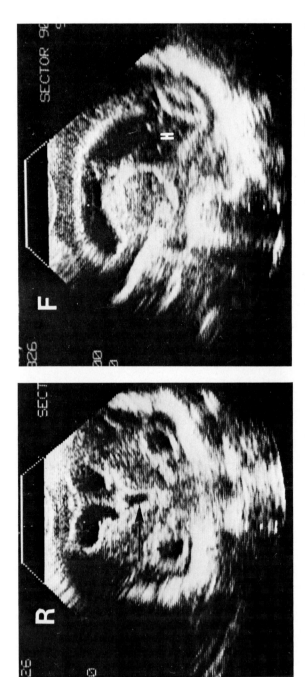

FIG. 17. continued. **Left:** By 20 days of age, the blood in the ventricles has been resorbed to a great extent, but there remain bilateral caudate nucleus densities and marked ventricular enlargement as seen on this coronal section just beyond the foramen of Monro. The third ventricle (*arrow*) is seen and therefore enlarged. **Right:** Sagittal section of the midline at the same time reveals blood in the occipital horn (H) as well as a dense caudate nucleus bleed.

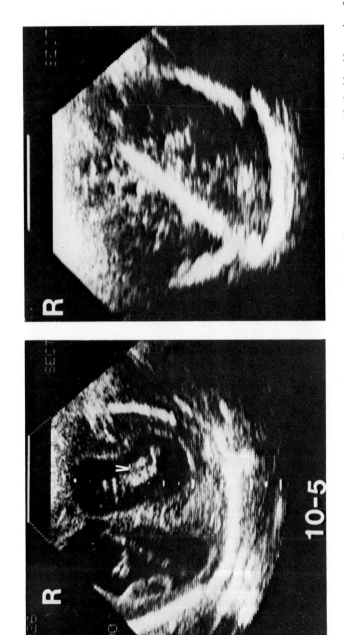

FIG. 17. continued. **Left:** By 1 month of age, there are septations within the ventricles and the halo-like appearance of intraventricular blood (*arrowhead*) on this coronal section. The ventricles continue to enlarge despite repeated lumbar punctures. **Right:** A shunt was placed as demonstrated on this coronal section, and the ventricles decreased rapidly to normal size.

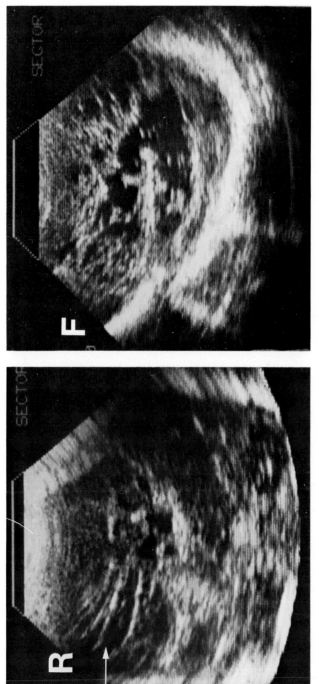

FIG. 18. Parenchymal lucencies at 1 month of age in the distribution of the middle cerebral artery. **Left:** Coronal section of the body of the lateral ventricles reveals multiple sonolucent areas in the right parietal hemisphere (*arrow*). **Right:** Sagittal section in this region reveals the liquefaction of the brain parenchyma.

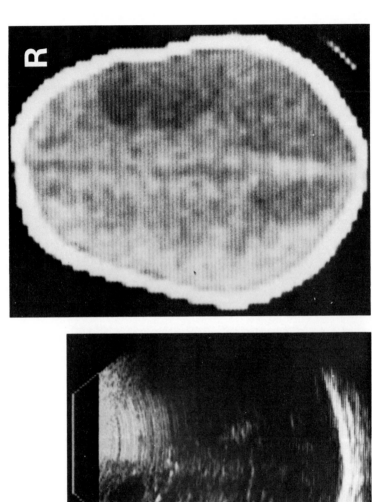

FIG. 18. continued. **Left:** Coronal section posteriorly in the occipital hemisphere reveals further sonolucent areas. **Right:** CT study shows the distribution of this disease in the region of the middle cerebral artery on the **right**. This most likely represents an infarction in the distribution of this vessel.

or parenchymal hypodense areas has been extensive (2,12,25,29,31,67). There is considerable debate with regard to the significance of these findings in a premature infant and little correlation with either neurological sequelae or autopsy results. If, however, with improved resolution of CT, ischemic lesions of the brain with prognostic significance are recognizable, the sonographic examination *alone* will not suffice. We have not reached this stage in premature infants, but in term babies the role of CT is quite different (*see below*).

The sonographic and CT findings of ventricular dilatation occur several weeks before actual increase in the infant's head circumference (42,79,84). This is because there is an abundant amount of water in the cerebral white matter and very little myelin. After the insult, the ventricles start to dilate for any of the reasons described above, and the water is "squeezed out of the brain." Because this is a pliable system, and there is not much myelin, the brain tissue is compressed, and it is not until late in the course, approximately 2 weeks after ventricular dilatation, that the head circumference starts to increase. It is for these reasons that an imaging procedure is necessary in the high-risk group, as the clinical anticipation of PVH-IVH and its sequelae is not sensitive. Ventricular enlargement may occur any time during the first 2 weeks. Progression on two examinations is defined as posthemorrhagic hydrocephalus. However, spontaneous cessation and eventual return to normal ventricular size frequently occurs (Fig. 19). If the ventricles continue to enlarge, intervention may be necessary. At this juncture, the indications for intervention and the efficacy of various interventional procedures have not been subject to large, controlled studies (54,59). In fact, only now are control studies in progress to determine if drug intervention at birth can prevent the occurrence of bleeding (17,21).

For all these reasons, the following scheme for performance of ultrasonic evaluation in premature infants for detection of bleeding and hydrocephalus is suggested:

(a) An initial scan, sometime between 3 and 5 days. This will detect virtually all hemorrhages in the high-risk group.

(b) If the child is normal on the first examination, a second study may be performed at 2 weeks of age. This is an optional examination but certainly allows one to declare the child normal in terms of this complication with a great deal of certainty.

(c) If a bleed is detected on the first examination, studies are conducted every 3 days until 2 weeks of age. If there has been no ventricular enlargement at this point, the scans can be done intermittently as progression of hemorrhage and initial ventricular enlargement occur before 2 weeks of age in the vast majority of infants. Subependymal bleeds may remain recognizable sonographically from weeks to several months before resolving.

(d) When ventricular enlargement is found, scans are done every 3 days to 2 weeks of age, then weekly until the ventricles stabilize. If therapeutic

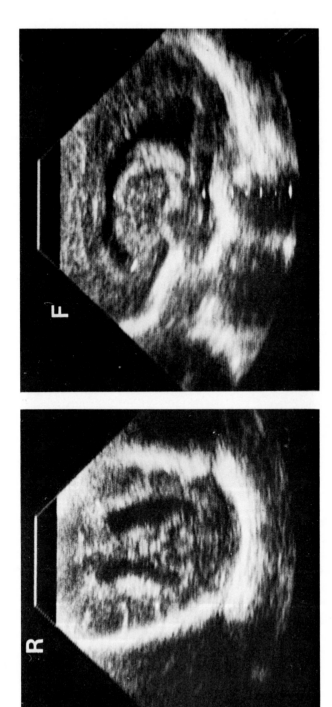

FIG. 19. Resolution of posthemorrhagic hydrocephalus *without* therapeutic intervention. **Left:** Coronal section at 2 weeks of age in this premature infant reveals the enlarged occipital portion of the lateral ventricles and some dense blood in the lateral aspect of the right ventricle. **Right:** Sagittal section at the same time shows generalized ventricular enlargement.

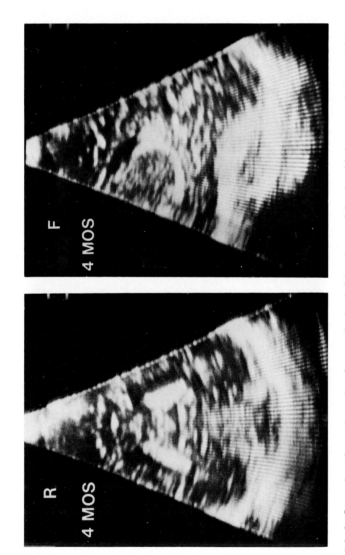

FIG. 19, continued. Left: Coronal section at 4 months of age at the level of the glomus of the choroid plexus reveals the ventricles now to be normal. Right: Sagittal section at the same time confirms the normalcy of these ventricles.

measures such as lumbar puncture or ventriculotomy have been performed, more frequent scans may be necessary. In this group, it is difficult to be sure of the endpoint, and this is especially true following the report of Hill et al. (37), who noted 11 infants whose enlarged ventricles stabilized but then had progressive ventricular dilatation between 12 and 84 days. Nevertheless, once stability of ventricular size has been achieved and no further therapeutic measures are ongoing, the examinations can be tailored precisely to the patient.

A schema for evaluation of term infants is somewhat different. As described above, these infants suffer from subdural hematomas and primary subarachnoid bleeding. Recent work has expanded the spectrum of bleeding to now include intracerebral parenchymal hemorrhage, intraventricular hemorrhage without periventricular hemorrhage, and posterior fossa subdural hemorrhage (16,19,55,68). These children present with seizures or other neurological abnormalities. Surprisingly, many of these infants have a negative history for ischemic-hypoxic episodes and birth trauma. In a large CT study, extensive areas of hypodensity occurred in children who were neurologically impaired on follow-up (29).

Since ultrasound cannot detect primary subarachnoid hemorrhage or parenchymal hypodense areas, the sonographic examination is merely used as a preliminary test in these term infants. If negative, term babies with ischemic-hypoxic histories, seizures, or neurological findings should have a CT scan.

CONGENITAL ABNORMALITIES

It is clear that enlarged ventricles are easily detected by the sonographic technique (39,48,56,70–72). This method is accurate in following the course and/or results of surgical intervention. When the degree of hydrocephalus is great and there are no unusual sonographic findings, the neurosurgeon may act on this imaging procedure alone. However, when complex abnormalities are detected, i.e., holoprosencephaly, porencephalic changes, Dandy-Walker malformation, etc., the sonogram is followed by CT for added information.

The classic sonographically detected abnormalities are "fluid-filled"— sonolucent (23,64,65). Ultrasonic detection of the Dandy-Walker syndrome (as opposed to other posterior fossa abnormalities) is particularly good, as the sagittal view demonstrates the aqueduct extending into a very dilated fourth ventricle (4). The hypoplasia of the cerebellum is striking (Fig. 20). Holoprosencephaly—the absence of a central ventricular septum—can also be seen early (Fig. 24).

In porencephaly, the sonogram can demonstrate the large fluid collection adjacent to the ventricles; however, the added dimension of a CT scan helps with the detection of ventricular communication (Fig. 22).

Arachnoid cysts are fluid collections that do not communicate with the ventricular system. These are found in various locations and are firmly

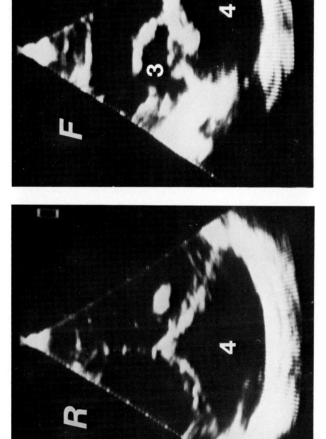

FIG. 20. Dandy-Walker syndrome. **Left:** Coronal section shows enlarged lateral ventricles and a large sonolucent region below the tentorium (4), the enlarged fourth ventricle or Dandy-Walker cyst. **Right:** Sagittal section shows a gaping foramen of Monro, the large third ventricle (3), the aqueduct, and massive fourth ventricle (4). (From ref. 74, with permission.)

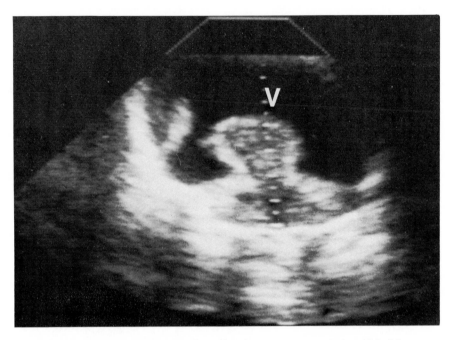

FIG. 21. Holoprosencephaly, coronal sections. Note large common central ventricle (V).

diagnosed only after multiple imaging procedures—sonograms, CT—with and without contrast. If after these studies there is still confusion, an arteriogram may be done.

Less frequently, "solid"—echogenic—lesions have been detected (65,73,74). These would include hemorrhage with parenchymal extension, vascular abnormalities in which there is so much vascular tissue that it appears echogenic, and brain tumors (Fig. 23). In addition, there are now reports of lipomas of the corpus callosum and visualization of calcification in congenital intrauterine infections. Ventriculitis, secondary to acquired meningitis, has also been described (8,38,65).

There is a major vascular lesion, the vein of Galen malformation, that is quite diagnosable with the real-time ultrasonic technique (Fig. 24). This is an important diagnosis to make, as the child may present clinically in congestive heart failure with cardiomegaly and there may be little clue to the etiology of the high-output failure. The sonographic appearance of these lesions in coronal sections is a large lucency in the vicinity of the third ventricle. It may have multiple feeders. On sagittal sections, the lesion is seen behind the fluid-filled third ventricle and is noted to drain, in the majority of cases, into the straight sinus. There may be ventricular dilatation and certainly deviation of the third ventricle, secondary to the mass effect of this lesion.

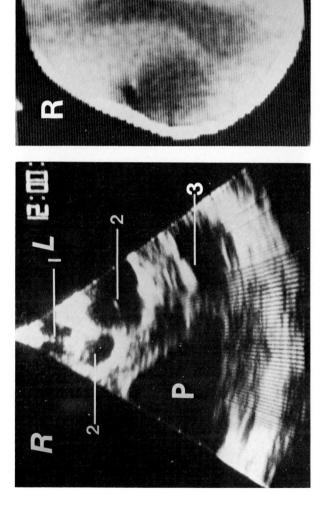

FIG. 22. Porencephaly. **Left:** Coronal section shows the anterior longitudinal cerebral fissure (1) and the body of the lateral ventricles (2). The right lateral ventricle is being compressed by a large porencephalic cyst (P). A dilated temporal horn (3) is seen on the left. **Right:** CT reversed so the patient's right is on the viewer's left to correspond with the ultrasonic evaluation confirms these findings. (Modified and reprinted from ref. 73, with permission.)

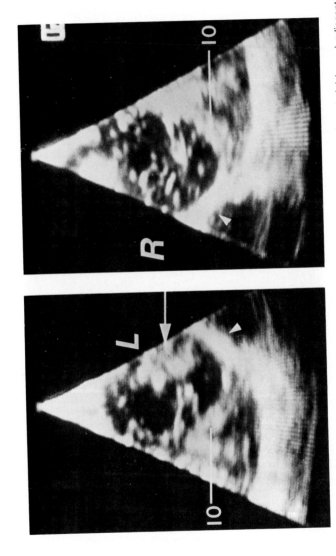

FIG. 23. Parenchymal lesions. **Left:** Coronal section of the left temporal-parietal region reveals echogenic material (*arrow*) adjacent to the temporal-parietal bones (*arrowhead*) and above the tentorium (10). **Right:** Similar section on the *right* side shows normal parenchyma.

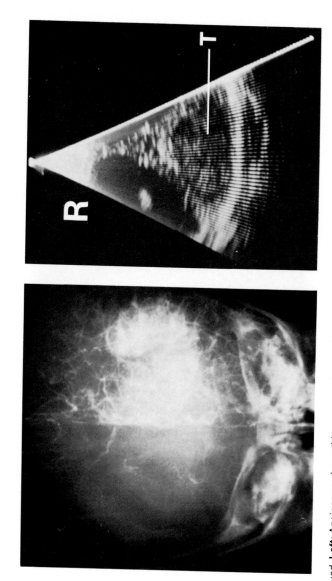

FIG. 23. continued. **Left:** Angiogram shows this racemose hemangioma in this region. (**A**, **B**, and **C** from ref. 73, with permission.) **Right:** Coronal section reveals dense echogenic material (T) in the posterior fossa and dilated lateral ventricles.

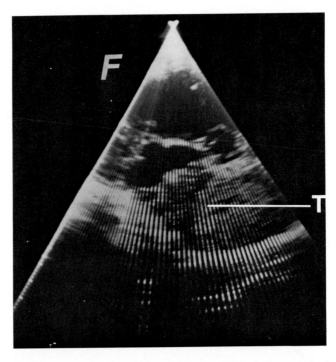

FIG. 23. continued. Sagittal section of this infant reveals the density in the posterior fossa (T) and the enlarged third and lateral ventricles. This proved to be a primitive neuro-ectodermal tumor. (**D** and **E** from ref. 74, with permission.)

Babcock and Han (6) have presented distinctive changes in myelomeningocele patients. These include widening of the intrahemispheric fissure, large massa intermedia with large third ventricle, and low position of the tentorium with a small posterior fossa (Fig. 25) (other changes have been described with axial scanning).

SUMMARY

Real-time sector scanning for the evaluation of neonatal intracranial contents is an easy, definitive procedure. This chapter outlines some of the techniques, advantages, and limitation of this modality. A practical approach to management of these infants is presented.

ACKNOWLEDGMENTS

I express special thanks to Jacqueline Roskamp, M.D. and Chung Ho-Chang, M.D., pathologists at Children's Hospital of Michigan, who provided guidance and helped during the pathological evaluations.

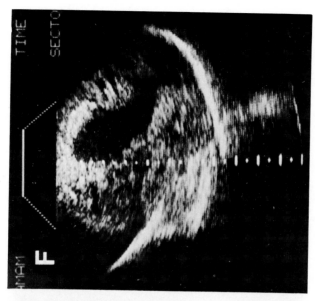

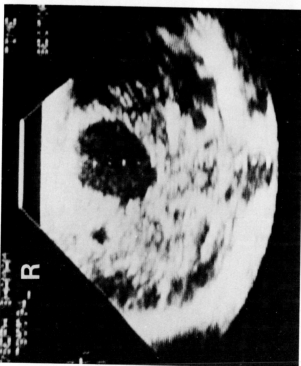

FIG. 24. Vein of Galen malformation. **Left:** Coronal section at the level of the body of the lateral ventricles reveals a large anechoic structure in the middle of the brain. **Right:** Sagittal section in the midline at the same level reveals this anechoic structure to be behind the third ventricle and to drain back into the straight sinus.

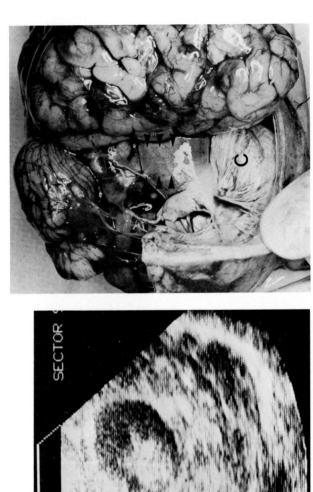

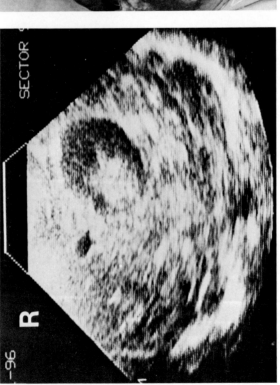

FIG. 24. continued. **Left:** Coronal section at autopsy reveals a large, echogenic clot within this venous aneurysm. **Right:** Pathological specimen shows the large vein of Galen malformation (*arrowheads*) on this view of the inferior surface of the brain. C, Cerebellum.

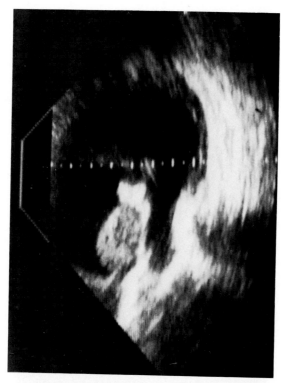

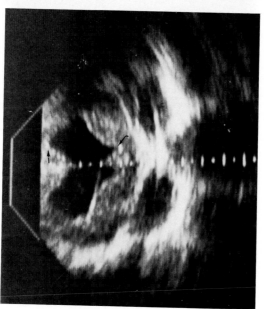

FIG. 25. Arnold-Chiari malformation. **Left:** Coronal section. Large lateral ventricles with angled superior and inferior margins (*arrows*). Septum pellucidum is absent. **Right:** Sagittal section off the midline shows the ventricular dilatation.

REFERENCES

1. Ahmann, P. A., Lazzara, A., Dykes, F. D., Brann, A. W., Jr., and Schwartz, J. F. Intraventricular hemorrhage in the high-risk preterm infants: Incidence and outcome. *Ann. Neurol.*, 7:118–124, 1980.
2. Albright, L., and Fellows, R. Sequential CT scanning after neonatal intracerebral hemorrhage. *AJNR*, 2:133–137, 1981.
3. Allan, W. C., Roveto, C. A., Sawyer, L. R., and Courtney, S. E. Sector scan ultrasound imaging through the anterior fontanelle. *Am. J. Dis. Child.*, 134:1028–1031, 1980.
4. Archer, C. R., Darwish, H., and Smith, K., Jr. Enlarged cisternae magnae and posterior fossa cysts simulating Dandy-Walker syndrome on computed tomography. *Radiology*, 127:681–686, 1978.
5. Babcock, D. S., Han, B. K., and LeQuesne, G. W. B-mode gray scale ultrasound of the head in the newborn and young infant. *AJR*, 134:457–468, 1980.
6. Babcock, D. S., and Han, B. K. Cranial sonographic findings in meningomyelocele. *AJNR*, 1:493–499, 1980.
7. Babcock, D. S., and Han, B. K. The accuracy of high resolution, real-time ultrasonography of the head in infancy. *Radiology*, 139:665–676, 1981.
8. Babcock, D. S., and Han, B. K. *Cranial Ultrasonography of Infants*. Williams & Wilkins, Baltimore, 1981.
9. Bejar, R., Curbelo, V., Coen, R. W., Leopold, G., James, H., and Gluck, L. Diagnosis and follow-up of intraventricular and intracerebral hemorrhages by ultrasound studies of infant's brain through the fontanelle and sutures. *Pediatrics*, 66:661–673, 1980.
10. Ben-Ora, A., Eddy, L., Hatch, G., and Solida, B. The anterior fontanelle as an acoustic window to the neonatal ventricular system. *J. Clin. Ultrasound*, 8:65–67, 1980.
11. Binder, G. A., Haughton, V. M., and Ho, K-C. *Computed Tomography of the Brain in Axial, Coronal and Sagittal Planes*. Little, Brown, Boston, 1979.
12. Brant-Zawadzki, M., and Enzmann, D. R. Using computed tomography of the brain to correlate low white-matter attentuation with early gestational age in neonates. *Radiology*, 139:105–108, 1981.
13. Brinker, R. A., and Taveras, J. M. Ultrasound cross-sectional pictures of the head. *Acta Radiol. [Diagn.] (Stockh.)*, 5:745–753, 1966.
14. Burstein, J., Papile, L., and Burstein, R. Subependymal germinal matrix and intraventricular hemorrhage in premature infants: Diagnosis by CT. *Am. J. Roentgenol.*, 128:971–976, 1977.
15. Burstein, J., Papile, L., and Burstein, R. Intraventricular hemorrhage and hydrocephalus in premature newborns: A prospective study with CT. *AJR*, 132:631–635, 1979.
16. Cartwright, G. W., Culbertson, K., Schreiner, R. L., and Garg, B. P. (1979): Changes in clinical presentation of term infants with intracranial hemorrhage. *Develop. Med. Child. Neurol.*, 21:730–737, 1979.
17. Cooke, R. W. I., Morgan, M. E. I., and Massey, R. F. Phenobarbitone to prevent intraventricular haemorrhage (letter). *Lancet*, ii:414–415, 1981.
18. Dewbury, K. C., and Aluwihare, A. P. R. The anterior fontanelle as an ultrasound window for study of the brain: A preliminary report. *Br. J. Radiol.*, 53:81–84, 1980.
19. Donat, J. F., Okazaki, H., Kleinberg, F., and Reagan, T. J. Intraventricular hemorrhages in full-term and premature infants. *Mayo Clin. Proc.*, 53:437–441, 1978.
20. Donat, J. F., Okazaki, H., and Kleinberg, F. Cerebellar hemorrhages in newborn infants. *Am. J. Dis. Child.*, 133:441, 1979.
21. Donn, S. M., Roloff, D. W., and Goldstein, G. W. A controlled trial of phenobarbital for prevention of intraventricular hemorrhage in preterm infants. *Ann. Neurol.*, 10:295, 1981.
22. Dykes, F. D., Lazzara, A., Ahmann, P., Blumenstein, B., Schwartz, J., and Brann, A. W. Intraventricular hemorrhage: A prospective evaluation of etiopathogenesis. *Pediatrics*, 66:42–49, 1980.
23. Edwards, M. K., Brown, D. L., Muller, J., Grossman, C. B., and Chua, G. T. Cribside neurosonography: Real-time sonography for intracranial investigation of the neonate. *AJNR*, 1:501–505, 1980.
24. Enzmann, D. R., Britt, R. H., Lyons, B. E., Buxton, J. L., and Wilson, D. A. Natural history of experimental intracerebral hemorrhage: Sonography, computed tomography and neuropathology. *AJNR*, 2:517–526, 1981.

25. Estrada, M. El-Gammel, T., and Dykes, P. R. Periventricular low attenuations—A normal finding in computerized tomographic scans of neonates? *Arch. Neurol.*, 37:754–756, 1980.
26. Farruggia, S., and Babcock, D. S. The cavum septi pellucidi: Its appearance and incidence with cranial ultrasonography in infancy. *Radiology*, 139:147–150, 1981.
27. Fiske, C. E., Filly, R. A., and Callen, P. W. Sonographic measurement of lateral ventricular width in early ventricular dilation. *J. Clin. Ultrasound*, 9:303–307, 1981.
28. Fiske, C. E., Filly, R. A., and Callen, P. W. The normal choroid plexus: Ultrasonographic appearance of the neonatal head. *Radiology*, 141:467–471.
29. Fitzhardinge, P. M., Flodmark, O., Fitz, C. R., and Ashby, S. The prognostic value of computed tomography as an adjunct to assessment of the term infant with postasphyxial encephalopathy. *J. Pediatr.*, 99:777–781, 1981.
30. Fleischer, A. C., Hutchinson, A. A., Kirchner, S. G., James, A. E., Jr. Cranial sonography of the preterm neonate. *Diagn. Imaging*, 3:20–28, 1981.
31. Flodmark, O., Becker, L. E., Harwood-Nash, D. C., Fitzhardinge, P. M., Fitz, C. R., and Chuang, S. H. Correlation between computed tomography and autopsy in premature and full-term neonates that have suffered perinatal asphyxia. *Radiology*, 137:93–103, 1980.
32. Garrett, W. J., Kossoff, G., and Warren, P. S. Cerebral ventricular size in children. *Radiology*, 136:711–715, 1980.
33. Grant, E. G., Schellinger, D., Borts, F. T., McCullough, D. C., Friedman, G. R., Sivasubramanian, K. N., and Smith, Y. Real-time sonography of the neonatal and infant head. *AJNR*, 1:487–492, 1980.
34. Grant, E. G., Borts, F., Schellinger, D., McCullough, D. C., and Smith, Y. Cerebral intraparenchymal hemorrhage in neonates: Sonographic appearance. *AJNR*, 2:129–132, 1981.
35. Haber, K., Wachter, R. D., Christenson, P. C., Vaucher, Y., Sahn, D. J., and Smith, J. R. Ultrasonic evaluation of intracranial pathology in infants: A new technique. *Radiology*, 134:173–178, 1980.
36. Hambleton, G., and Wigglesworth, J. S. Origin of intraventricular haemorrhage in the preterm infant. *Am. J. Dis. Child.*, 51:651–659, 1976.
37. Hill, A., and Volpe, J. J. Normal pressure hydrocephalus in the newborn. *Pediatrics*, 68:623–629, 1981.
38. Horbar, J. Persistent ventriculitis. *Hosp. Pract.*, 5:96–97, 1980.
39. Horbar, J. D., Walters, C. L., Philip, A. G. S., and Lucey, J. F. Ultrasound detection of changing ventricular size in posthemorrhagic hydrocephalus. *Pediatrics*, 66:674–678, 1980.
40. Johnson, M. L., Mack, L. A., Rumack, C. M., Frost, M., and Rashbaum, C. B-mode echoencephalography in the normal and high risk infant. *AJR*, 133:375–381, 1979.
41. Johnson, M. L., Dunne, M. G., Mack, L. A., and Rashbaum, C. L. Evaluation of fetal intracranial anatomy by static and real-time ultrasound. *J. Clin. Ultrasound*, 8:311–318, 1980.
42. Korobkin, R. The relationship between head circumference and the development of communicating hydrocephalus in infants following intraventricular hemorrhage. *Pediatrics*, 56:74–77, 1975.
43. Kossoff, G., Garrett, W. J., and Radavanovich, G. Ultrasonic atlas of normal brain of infant. *Ultrasound Med. Biol.*, 1:259–266, 1974.
44. Krishnamoorthy, K. S., Fernandez, R. A., Momose, K. J., DeLong, G. R., Moylan, F. M. B., Todres, I. D., and Shannon, D. C. Evaluation of neonatal intracranial hemorrhage by computerized tomography. *Pediatrics*, 59:165–172, 1977.
45. Krishnamoorthy, K. S., Shannon, D. C., DeLong, G. R., Todres, I. D., and Davis, K. R. Neurologic sequelae in the survivors of neonatal intraventricular hemorrhage. *Pediatrics*, 64:233–237, 1979.
46. Larroche, J-C. Post-haemorrhagic hydrocephalus in infancy anatomical study. *Biol. Neonate*, 20:287–299, 1972.
47. Lazzara, A., Ahmann, P., Dykes, F., Brann, A. W., Jr., and Schwartz, J. Clinical predictability of intraventricular hemorrhage in preterm infants. *Pediatrics*, 65:30–34, 1980.
48. Lees, R. F., Harrison, R. B., and Sims, T. L. Gray scale ultrasonography in the evaluation of hydrocephalus and associated abnormalities in infants. *Am. J. Dis. Child.*, 132:376–378, 1978.
49. Leech, R. W., and Kohnen, P. Subependymal and intraventricular hemorrhages in the newborn. *Am. J. Pathol.*, 77:465–476, 1974.

50. Leksell, L. Echoencephalography: Detection of the intracranial complications following head injury. *Acta Chir. Scand.*, 110:301–315, 1956.
51. Lipscombe, A. P., Blackwell, R. J., Reynolds, E. O. R., Thornburn, R. J., Cusick, G., and Pape, K. E. Ultrasound scanning of brain through anterior fontanelle of newborn infants (letter). *Lancet*, ii:39, 1979.
52. Lipscomb, A. P., Reynolds, E. O. R., Blackwell, R. J., Thornburn, R. J., Stewart, A. L., Cusick, G., and Whitehead, M. D. Pheumothorax and cerebral haemorrhage in preterm infants. *Lancet*, i:414–416, 1981.
53. London, D. A., Carroll, B. A., and Enzmann, D. R. Sonography of ventricular size and germinal matrix hemorrhage in premature infants. *AJR*, 135:559–564, 1980.
54. Mantovani, J. F., Pasternak, J. F., Mathew, O. P., Allan, W. C., Mills, M. T., Casper, J., and Volpe, J. J. Failure of daily lumbar punctures to prevent the development of hydrocephalus following intraventricular hemorrhage. *J. Pediatr.*, 97:278–281, 1980.
55. Mitchell, W., and O'Tuama, L. Cerebral intraventricular hemorrhages in infants: A widening age spectrum. *Pediatrics*, 65:35–39, 1980.
56. Morgan, C. L., Trought, W. S., Rothman, S. J., and Jimenez, J. P. Comparison of gray-scale ultrasonography and computed tomography in the evaluation of macrocrania in infants. *Radiology*, 132:119–123, 1979.
57. Pape, K. E., and Wigglesworth, J. S. *Haemorrhage, Ischemic and the Perinatal Brain.* William Heinemann, London; J. P. Lippincott, Philadelphia, 1979.
58. Papile, L., Burstein, J., Burstein, R., and Koffler, H. Incidence and evolution of subependymal and intraventricular hemorrhage: A study of infants with birth weights less than 1,500 grams. *J. Pediatr.*, 92:529–534, 1978.
59. Papile, L., Burstein, J., Burstein, R., Koffler, H., Koops, B. L., and Johnson, J. D. Posthemorrhagic hydrocephalus in low-birth-weight infants: Treatment by serial lumbar punctures. *J. Pediatr.*, 97:273–277, 1980.
60. Rom, S., Serfontein, G. L., and Humphreys, R. P. Intracerebellar hematoma in the neonate. *J. Pediatr.*, 93:486–488, 1978.
61. Rumack, C. M., McDonald, M. M., O'Meara, O. P., Sanders, B. B., and Rudikoff, J. C. CT detection and course of intracranial hemorrhage in premature infants. *Am. J. Roentgenol*, 131:493–497, 1978.
62. Rumack, C. M., Johnson, M. L., McDonald, M. M., Hathaway, W. L., Koops, B., and Guggenheim, M. A. Timing and course of neonatal intracranial hemorrhage on real-time and static ultrasound. *Presented to Society of Pediatric Radiology, San Francisco*, March, 1981.
63. Rumack, C. M., Guggenheim, M. A., Rumack, B. H., Peterson, R. G., Johnson, M. L., and Braithwaite, W. R. Neonatal intracranial hemorrhage and maternal use of aspirin. *Obstet. Gynecol.*, 58:525–565, 1981.
64. Sauerbrei, E. E., Harrison, P. B., Ling, E., and Cooperberg, P. L. Neonatal intracranial pathology demonstrated by high-frequency linear array ultrasound. *J. Clin. Ultrasound*, 9:33–36, 1981.
65. Sauerbrei, E. E., and Cooperberg, P. L. Neonatal brain: Sonography of congenital abnormalities. *AJNR*, 2:125–128, 1981.
66. Sauerbrei, E. E., Digney, M., Harrison, P. B., and Cooperberg, P. L. Ultrasonic evaluation of neonatal intracranial hemorrhage and its complications. *Radiology*, 139:677–685, 1981.
67. Schrumpf, J. D., Sehring, S., Killpack, S., Brady, J. P., Hirata, T., and Mednick, J. P. Correlation of early neurologic outcome and CT findings in neonatal brain hypoxia and injury. *J. Comput. Assist. Tomogr.*, 4:445–450, 1980.
68. Serfontein, G. L., Rom, S., and Stein, S. Posterior fossa subdural hemorrhage in the newborn. *Pediatrics*, 65:40–43, 1980.
69. Shankaran, S., Slovis, T. L., Bedard, M. P., and Poland, R. L. Sonographic classification of intracranial hemorrhage. A prognostic indicator of mortality, morbidity and short-term neurologic outcome. *J. Pediatr.*, 100:469–475, 1982.
70. Shkolnik, A., and McLone, D. G. Intraoperative real-time ultrasonic guidance of ventricular shunt placement in infants. *Radiology*, 141:515–517, 1981.
71. Silverboard, G., Horder, M. H., Ahmann, P. A., Lazzara, A., and Schwartz, J. F. Reliability of ultrasound in diagnosis of intracerebral hemorrhage and posthemorrhagic hydrocephalus: Comparison with computed tomography. *Pediatrics*, 66:507–514, 1980.

72. Skolnick, M. L., Rosenbaum, A. E., Matzuk, T., Guthkelch, A. N., and Heinz, E. R. Detection of dilated cerebral ventricles in infants: A correlative study between ultrasound and computed tomography. *Radiology*, 131:447–451, 1979.
73. Slovis, T. L., and Kuhns, L. R. Real-time sonography of the brain through the anterior fontanelle. *AJR*, 136:277–286, 1981.
74. Slovis, T. L. Real-time ultrasound of the intracranial contents. In: *Clinics in Diagnostic Ultrasound, Vol. 8: Ultrasound in Pediatrics*, edited by J. O. Haller and A. Shkolnik, pp. 13–27, 1981. Churchill, Livingstone, New York.
75. Slovis, T. L., Kelly, J. K., Eisenbrey, A. E., Quiroga, A., and Greininger, N. The detection of extracerebral fluid collections by real-time sector scanning through the anterior fontanelle. *J. Ultrasound Med.*, 1:41–42, 1982.
76. Slovis, T. L., Poland, R. L., and Shankaran, S. Normal values for ventricular size as determined by real time sonographic techniques. *(In preparation)*.
77. Tenner, M. S., and Wodraska, G. M. *Diagnostic Ultrasound in Neurology*. John Wiley & Sons, New York, 1975.
78. Vlieger, M. Evaluation of echoencephalography. *J. Clin. Ultrasound*, 8:39–47, 1980.
79. Volpe, J. J., Pasternak, J. F., and Allan, W. C. Ventricular dilation preceding rapid head growth following neonatal intracranial hemorrhage. *Am. J. Dis. Child.*, 131:1212–1215, 1977.
80. Volpe, J. J. Neonatal periventricular hemorrhage: Past, present, and future. Editor's column. *J. Pediatr.*, 92:693–696, 1978.
81. Volpe, J. J. Intracranial hemorrhage in the newborn: Current understanding and dilemmas. *Neurology*, 29:632–635, 1979.
82. Volpe, J. J. Evaluation of neonatal periventricular-intraventricular hemorrhage. A major advance. *Am. J. Dis. Child.*, 134:1023–1025, 1980.
83. Volpe, J. J. Neonatal intraventricular hemorrhage. *N. Engl. J. Med.*, 304:886–891, 1981.
84. Volpe, J. J. *Neurology of the Newborn*. W. B. Saunders Co., Philadelphia, 1981.
85. Wells, P. N. T. Real-time scanning systems. In: *Clinics in Diagnostic Ultrasound, Vol. 5*, edited by P. N. T. Wells and M. C. Ziskin, pp. 69–84, 1980, Churchill Livingstone, New York, 1980.
86. Wigglesworth, J. S., and Pape, K. E. An integrated model for haemorrhagic and ischaemic lesions in the newborn brain. *Early Hum. Dev.*, 2:179–199, 1978.

Ultrasound Annual 1982,
edited by Roger C. Sanders.
Raven Press, New York © 1982.

Update on the Gallbladder

Hsu-Chong Yeh

Department of Radiology of The Mount Sinai Hospital and School of Medicine of the City University of New York, New York, New York 10029

The role of ultrasonography in the investigation of gallbladder disease has become increasingly important during the past few years as a result of improvement in gray-scale ultrasonic scanners and refinement of scanning technique. The accuracy of cholecystonography has increased to the degree that it is now almost the same, or in some cases surpasses, that of oral cholecystography. Further, as a result of (a) its noninvasive nature, (b) lack of ionizing radiation, (c) the fact that ingestion or injection of contrast material is not required, (d) visualization of the gallbladder is not dependent on the function of the gallbladder or the patency of biliary tract, and (e) the simultaneous visualization of other intraabdominal structures, ultrasonography has replaced oral cholecystography and intravenous cholangiography as the primary screening procedure for gallbladder or biliary tract disease in certain groups of patients, in particular, in those suspected of having acute cholecystitis or in patients with jaundice.

NORMAL ANATOMY

The gallbladder is a pear-shaped musculomembranous sac attached to the under surface of the liver in the gallbladder fossa; it may project beyond the inferior border of the liver. The anterior surface, which is adherent to the gallbladder fossa, is therefore nonperitonealized, i.e., a bare area. The rest of the gallbladder is peritonealized. The gallbladder fossa is located between the two physiological lobes of the liver. Since the shape of the liver may be quite variable, e.g., the right lobe of the liver may be quite elongated in craniocaudal dimension and the left lobe may be short, the "under surface" of the middle portion of the liver may actually rotate to face toward the left; the bare area of the gallbladder, in many cases, is actually on the lateral surface of the gallbladder.

The gallbladder fossa is best identified on the transverse scan as an indentation on the posterior surface of the liver. The gallbladder is located in the indentation and protrudes posteriorly outside the liver. It is usually located in the right upper quadrant of the abdomen, a few centimeters below the level of the porta hepatis, anterior or anterolateral to the right kidney. On occasion, it may be located on an extreme lateral or anterior location, immediately beneath the rib cage, and, therefore, be difficult to delineate by ultrasonography. Rarely, it is located medially anterior to the inferior vena cava or pancreas. When a mesentery is present, the gallbladder may be mobile and herniate through the foramen of Winslow into the lesser peritoneal sac to be located to the left of the midline (16).

On longitudinal scanning, the gallbladder is located beneath the inferior surface of the liver; the anterior gallbladder wall appears attached to the posterior liver surface. The gallbladder "fossa" or indentation of the liver is usually not seen in this scanning plane. However, in about 65% of patients, a linear echo may be seen extending from the superior end of the gallbladder to connect the right or main portal vein (4). This echogenic line represents a portion of a main lobar fissure of the liver. It may also be seen on the transverse scan (37) above the level of gallbladder running from the posterior surface of the liver anterolaterally in the same oblique orientation as the gallbladder.

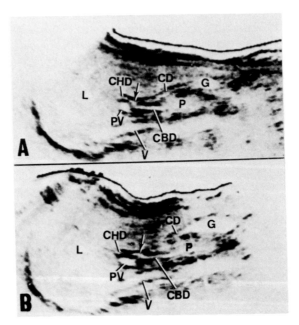

FIG. 1. Normal anatomy. Oblique view in the decubitus position, right side up. **A:** The gallbladder (G) is on the inferior surface of the liver. The cystic duct (CD) joins (*arrow*) the common hepatic bile duct (CHD) anterior to the right branch of the portal vein (PV) beyond which the common hepatic duct (CHD) becomes the common bile duct (CBD). P, Pancreas; V, inferior vena cava. **B:** On scanning slightly more laterally, the main portion of the gallbladder (G) is seen. The common bile duct enters the pancreas posteriorly. (From ref. 53, with permission.)

An oblique scan from superolateral to inferomedial on semidecubitus position, right side up, will usually delineate the gallbladder along its long axis (Fig. 1). The common bile duct is usually seen in a scanning plane medial to the gallbladder. The cystic duct is infrequently visualized, but occasionally may be seen to join the common bile duct slightly inferior to the right branch of portal vein (53) (Fig. 1).

The gallbladder is usually divided, without clear anatomical landmarks, into three parts: the fundus, which represents the blind end portion; the neck, a narrow tube-like structure that tapers into the cystic duct; and the body, which represents the remaining portion between the neck and the fundus.

The size of the gallbladder is quite variable. Considerable overlap exists between an early dilated gallbladder and a distended normal gallbladder (29). Therefore, strict criteria for the upper limit of normal size have not been established. Normally, it measures about 5 to 7 cm × 2 to 3 cm. When the largest transverse diameter (or width) is more than 3.5 to 4 cm, dilatation may be suggested (16). If the gallbladder contracts after a fatty meal or after injection of cholecystokinin and the common bile duct remains normal-sized as before, the apparently large gallbladder is probably not a result of obstruction. In some patients with common bile duct obstruction but without significant dilatation of the duct, a fatty meal may cause the gallbladder to contract and the common bile duct to become dilated. If the gallbladder remains unchanged, obstruction of cystic duct or common bile duct may be suggested. However, some normal gallbladders may not contract after a fatty meal.

SCANNING TECHNIQUE

As with oral cholecystography, the examination of the gallbladder with ultrasonography is best accomplished when the gallbladder is distended with bile, i.e., after 6 to 12 hr of fasting. Both transverse and longitudinal scans in a supine position should be done, with systematic 1-cm increments to cover the upper abdomen. More detailed closed-space scanning in 3- to 5-mm increments may be done over the gallbladder to ensure not missing a tiny stone when no stone is found.

Real-time scanning is very useful in examination of the gallbladder because it is fast, convenient, and accurate. Sector real-time scanning is more versatile than a linear array because it can be used through an intercostal space. However, high-resolution linear scanners are also excellent for gallbladder examination in most patients (26), except when the gallbladder is high and completely within the rib cage, when it may be largely obscured by rib shadows. However, because of larger rectangular-shaped scanning field, it is usually easier to find the gallbladder with a linear scanner.

A sustained deep inspiration is usually helpful in bringing down and stabilizing the gallbladder so that a steady sector or sector-linear sweep can be performed to obtain maximal detail. To delineate the smooth posterior wall of the gallbladder so that a tiny stone can be clearly seen, one should direct the ultrasound beam as perpendicular to the posterior gallbladder wall as possible. Sometimes a sector scanning through an interosteal space may be necessary for this purpose. On the other hand, to split up a combined linear echo derived from a "layer of stones" on the posterior gallbladder wall so that individual stones can be recognized, scanning from the fundus toward the posterior wall may be necessary. This is discussed later.

Scanning in a decubitus, right side up, or upright position should be done on most patients for the following purposes:

(a) It will demonstrate gravitational shift of gallstone.

(b) A small stone hidden in the neck of the gallbladder may move down and be detected.

(c) Some high-positioned gallbladders may move from within the rib cage and become easier to examine.

(d) A gas-filled bowel loop indenting the gallbladder and simulating a stone may move away and become separate from the gallbladder (6).

(e) An anteroposteriorly oriented gallbladder may become more horizontally oriented so that the ultrasound beam can be directed more perpendicular to the gallbladder wall to obtain maximum detail.

(f) Oblique scanning in a decubitus position may improve the delineation of the common bile duct.

With a manual scanner, decubitus scanning is usually easier and more rewarding than scanning in the upright position. Real-time scanning can be done easily in either a decubitus or an upright position.

Since the gallbladder may not be responsible for the clinical symptoms, structures other than gallbladder should be included in the examination, in particular, the pancreas, liver, the right kidney, the right subphrenic space, and Morrison's pouch.

CONGENITAL ANOMALIES

It is difficult to diagnose agenesis or hypoplasia of gallbladder by ultrasonography, since some small "chronically diseased" contracted gallbladders are also not visualized. A double gallbladder may appear as two parallel pear-shaped cystic structures in the gallbladder fossa (Fig. 2). An angulated gallbladder may simulate a double gallbladder (Fig. 3). A prominent inferior vena cava may also simulate a second gallbladder in a decubitus position when the inferior vena cava is sectioned obliquely. Therefore, demonstrating the whole course of a structure by scanning in multiple planes or demonstrating the pulsation by real-time scanning is important in making such a distinction. An elongated gallbladder with segmental involvement by adenomyomatosis in its middle portion may simulate a double gallbladder, one segment being superior

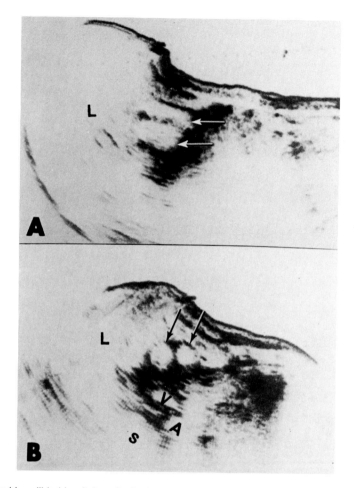

FIG. 2. Double gallbladder. **A:** Longitudinal scan 7 cm to the right of the midline. Two gallbladders (*arrows*) are seen. Some weak echoes in the posterior gallbladder represent sludge of bile. L, Liver. **B:** Transverse scan in decubitis position, right side up. Both gallbladders (*arrows*) are clearly within the gallbladder fossa. V, Inferior vena cava; A, aorta, S, spine.

to the other. It may not be difficult to diagnose an intrahepatic gallbladder if it retains the shape of the gallbladder (Fig. 4) and contracts after a fatty meal. Otherwise, differentiation from a liver cyst may be difficult and oral cholecystography or correlation with computed tomography (CT) after ingestion of telepaque may be necessary in the diagnosis.

Angulation or kinking of the gallbladder may result in a Phrygian cap deformity (Fig. 5). Sonographically, a septum may be seen within the gallbladder and the gallbladder may appear bent.

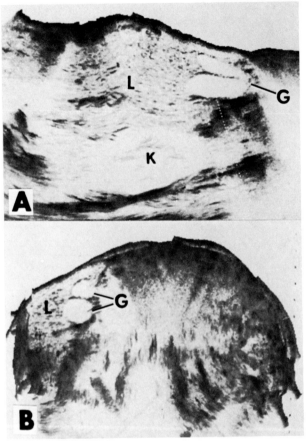

FIG. 3. Markedly angulated gallbladder. **A:** Longitudinal scan 5 cm to the right of midline. The gallbladder (G) is folded at its middle portion. If the angulated portion is not delineated or obscured by bowel gas, this may be mistaken for double gallbladder. L, Liver. **B:** Transverse scan 4 cm above umbilicus.

GALLSTONES

With improvement in gray-scale ultrasound instruments, both manual and real-time scanners, sonographic examination for gallstones has become highly accurate. Accuracy ranges from 91 to 96% (2,8,30,34), which is close to that of oral cholecystography. Although accuracy for oral cholecystography in well-opacified positive cases is as high as 98% (1), the overall accuracy is not as high, since a negative diagnosis is not quite as reliable as a positive one (35). The main errors in oral cholecystography (35) are: (a) small lucent stones

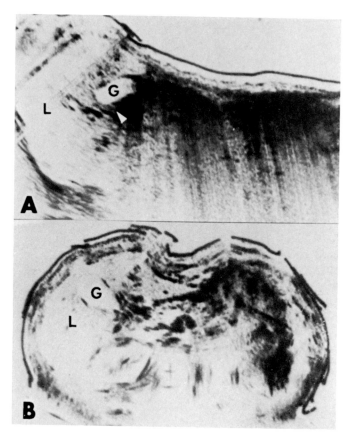

FIG. 4. Partially intrahepatic gallbladder. The gallbladder (G) is surrounded by the liver on both longitudinal (A) and transverse (B) scannings except at the fundal portion on a longitudinal scan. L, Liver; *arrowhead*, portion of liver posterior to the gallbladder.

seen only on films with horizontal beam as a lucent line. It is even easier to miss one or two small lucent stones. (b) A tiny calcified stone may be obscured by contrast medium. Ultrasonography may show these stones. In patients suspected of having acute cholecystitis, ultrasonography is usually more helpful than intravenous cholangiography (46).

Since ultrasonography is not restricted by the function of the liver or gallbladder or by patency of the biliary duct, it can be done on most patients with acute cholecystitis. Such patients are usually anorexic and not eating; therefore the gallbladder usually can be seen on emergency examination. Although gas distention of the abdomen as a consequence of acute abdomen may cause some difficulty in visualizing the gallbladder from below the costal margin, this can usually be overcome by scanning through an intercostal

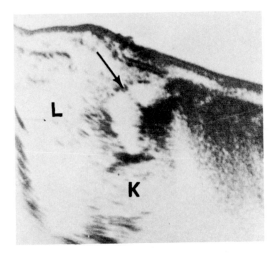

FIG. 5. Phrygian cap deformity of the gallbladder. Longitudinal scan 6 cm to the right of the midline. The gallbladder is angulated and a septum (*arrow*) is seen. L, Liver; K, kidney.

space. Intravenous cholangiography, on the other hand, is limited by jaundice (i.e., not applicable in patients with a serum bilirubin level higher than 4 mg/dl as in about 20% of the patients) (46). A history of allergy to the contrast material with possible serious complications is not uncommon and also needs to be considered.

The ultrasonographic features of gallstones may be categorized into four types:

Type I: This is a discrete highly echogenic lesion or lesions in the dependent portion of an echo-free gallbladder lumen (Fig. 6). The lesion casts an acoustic shadow and moves with gravity as the patient changes position. When these features are present, one can be 100% certain that there is a stone (or stones) (7). A stone impacted in the neck of the gallbladder (Fig. 7), or small stones adherent to the gallbladder wall (Fig. 8), will not move with gravity. When a slit of gallbladder lumen is visualized, however, the presence of a gallstone (or gallstones) is established. Such a narrow gallbladder lumen appears as double curves ["double-arc-shadow sign" (40)] or as a crescent-shaped echo-free space ("crescent sign") (Fig. 9B).

Type II: The gallbladder is not visualized but strong echoes with an acoustic shadow are seen in the gallbladder fossa. The accuracy of predicting a gallstone (or stones) by this feature is also very high, approaching 96% (7).

Type III: Multiple weak echoes in the dependent portion of the gallbladder form a level. These echoes move with gravity. No acoustic shadow is seen. The cause of these echoes is difficult to determine in most instances, and a variable number of patients prove to have small stones.

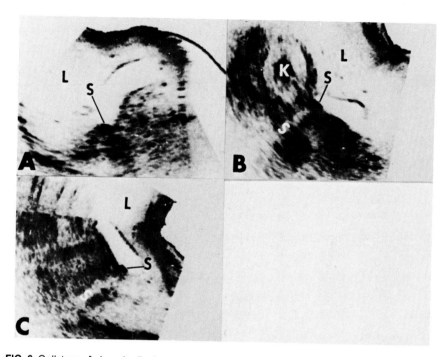

FIG. 6. Gallstone. A: Longitudinal scan 6 cm to the right of midline. The stone (S) is in the neck of the gallbladder, which is markedly posteriorly located in this patient and, therefore, not well seen. The acoustic shadow is also poorly seen. L, Liver. B: Oblique scan in decubitus position, right side up. The stone (*black* S) and the acoustic shadow (*white* S) are better seen. Although the neck of the gallbladder is higher than the fundus, the stone did not move. C: After the patient was turned to prone position for a few minutes and rescanned in decubitus position, the stone moved to the fundus of the gallbladder.

Type IV: This is a discrete echogenic focus that does not cause an acoustic shadow but moves with gravity. If the lesion is a small one and highly echogenic, it may represent a small stone. If the lesion is relatively large and poorly echogenic, it may represent a stone (or stones), a ball of sludge of bile, or a blood clot or parasitic cyst.

Since the demonstration of an acoustic shadow allows one to make a definite diagnosis of gallstones, there is a considerable literature on formation or absence of the acoustic shadow. Filly et al. (13) have demonstrated by *in vitro* scanning that all gallstones will cast an acoustic shadow regardless of the size, shape, surface characterization, composition, or calcium content of the stone. Their specimens included stones of 1 to 40 mm in size as well as very small formed granules or sand. The sand was scanned *in toto* and also cast a shadow. The sound is mainly a result of absorption of sound energy by the stone (50) as well as a severe acoustic mismatch impedance between the bile

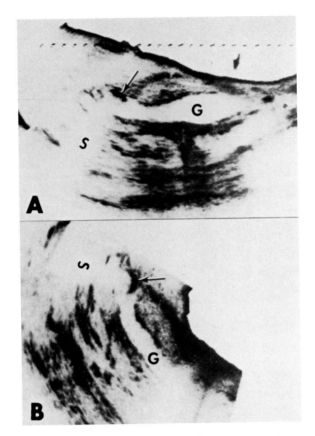

FIG. 7. Large stone impacted in the gallbladder neck. **A:** Scanning on decubitus position, right side up. The gallbladder (G) is long and enlarged. A large stone (*arrow*) is in the neck of the gallbladder, casting an acoustic shadow(s). **B:** On scanning in semiprone position, the stone (*arrow*) remained high in the gallbladder neck and did not move with gravity.

and the stone, which causes reflection and scattering of most or all of the incident ultrasound beam. The factors responsible for nonshadowing of some stones at ultrasonography are:

(a) The size of the stone is small in comparison with the beam width (13), allowing sound beam to pass alongside the stone, i.e., the stone does not block completely the beam path. Depending on the size of the stone in relation

FIG. 8. Stones cling on to the gallbladder wall simulating calcification of gallbladder wall. **A:** Transverse scan 8 cm above umbilicus. An echogenic focus (*arrow*) on the lateral wall of the gallbladder (G) casting an acoustic shadow(s) represents several small stones. Note the gallbladder wall is thickened as a result of cholecystitis. K, Kidney. **B:** Oblique scan in decubitus position, right side up. The stones still cling on to the gallbladder wall. **C:** Scan in semiprone position, the stones remain high.

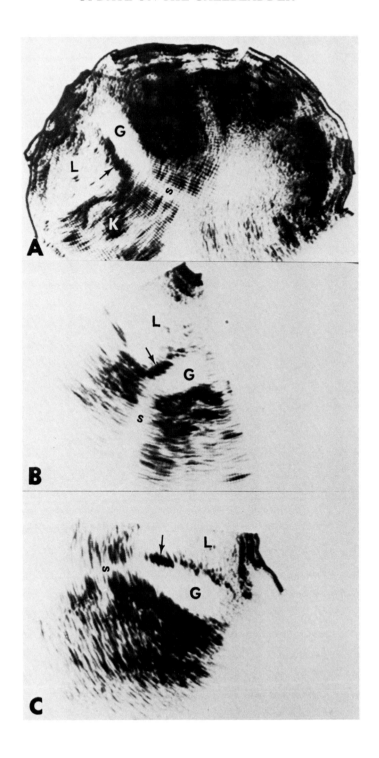

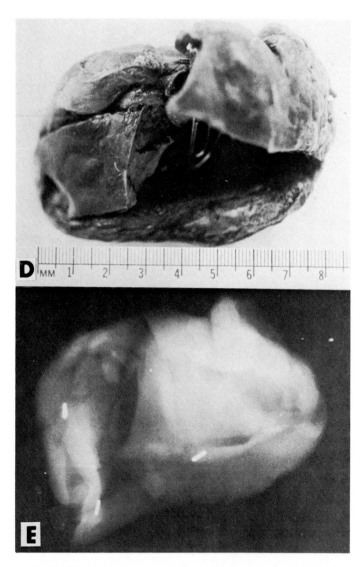

FIG. 8. continued. **D:** Specimen shows markedly thickened wall of gallbladder. The wall is not uniformly thickened. There are also free small stones (*not shown*). **E:** Radiograph of the specimen shows no evidence of calcification in the gallbladder wall. (A tiny white dot is an artifact.) Three surgical clips are seen.

to the beam width, the acoustic shadow may be complete, partial, or undiscernible (i.e., nonshadowing). Therefore, to detect (or produce) an acoustic shadow from a small stone, a focused or high-frequency transducer should be used. One should select a transducer with an appropriate focal length so that

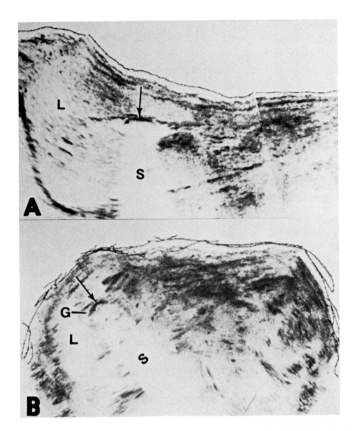

FIG. 9. Large gallstone. **A:** Longitudinal scan 7 cm to the right of midline. Only the anterior surface of the stone (*arrow*) is delineated as a curvilinear echo because of large strong acoustic shadow(s). L, Liver. **B:** Transverse scan 6 cm above umbilicus. The diameter of the gallstone (*arrow*) is almost as large as that of the gallbladder. Therefore only a small crescent rim of gallbladder lumen (G) is visualized, i.e., crescent sign. When the gallbladder is small and not well seen on both longitudinal and transverse scans, this sign will help to identify the stone with a high degree of certainty.

the gallbladder will fall within the focal zone. Although a single small stone may not cause shadowing, an aggregation of such small stones will cast shadows (13,20).

(b) Intrinsic ultrasonic attenuation or absorption of sound energy by the gallstone. The more attenuation, the stronger the shadowing. More highly attenuating stones tend to contain a large percentage of crystalline material (38).

(c) The orientation of the faceted stone to the sound beam affects the shadowing. A faceted stone that rotates within the sound beam will cause the magnitude of the acoustic shadow to change (18). When the flat surface of the stone is perpendicular to the sound beam, maximum shadowing can be observed because maximum reflection of the sound beam occurs. When the flat surface is not perpendicular to the sound beam, the shadow may be partial (weak) or undiscernible. Therefore, to bring out a shadow, one should perform multiple scans with different transducer positions and angulations.

(d) A weak shadow may become undiscernible if the gain setting is too high because the shadow is flooded with echoes. An optimal gain setting is important to demonstrate the acoustic shadow.

(e) The use of compression amplification signal processing owing to inaccuracies in the dynamic range of TV display units may preclude perception of a weak shadow (51). An optimal pre- or postprocessing mechanism for a digital scanner may bring out a weak shadow.

(f) Diffraction around the stone. A large amount of diffraction may explain a decrease in the size of the shadow. This has been demonstrated by *in vitro* Schlieren photography of gallstones that were insonated under an ultrasound beam (18).

The acoustic shadow from gallstones is usually different from that of bowel gas (50). The gas shadow usually contains numerous small echoes that gradually fade out distally. These echoes are caused by the reverberation of sound beams consequent upon the near-total reflection of sound energy at tissue-gas interfaces. The stone shadow is, however, usually "echo-free," i.e., appears "clean" as a result of absorption of sound energy. Although this is not always reliable in differentiating a gallstone from a gas-containing bowel loop, when a "clean" acoustic shadow is observed originating from the gallbladder fossa, the sign adds confidence in the diagnosis of a gallstone.

Crade reported a 96% incidence of gallstone in a group of patients who were scanned in a fasting state but no gallbladder could be found on sonography (7,8). Leopold et al. (30) reported a similar experience.

Gallstones may have many acoustic features. A relatively small stone appears as a high echogenic round lesion with an acoustic shadow. Only the anterior surface of a larger-size stone may be seen as a curvilinear echo (Fig. 9). All the rest of the stone is obscured by the acoustic shadow. Faceted stone may show a flat surface or a strong linear horizontal linear echo with a shadow behind it. When the sound beam is less attenuated, echoes within the stone can be seen and the whole stone rather than its anterior surface can be delineated. A coarse line of irregular strong echoes with a single large shadow may represent multiple stones that lie adjacent to each other. When these stones are small or faceted, the "line" of the echogenic structure may be smooth and sharp, simulating the normal smooth posterior wall of the gallbladder. The only evidence of the presence of the stones is the acoustic shadow behind the "line" of echoes. However, a gas-filled bowel loop posterior to the gallbladder may have similar features. Although a "clean" acoustic shadow is usually attributable to gallstones (50), this sign is not entirely reliable and diagnosis of gallstones will not be satisfactory if based solely on this feature. To split up the echoes to demonstrate individual stones, one should scan from different directions either by tilting the transducer at a different angle (Fig. 10) or by changing the patient's position.

There are many conditions that may mimic gallstones on sonography. The most common one is a gas-filled bowel loop indenting the gallbladder, as bowel gas also appears as a strongly echogenic structure with acoustic shadow. By changing the patient's position (e.g., decubitus) or during deep inspiration or expiration the bowel may move away from the gallbladder. Peristaltic movement within the bowel loop may be observed on real-time scanning. Heister's valve or fibrous tissue in the periportal area may also be echogenic and associated with shadowing. Shadowing from this area may be mistaken for a small stone in the neck of the gallbladder or common bile duct. With an awareness of such anatomic variations, scanning in different positions to separate the valve of Heister from the gallbladder may solve the problem.

Food particles and air may enter the gallbladder after cholecystoenterostomy or other forms of bypass surgery in anomalies with a biliary fistula. Strong echoes and acoustic shadowing may then appear within the gallbladder, simulating gallstones (19).

In patients with a previous history of cholecystectomy, strong echoes with an acoustic shadow are seen frequently in the gallbladder fossa simulating

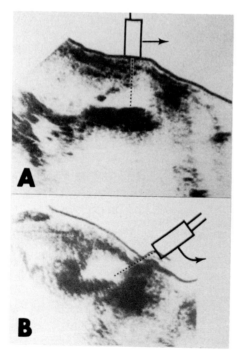

FIG. 10. Multiple gallstones. A: Oblique scan in semidecubitus position, right side up. Scanning was done with the transducer vertically oriented, i.e., perpendicular to the posterior wall of the gallbladder. A sharp smooth posterior echogenic surface simulates a normal posterior wall. A large acoustic shadow posteriorly may be attributable to multiple stones or a gas-filled bowel loop posterior to the gallbladder. B: Repeat the scanning with transducer directed from medially. Multiple stones are now clearly seen.

gallstones. In a majority of these patients, postoperative scars of the gallbladder are responsible but echoes may also be caused by surgical clips; only occasionally a common bile duct stone may be present (39).

There is a high incidence of cholelithiasis in patients with sickle-cell anemia, even in children (28,45). Since ultrasound is simple and accurate for detecting gallstones and involves no ionizing radiation, it is an ideal method of screening patients with sickle-cell anemia. Sarnaik et al. (45) used ultrasonography to screen 234 pediatric patients with sickle-cell anemia and detected gallstones in 23%. The incidence ranged from 8.5% at 2 to 4 years of age to 35% at 15 to 18 years of age. The majority of patients had no classic symptoms of gallbladder disease. On the other hand, Harned and Babbitt (22) reviewed 367 cases of cholelithiasis in children in the literature with his own cases; 81% were without evidence of hemolytic anemia. The etiology of stones in most of these patients was not clear; many were associated with chronic cholecystitis. Oral cholecystography was diagnostic of stone in only 70%. In the remaining 30%, the gallbladder was not visualized.

NONSHADOWING INTRALUMINAL ECHOES

Nonshadowing echoes are frequently observed in the gallbladder. There are three varieties:

(a) Diffuse low-amplitude echoes in the whole gallbladder or in the dependent portion of the gallbladder forming a fluid level that moves with gravity. This is mostly caused by biliary sludge, i.e., inspissated bile or thick viscid bile. Sludge may occur in those bedridden for long periods of time, after prolonged fasting, hyperalimentation, liver cirrhosis, or obstruction of the common bile duct (48), or in association with sepsis. In some cases small stones may be found with sludge. Pus, cholesterol crystals (5), or blood (Fig. 11) may also cause echoes within the gallbladder. Some insoluble mineral precipitation or abnormal mucoprotein production by the gallbladder may also cause such echoes.

Conrad et al. (5) showed that viscous desiccated bile appeared echogenic. In their experiments with dogs, they suggested that highly desiccated bile may be as echogenic as liver parenchyma and the gallbladder was therefore not imaged. In my experience, however, the gallbladder can still be seen, for one can see the ring-shaped wall (Fig. 12) defining the margin of the liver. Filly (12) et al. filtered the echogenic component of biliary sludge. The bile filtrate became echo-free. They observed, under microscopy, that the filtration residue is particulate matter, predominantly calcium bilirubinate granules. Cholesterol crystals were found in one of five patients investigated.

If sludge echoes are due to prolonged fasting, they will disappear on repeat examination after the patient begins food intake by mouth. The echoes in these patients are, therefore, most likely of no clinical significance. In patients

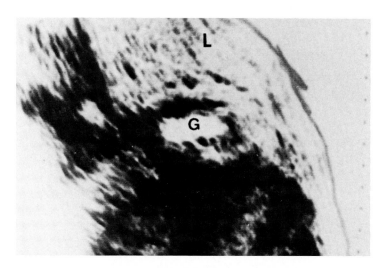

FIG. 11. Blood in the gallbladder in a patient who was on anticoagulant therapy for his cardiac condition. Oblique scan in decubitus position, right side up. Multiple echoes in the dependent portion of the gallbladder (G) represent blood.

without prolonged fasting, however, a finding of weak echoes should be considered abnormal although the clinical significance is as yet uncertain (12).

(b) One or more discrete nonshadowing echoes measuring less than 5 mm. Whether or not the lesion moves with gravity, it most likely represents a stone. But if it is constant in location, the possibility of a polyp should also be considered. Demonstration of the motility of a small stone may require strenuous effort. One may have to turn the patient into a prone position for several minutes and then turn the patient back to a decubitus position and repeat the examination to demonstrate movement. Since a small stone may be purely cholesterol and of low specific gravity, it may cling to the gallbladder wall in viscous bile and not readily sink into the dependent portion of the gallbladder. Simeone et al. found an 81% overall incidence of stones in patients with discrete small echoes in the gallbladder whether mobile or immobile (48). Although McIntosh et al. (7) found polyps in 12 out of 13 patients who had a constant immobile echogenic nonshadowing focus, a gallbladder polyp is a rare lesion (42), such an ultrasound feature occurs quite frequently with stones. Similarly a small fixed discrete echogenic shadowing focus on the gallbladder wall may also represent a stone (Fig. 13); calcification of the gallbladder wall can give the same appearance.

(c) A relatively large mass-like nonshadowing lesion. The nature of such lesions is frequently not apparent (25). However, if movement with gravity can be demonstrated, a carcinoma can be excluded. Again, strenuous effort may be necessary to demonstrate such movement, since the surrounding bile

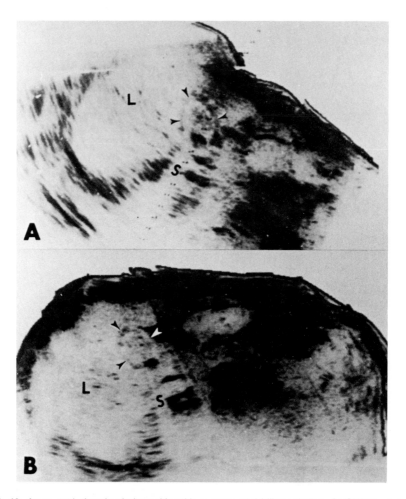

FIG. 12. Acute and chronic cholecystitis with a stone and biliary sludge. **A:** Oblique scan in decubitus position, right side up. **B:** Transverse scan 15 cm above umbilicus. The sludge causes diffuse weak echoes similar to those of the liver parenchyma. The gallbladder could be easily missed, especially on a transverse scan. The gallbladder is identified, however, because of a thickened hyperechoic wall (*arrowheads*). A stone is present in the dependent portion casting an acoustic shadow(s). L, Liver.

may be very viscous and the lesion may be composed of viscous material, e.g., a ball of biliary sludge. Further, the lesion may be of only slightly higher specific gravity than the surrounding bile, and therefore not easily moved. One may have to place the patient in a prone position to demonstrate movement. When the gallbladder is relatively small and globular, the movement of a relatively large lesion is even harder to demonstrate. The most common cause of round balls of echoes in the gallbladder is probably a ball of biliary

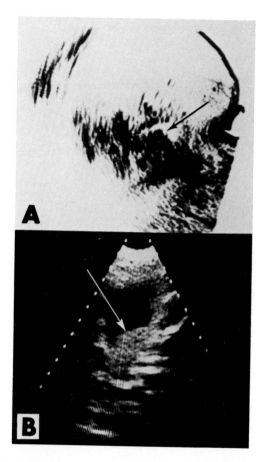

FIG. 13. Nonshadowing large gallstone. **A:** Transverse scan in decubitus position right side up. A large nonshadowing lesion (*arrow*) is seen in the dependent portion of the gallbladder. The scan was performed with a 2.25-mHz collimated transducer. **B:** Real-time sector scanning with 3.5-mHz transducer shows the same feature.

sludge (Fig. 14). This is an irregularly bordered poorly echogenic lesion without a discrete highly echogenic or sharp margin. It moves slowly with gravity and, at the same time, may change its shape. Some small strong echoes with or without an acoustic shadow may be seen within a lesion representing a small stone (or stones). The lack of a sharp echogenic margin separates it from a tumor.

Other possible causes of round echogenic nonshadowing gallbladder lesions are:

(a) a blood clot; (b) a ball of pus; (c) food content in a patient with a history of previous biliary bypass surgery or biliary fistula caused by chronic perforation; (d) parasitic cyst; (e) a fluid-filled bowel loop indenting on the gallblad-

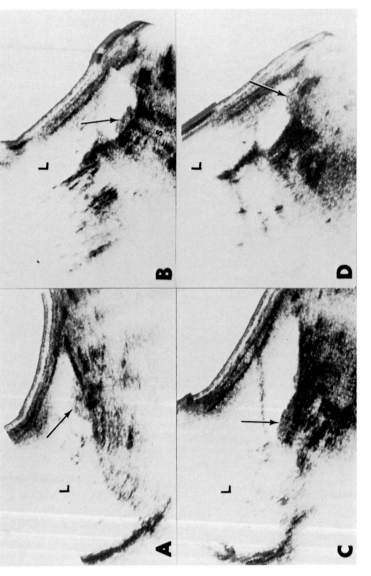

FIG. 14. Biliary sludge forming a nonshadowing echogenic lesion. **A:** Longitudinal scan 7 cm to the right of midline. A nonshadowing lesion (*arrow*) representing biliary sludge is seen on the posterior wall of the gallbladder. **B:** Transverse scan in decubitus position, right side up. The lesion appears somewhat irregular. A small highly echogenic shadowing focus within the lesion represents a small stone. **C:** Oblique scan in decubitus position. The lesion is near the neck of the gallbladder. **D:** Rescanning in a decubitus position after the patient was prone for about 10 min. The lesion has moved down to the fundus of the gallbladder.

der. When food particles are suspended in the fluid, weak echoes may be seen (Fig. 15). (f) An unusual type of stone (Fig. 13). Although a stone of such size almost always casts an acoustic shadow, on rare occasions an acoustic shadow cannot be demonstrated. The echoes are diffuse and of relatively low level compared with the usual stone, but of a higher level than those from sludge.

CHOLECYSTITIS

The ultrasonographic features of cholecystitis are: (a) increased gallbladder wall thickness. (b) Other abnormal ultrasonic gallbladder wall features such as an altered echogenicity of the thickened wall and irregularity or ill-defined outer margins. (c) Ultrasonic Murphy's sign. (d) Pericholecystic collections of fluid. Other findings that support the diagnosis of cholecystitis are: (a) the presence of gallstones; (b) numerous intraluminal weak echoes that may be in the dependent portion forming a fluid level or diffusely scattered throughout the gallbladder in a patient who is not fasting. These echoes may represent

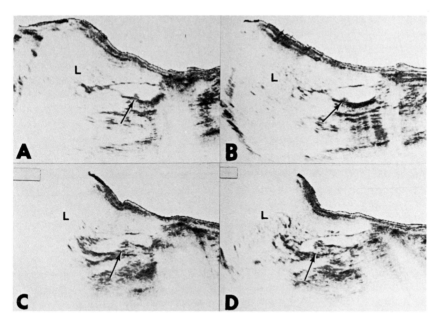

FIG. 15. Bowel loop indenting the gallbladder simulating an intraluminal nonshadowing lesion. **A:** Longitudinal scan 6 cm to the right of midline. A small echogenic nonshadowing lesion (*arrow*) on the posterior wall of the gallbladder actually represents a bowel loop indenting the gallbladder. L, Liver. **B:** Repeat scan with suspended deep inspiration. The "lesion" has moved cephalad in the gallbladder. This is actually caused by inferior movement of the gallbladder but not the bowel loop. **C:** Oblique scan in semidecubitus position, right side up. The "lesion" is now larger and appears to be part of an extrinsic structure, i.e., a bowel loop. **D:** Repeat scan as **C** but with suspended deep inspiration. The "lesion" again moves cephalad.

tissue debris or purulent gallbladder contents. Although none of the above features are pathognomonic of cholecystitis, when multiple features are seen in the clinical setting of cholecystitis, the diagnosis is almost certain. Some features are more suggestive of cholecystitis than others, and, therefore, are discussed in detail.

Gallbladder Wall: Thickness

The measurement of gallbladder wall thickness is most accurate in a region where the gallbladder wall is perpendicular to the ultrasound beam. However, since the gallbladder wall may be unevenly thickened, the thickness of other parts of the gallbladder should also be observed. The black echo line or lines are usually included in the measurement. Since the posterior wall is usually more difficult to measure than the anterior wall because of the effect of enhancement of the sound beam behind the "echo-free" bile, most authors (14,47) measure the anterior wall or the wall adjacent to the liver, i.e., the bare area of the gallbladder. By correlation with surgical specimens, the accuracy of the measurement has been found to be within a millimeter in 90% of cases (14). In fact, when gain settings are adjusted correctly for optimum visualization of the posterior wall, that wall can also be measured accurately. In examination of surgical specimens, Engel et al. (11) suggested that measurement of posterior wall is preferable to measurement of the anterior wall (bare area), since the latter was generally disrupted during surgical procedures. They obtained an accuracy of within 1 mm in 93% of patients and within 1.5 mm in 100% of patients by this method.

Generally, the thickness of the normal gallbladder wall in fasting patients is less than 3 mm (11,14,43), with the majority less than 2 mm (14) or 2.5 mm (11). Occasionally (2%) it may be as thick as 4 to 5 mm (14). Although initial experience suggested that a thickened gallbladder wall indicates gallbladder disease (33), and on pathologic examination the gallbladder wall in patients with chronic cholecystitis is almost always thickened, Sanders (43) found that the gallbladder wall was thickened in 45% of the patients with acute cholecystitis and in only 10% of the patients with chronic cholecystitis. Further, more thickening of the gallbladder wall can be associated with many conditions other than cholecystitis (10,52). It occurs in normal gallbladders that contract after meals. Fiske et al. (15) found marked hypoalbuminemia (average 2.7%/100 ml) in 40 out of 42 patients with a thickened gallbladder wall. Twenty-seven of these patients had chronic alcoholic liver disease. In 20 patients with a thickened wall, Sklaer et al. (47) found 12 without cholecystitis, but associated with hepatitis, alcoholic liver disease with a low serum albumin, heart failure, renal disease, or multiple myeloma. The gallbladder wall may appear artifactually thickened when surrounded by ascites (15). Diffuse infiltrative carcinoma (52) or lymphoma may also cause thickening of the gallbladder wall. The gallbladder wall may not be uniformly thickened in cholecystitis, whether acute or chronic (Fig. 16).

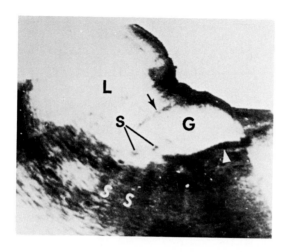

FIG. 16. Acute and chronic cholecystitis with stones. Oblique scan in semidecubitus position. The gallbladder wall is hyperechoic, the posterior wall (*white arrowhead*) is much thicker than the anterior one (*arrow*). There are stones (*black* S) above weak echoes caused by sludge. S, Shadow.

Ultrasonographic Features of the Thickened Gallbladder Wall

In acute cholecystitis the thickened gallbladder wall is generally poorly echogenic (15,44). Low-level echogenicity may involve only the inner layer (14) or the whole thickness of the wall (Fig. 17). This thickening is caused by edema and inflammatory changes that involve either the mucosa or all layers of the wall. The wall of a normal but contracted gallbladder may also be poorly echogenic. In acute cholecystitis the outer margins of the wall may be ill-defined or slightly irregular. In some patients, however, the gallbladder wall may be partly or totally hyperechoic (Fig. 18). Marchal et al. (32) described another feature of the thickened gallbladder wall in 11 patients that they considered characteristics for acute cholecystitis: a thickened wall consisted of a hyperechoic inner and outer layers and a sonolucent middle layer. They found one surgical specimen in which the mucosa was completely exfoliated; the inner hyperechoic layer represents necrotic lamina propria and muscularis. The middle anechoic (or hypoechoic) layer represents edematous thickened subserosa. The outer irregular hyperechoic layer represents edema and cellular infiltration of subserosa and adjacent liver parenchyma. These findings represent certain stages of acute cholecystitis, and may not be seen in every case of acute cholecystitis.

In chronic cholecystitis, the thickened walls are generally of high echogenicity, probably as a result of fibrotic changes. The echogenicity is, however, variable. In some instances, the echoes are similar to liver parenchyma, and are not, therefore, readily apparent (Fig. 19).

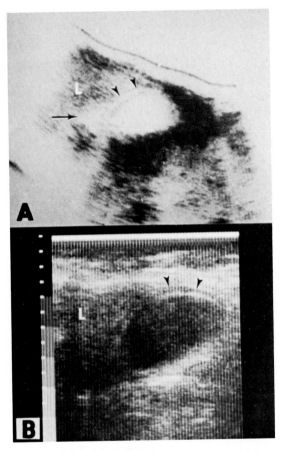

FIG. 17. Acute and chronic cholecystitis. **A:** Oblique scan in semidecubitus position, right side up. The anterior wall (*arrowheads*) of the gallbladder is thickened and hypoechoic. An ill-defined hypoechoic area (*arrow*) near the neck of the gallbladder is due to inflammatory changes in the adjacent liver parenchyma and not an abscess. L, Liver. A stone in the cystic duct was not visualized. **B:** Real-time scan of the same structure.

In the presence of ascites, the thickened gallbladder wall usually is hyperechoic. However, a double hyperechoic line and hypoechoic middle layer may also be seen.

Ultrasonographic Murphy's Sign

Since the location of the gallbladder is quite variable, the relationship of the point of maximum pain or tenderness to the actual site of gallbladder may not be quite clear on palpation alone. With the aid of ultrasonography, the exact

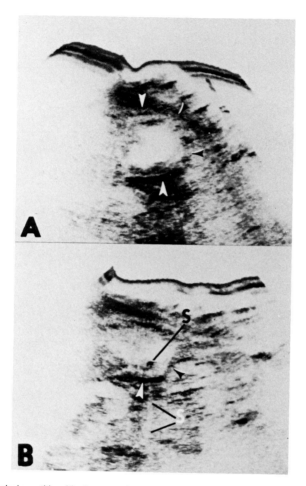

FIG. 18. Acute cholecystitis with stones and pericholecystic inflammatory mass. **A:** Oblique scan in semidecubitus position, right side up. The gallbladder wall is thickened. The thickened gallbladder wall is mainly hypoechoic except for the hyperechoic posterior wall. A hyperechoic band (*arrowhead*) surrounding the gallbladder represents a pericholecystic inflammatory mass. This together with gallbladder wall form a "layered" wall appearance. **B:** Oblique scan slightly lateral than **A**. A stone (*black* S) is seen within the gallbladder, casting an acoustic shadow (*white* S).

location of the gallbladder can be pinpointed on the patient. Hence whether the pain or tenderness is related to the site of the gallbladder can be more precisely evaluated. Therefore, the Murphy's sign demonstrated by ultrasonography is more accurate and of greater clinical significance than one demonstrated by simple physical examination.

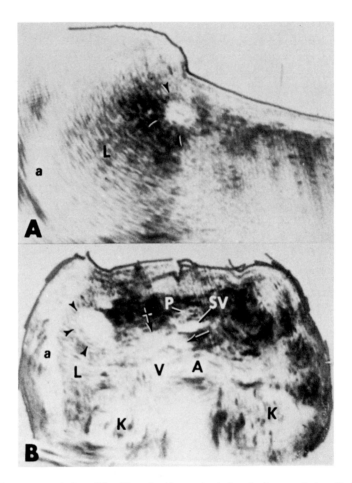

FIG. 19. Gangrenous cholecystitis with perforation and subphrenic abscess. **A:** Longitudinal scan 9 cm to the right of midline. The gallbladder wall (*arrowhead*) is thickened as a result of necrosis, hemorrhage, and inflammatory changes superimposed on fibrosis. The echo pattern of the gallbladder wall is similar to that of the adjacent liver parenchyma, hence not well seen. Some weak echoes are present within the gallbladder caused by blood and necrotic tissue debris. A subphrenic abscess (a) is seen. L, Liver. **B:** Transverse scan 9 cm above the umbilicus. The thickened lateral wall (*arrowheads*) of the gallbladder is hypoechoic. A subphrenic abscess (a) extends down lateral to the liver. Nodules (*arrow and broken arrow*) posterior to splenic vein (SV) and pancreas (P) represent metastatic adenocarcinoma to lymph nodes. The primary carcinoma was not found. V, Inferior vena cava; A, aorta; K, kidney.

Pericholecystic Collection of Fluid and Pericholecystitis

In 5 to 10% of patients with acute cholecystitis, the gallbladder may perforate. There are three types of perforation: (a) Acute free perforation; this may cause bile peritonitis and may have a catastrophic outcome. (b) Subacute perforation; this may result in walled-off abscess or pericholecystitic abscess. The pericholecystitic collection of fluid may also be caused by exudate from gangrenous or severe acute cholecystitis. (c) Chronic perforation; this may result in internal biliary fistula.

A pericholecystic collection of abscess or fluid may appear as an echo-free band partly or totally around the gallbladder. Since the fluid (or pus) may contain necrotic tissue debris or particles, a band of low-level echoes may be seen (3,21). This, together with the thickened gallbladder wall, may be mistaken for a markedly thickened layered gallbladder wall. The collection of fluid may be ill-defined, irregular, and/or multiloculated. It may have a complex echo-pattern with a predominantly echogenic or a predominantly anechoic area (31). Pericholecystic inflammatory tissue without actual fluid (or pus) may show similar features (Fig. 18). A liver abscess may be seen. However, inflammation of adjacent liver parenchyma without the formation of an abscess may appear as an ill-defined poorly echogenic area similar to an abscess (Fig. 17). A right subphrenic abscess is not unusual (Fig. 19).

Size of Gallbladder

In acute cholecystitis, the gallbladder may be either normal in size or enlarged. In chronic cholecystitis, the size of the gallbladder may be normal, large, or small. When it is small, the gallbladder may not be seen on the sonogram, especially if stones are present in the gallbladder. As previously mentioned, a nonvisualized gallbladder in a fasting patient is considered to be highly suggestive of gallbladder disease by many authors (14,29).

EMPHYSEMATOUS CHOLECYSTITIS AND PORCELAIN GALLBLADDER

These two entirely different entities have similar ultrasonographic features, i.e., a thick hyperechoic curvilinear structure with an acoustic shadow represents the anterior wall of the gallbladder (Fig. 20). The posterior wall may be completely or partially obscured by the acoustic shadow. The hyperechoic thick wall is either due to gas bubbles within an acutely swollen gallbladder wall in the case of emphysematous cholecystitis or to the calcification within a chronically fibrotic thickened wall in porcelain gallbladder. The gas bubbles in the gallbladder wall and/or within the gallbladder lumen are responsible for the acoustic shadow in the former (23), and the calcification causes shadowing in the latter. In emphysematous cholecystitis, the incidence of perforation of gallbladder is five times greater than that in acute cholecystitis, and therefore it is a surgical emergency. It may result from ischemia following an invasion of the gallbladder wall by gas-forming microorganisms. Gallstones are not usually associated with emphysematous cholecystitis. In porcelain gallbladder, there is a high incidence of gallbladder carcinoma. As a result of the acoustic

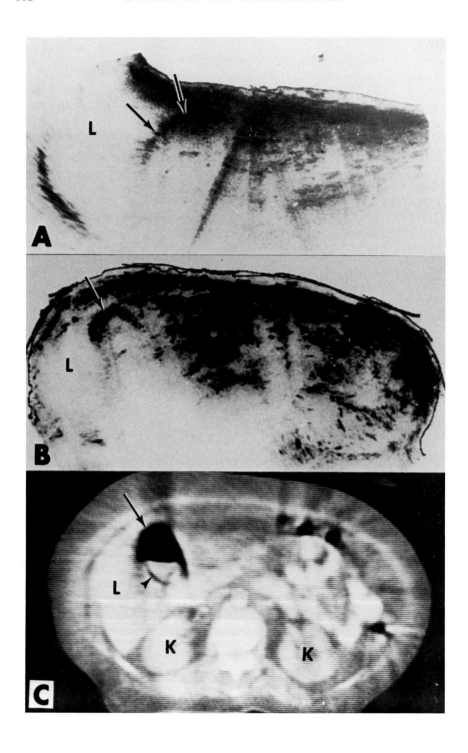

shadow, the tumor may be partly or completely obscured. A careful examination in different scanning planes should be attempted to try to exclude carcinoma.

Ultrasonographic differentiation between the two entities may be difficult. Although the acoustic shadow from gas in the emphysematous cholecystitis may contain more reverberation echoes than the acoustic shadow from calcification in porcelain gallbladder, this difference may not be very apparent. However, plain radiography will readily reveal the abnormal gas or calcification in the gallbladder wall. Therefore, radiographs should be performed when either of these entities is suspected. A CT scan may be helpful in demonstrating gas (Fig. 20) or calcification in the wall and tumor mass.

TUMOR OF THE GALLBLADDER

Ultrasonography is one of the most important imaging modalities in the diagnosis of carcinoma of the gallbladder. Although the ultrasonographic and CT features of the carcinoma of gallbladder have not often been described in the literature (9,24,36,52,54), ultrasonography and CT provide the only potential means of preoperative diagnosis, since, in most patients, a gallbladder carcinoma is not visualized by oral cholecystography, intravenous cholangiography, or percutaneous transhepatic cholangiography. A preoperative diagnosis was very rarely made before ultrasonography and CT were available.

The ultrasonographic features of carcinoma of the gallbladder can be grouped into four types (Fig. 21). The most common (Type I) is a poorly echogenic mass in the gallbladder fossa representing the gallbladder full of tumor. A variation (Type Ib) is a mass containing a highly echogenic focus that represents gallstone (or stones), i.e., "constrained stone sign" (Fig. 22), or a necrotic center in the mass or residual lumen of gallbladder. The second most common type (Type II) is a diffusely or focally thickened gallbladder wall caused by carcinomatous infiltration. The thickened gallbladder wall is poorly echogenic, similar to the findings in acute cholecystitis. Type III is an irregular hypoechoic mass protruding from the gallbladder wall into the lumen representing a fungating mass (Fig. 23). In Type IV, a fungating mass arises from an area of localized thickening of the gallbladder wall as a result of infiltration by the tumor (Fig. 24). The fungating mass (Type III) and well-localized wall thickening (localized Type II) without metastatic lesion may

---

FIG. 20. Emphysematous cholecystitis. **A:** Longitudinal scan 9 cm to the right of midline. A thickened hyperechoic anterior gallbladder wall (*arrows*) casts an incomplete acoustic shadow, so that the posterior wall is partly seen. L, Liver. **B:** Transverse scan 10 cm above umbilicus. A thick echogenic anterior gallbladder wall (*arrow*) is again seen. **C:** CT corresponding to **B**. The gas-filled anterior gallbladder wall (*arrow*) and intraluminal gas are seen together as a single gas bubble that forms a gas fluid level. The gas-filled posterior wall (*arrowheads*) is, however, clearly visualized. K, Kidney.

TYPE

Ia

b

II

III

IV

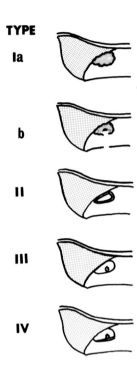

FIG. 21. Ultrasonographic features of gallbladder carcinoma. Type I: Gallbladder full of mass. Type Ia: Diffuse weak echoes or irregular echoes in the gallbladder. Type Ib: Strong central echoes in addition to diffuse weak echoes, i.e., a "constrained stone sign." Type II: Thickened gallbladder wall caused by infiltrating carcinoma. Type III: A fungating mass on the gallbladder wall. Type IV: A fungating mass with infiltrated thickened wall. (Modified from ref. 52, with permission.)

represent early stages of detectable carcinoma. Unfortunately, metastatic lesions are usually present when gallbladder carcinoma is first detected.

The features that distinguish Type I from empyema of the gallbladder or a gallbladder full of sludge of bile are (a) In carcinoma the contour of the gallbladder is usually irregular and sometimes ill-defined. (b) The internal echoes are irregular in size and distribution in carcinoma whereas they are uniform in size and homogeneous in empyema. (c) The central strong echoes in Type Ib usually is not seen in empyema. However, on rare occasion, a light stone may float in a thick bile (or pus) and cause such an appearance. If movement of the stone cannot be demonstrated, carcinoma should be suggested. A stone encased within the gallbladder (constrained stone sign) (Fig. 22) is almost a pathognomonic sign for Type Ib carcinoma. The Type II carcinomas are similar to acute cholecystitis in that the gallbladder wall may be uniformly or irregularly thickened. Although thickening confined to a relatively small area may be more suggestive of tumor, thickening of only the posterior or anterior wall may occur in cholecystitis (Fig. 16). Evidence of invasion of the liver or metastases to the peripancreatic or other retroperitoneal nodes are important findings that support a diagnosis of malignant disease. Correlation with clinical features may also help one to arrive at a correct diagnosis.

Type III carcinoma may resemble unusually large relatively hypoechoic nonshadowing gallstones (Fig. 13), a blood clot, or a ball of sludge of bile. If

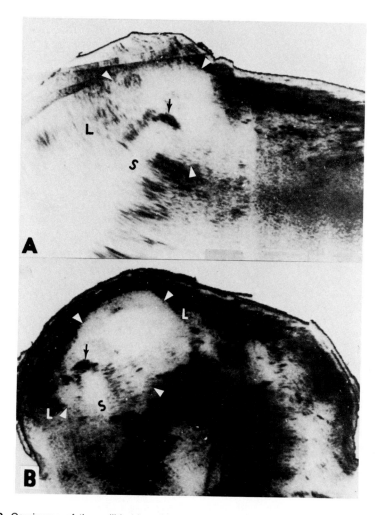

FIG. 22. Carcinoma of the gallbladder with a stone—"constrained stone sign" (Type lb). **A:** Longitudinal scan 7 cm to the right of midline. A large mass (*white arrowheads*) is located below the right lobe of the liver (L). The mass contains a stone (*arrow*) that casts an acoustic shadow (S). The mass represents gallbladder full of carcinoma that encases a stone. **B:** Transverse scan 10 cm above umbilicus. The stone (*arrow*) is located eccentrically within the gallbladder tumor (*white arrowheads*). The stone did not move in the decubitis position (*not shown*).

movement with gravity can be demonstrated, carcinoma can be excluded. An irregular but sharply defined surface to the lesion is more suggestive of carcinoma than the other entities. Local thickening of the underlying gallbladder wall (i.e., Type IV masses) further suggests the possibility of carcinoma. Again, evidence of invasion of the liver or metastasis will confirm the carcinoma.

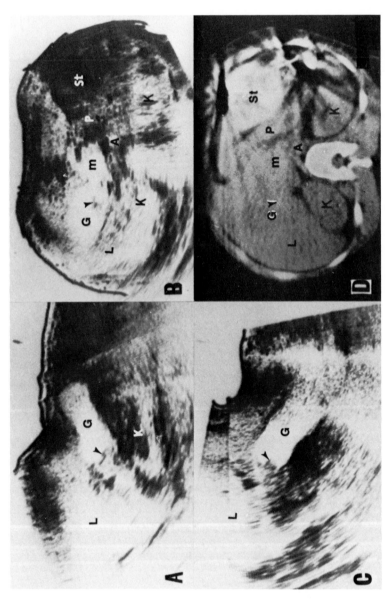

FIG. 23. Carcinoma of the gallbladder (Type II). An irregular fungating mass (*arrowhead*) is located medially near the gallbladder neck. The lesion did not move in the upright position (C). A metastatic lesion (m) is posterior to the splenic vein. A, Aorta; G, gallbladder; K, kidney; L, liver; P, pancreas; st, stomach. **A:** Longitudinal scan 7 cm to the right of midline. **B:** Transverse scan 12.5 cm above umbilicus. **C:** Longitudinal scan with the patient sitting. **D:** CT corresponding to **B** (From ref. 52, with permission.)

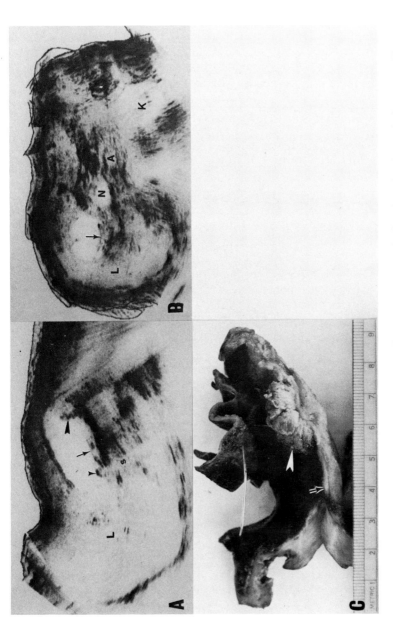

FIG. 24. Type IV carcinoma of the gallbladder in an 81-year-old woman. The gallbladder is markedly enlarged as the result of empyema. There is a fungating mass (*large arrowhead*) on the posterior wall of the gallbladder near the fundus. The posterior wall (*arrow*) is thickened as a result of infiltration by tumor. A stone (*small arrowhead*) with an acoustic shadow(s) is also seen. There is a metastasis to a retroperitoneal node (N) near the head of the pancreas. Note the irregular but sharp border to the mass (*large arrowhead*), which is different from that derived from a ball of biliary sludge. **A:** Longitudinal scan 4 cm to the right of the midline. **B:** Transverse scan 13 cm above the umbilicus. **C:** Surgical specimen. (From ref. 52, with permission.)

Carcinomas of the gallbladder are frequently associated with other features such as: (a) gallstones (in 65 to 95% of the patients), (b) chronic cholecystitis or empyema of the gallbladder, and (c) dilatation of biliary duct due to obstruction by direct invasion or compression by metastatic mass of the common bile or common hepatic duct. Although primary carcinomas of the gallbladder are usually hypoechoic, secondary lesions in the liver may be either hypoechoic or hyperechoic (52).

ADENOMYOMATOSIS

The ultrasonographic features of this entity have not been described much in the literature (41,43). The characteristic appearance is a localized solid lesion with small echo-free pockets that represent Rokitansky-Ashoff sinuses (41). When these echo-free pockets are too small to be delineated, a localized mass lesion may be seen (43). This may be difficult to differentiate from a carcinoma. I have seen a case with segmental involvement of the middle portion of gallbladder. The gallbladder was divided into two compartments, the proximal one being similar to a normal gallbladder. The distal one contained two large stones. The involved middle segment appeared as a solid mass. The lumen of the middle segment and Rokitansky-Aschoff sinuses were too small to be seen on the sonogram.

HYDROPS (MUCOCELE) OF THE GALLBLADDER

The gallbladder is markedly dilated as a result of obstruction of the cystic duct or neck of the gallbladder, usually by a stone, but less frequently by neoplasia within the cystic duct or kinking of the duct and sometimes by edema of the cystic duct or compression of the cystic duct by enlarged lymph nodes in the mucocutaneous lymph node syndrome. The entrapped bile is resorbed and the gallbladder becomes filled with a clear mucinous secretion from the gallbladder wall (42). Ultrasonography shows a large echo-free gallbladder that does not contract after a fatty meal or injection of cholecystokinin. Stones in the gallbladder neck may be demonstrated, whereas stones in the cystic duct may be difficult to visualize.

Mucocutaneous lymph node syndrome (Kawasaki disease) is a disease mostly seen in children that is prevalent in Japan but rare in the United States (17). Clinical features consist of prolonged fever, conjunctivitis, red mouth and lips, red swollen palms and soles followed by peeling, generalized rashes, nonsuppurative cervical lymphadenopathy, and hydrops of the gallbladder (27). The etiology is unknown; both infections and toxic causes have been suggested. The gallbladder eventually returns to normal as the patient's condition improves (27,49). A change in the size of the gallbladder can easily be assessed by ultrasonography. Ultrasonography is not only important in making the diagnosis of hydrops, but it is also an aid in the follow-up of such patients (Fig. 25).

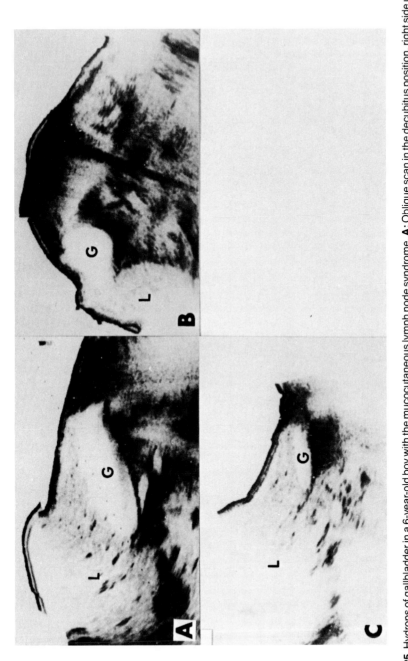

FIG. 25. Hydrops of gallbladder in a 6-year-old boy with the mucocutaneous lymph node syndrome. **A:** Oblique scan in the decubitus position, right side up: The gallbladder is markedly enlarged. A band of weak echoes in the gallbladder is a result of artifact. L, Liver. **B:** Transverse scan 6 cm above umbilicus. **C:** Same scan as **A**, repeated 1 month after medical treatment. The gallbladder (G) has decreased greatly in size.

REFERENCES

1. Baker, H. L., and Hodgson, J. R. Further studies on the accuracy of oral cholecystography. *Radiology*, 74:239–245, 1960.
2. Bartrum, R. J., Crow, H. C., and Foote, S. R. Ultrasonic and radiographic cholecystography. *N. Engl. J. Med.*, 296:538–541, 1977.
3. Bergman, A. B., Neiman, H. L., and Kraut, B. Ultrasonographic evaluation of pericholecystic abscess. *Am. J. Roentgenol.*, 132:201–203, 1979.
4. Callen, P. W., and Filly, R. A. Ultrasonographic localization of gallbladder. *Radiology*, 133:687–691, 1979.
5. Conrad, M. R., Janes, J. O., and Dietchy, J. Significance of low level echoes within the gallbladder. *Am. J. Roentgenol.*, 132:967–972, 1979.
6. Conrad, M. R., Leonard, J., and Landey, M. J. Left lateral decubitus sonography of gallstones in the contracted gallbladder. *Am. J. Roentgenol.*, 134:141–144, 1980.
7. Crade, M., Taylor, K. J. W., Rosenfield, A. T., and de Graaf, C. S. Pathologic correlation of cholecystosonography—a report of 145 cases. Surgical and pathological correlation of cholecystography and cholecystosonography. *Am. J. Roentgenol.*, 131:277–229, August 1978.
8. Crade, M., Taylor, K. J. W., Rosenfield, A. T., de Graaf, C. S. and Minihan, P. Surgical and pathologic correlation of cholecystosonography. *Am. J. Roentgenol.*, 131:227–229, 1978.
9. Crade, M., Taylor, K. J. W., Rosenfield, A. T., Ulreigh, S., Simeone, J., Sommer, G., and Viscomi, G. N. The varied ultrasonic character of gallbladder tumor. *JAMA*, 241:2195–2196, 1979.
10. Crade, M., Taylor, K. J. W., and Viscomi, G. N. Need for care in assessing gallbladder wall thickening (letters). *Am. J. Roentgenol.*, 135:423, 1980.
11. Engel, J. M., Deitch, E. A., and Sikkema, W. Gallbladder wall thickness: Sonographic accuracy and relation to disease. *Am. J. Roentgenol.*, 134:907–909, 1980.
12. Filly, R. A., Allen, B., Minton, M. J., Bernhaft, R., and Way, L. W. *In vitro* investigation of the origin of echoes within biliary sludge. *J. Clin. Ultrasound*, 8:193–200, 1980.
13. Filly, R. A., Moss, A. A., and Way, L. W. *In vitro* investigation of gallbladder stones shadowing with ultrasound tomography. *J. Clin. Ultrasound*, 7:255–262, 1979.
14. Finberg, H. J., and Birnholz, J. C. Ultrasound evaluation of the gallbladder wall. *Radiology*, 133:693–698, 1979.
15. Fiske, C. E., Laing, F. C., and Brown, T. W. Ultrasonographic evidence of gallbladder wall thickening in association with hypoalbuminemia. *Radiology*, 135:713–716, 1980.
16. Goldberg, B. B. Gallbladder and bile ducts. In: *Abdominal Gray Scale Ultrasonography*, edited by B. B. Goldberg, pp. 137–165, 1977. John Wiley, New York.
17. Goldsmith, R. W., Gribetz, D., and Strauss, L. Mucotaneous lymph node syndrome (MLNS) in the continental United States. *Pediatrics*, 57:431–434, 1976.
18. Gonzalez, L., and MacIntyre, W. J. Acoustic shadow formation by gallstones. *Radiology*, 135:217–218, 1980.
19. Gooding, G. A. Food particles in the gallbladder mimic cholelithiasis in a patient with a cholecystojejunostomy. *J. Clin. Ultrasound*, 9:346–347, 1981.
20. Grossman, M. Cholelithiasis and acoustic shadowing. *J. Clin. Ultrasound*, 6:182–184, 1978.
21. Handler, S. J. Ultrasound of gallbladder wall (letter). *Am. J. Roentgenol.*, 133:995, 1979.
22. Harned, R. K., and Babbitt, D. P. Cholelithiasis in children. *Radiology*, 117:391–393, 1975.
23. Hunter, N. D., and MacIntosh, P. K. Acute emphysematous cholecystitis, an ultrasonic diagnosis. *Am. J. Roentgenol.*, 134:592–593, 1980.
24. Itai, Y., Araki, T., Yashikawa, K., Furui, S., Yashiro, N., and Tasaka, A. CT of gallbladder carcinoma. *Radiology*, 137:713–718, 1980.
25. Jeanty, P., Gansbeke, D. V., Leclercq, F. R., Engelholm, L., Amman, W., Cooperberg, P., Gooding, G., and Kunstlinger, F. Unusual intraluminal masses of the gallbladder. *Presented at 67th Scientific Assembly and Annual Meeting of the Radiological Society of North America, Chicago, Illinois, Nov. 15–20, 1981.*
26. Krook, P. M., Allen, L. H., Bush, W. H., Malmer, G., and MacLean, M. D. Comparison of real-time cholecystosonography and oral cholecystography. *Radiology*, 135:145–148, 1980.

27. Kumari, S., Lee, W. J., and Baron, M. Hydrops of the gallbladder in a child: Diagnoses by ultrasonography. *Pediatrics*, 63:295–297, 1979.
28. Lachman, B. S., Lezerson, J., Starshek, R. J., Vaughters, F. M., and Werlin, S. L. The prevalence of cholelithiasis in sickle cell disease as diagnosed by ultrasound and cholecystography. *Pediatrics*, 64:601–603, 1979.
29. Leopold, G. R. Ultrasonography of jaundice. *Radiol. Clin. North Am.*, 17:127–136, 1979.
30. Leopold, G. R., Amberg, J., Gosink, B. B., and Mittelstaedt, C. Gray scale ultrasonic cholecystography: A comparison with conventional radiographic technique. *Radiology*, 121:445–448, 1976.
31. Madrazo, B. L., Francis, I., Hricak, H., Sandler, M. A., Hudak, S., and Kamidia, G. Gallbladder perforation: specific diagnosis by sonography. *Presented at 67th Scientific Assembly and Annual Meeting of the Radiological Society of North America, Chicago, Ill., Nov. 15–20, 1981.*
32. Marchal, G. J. F., Casaer, M., Baerl, A. L., Goddeeris, P. G., Kerremano, R., and Fevery, J. Gallbladder wall sonolucency in acute cholecystitis. *Radiology*, 133:429–433, 1979.
33. Marchal, G., Crolla, D., Baert, A. L., Fevery, J., and Kerremans, R. Gallbladder wall thickening: A new sign of gallbladder disease visualized by gray scale cholecystosonography. *J. Clin. Ultrasound*, 6:177–179, 1978.
34. Mittelstaedt, C., and Leopold, G. R. B-scan ultrasound of liver, gallbladder and pancreas. *Int. Surg.*, 62:277–281, 1977.
35. Ochsner, S. F. Performance and reliability of cholecystography. *South. Med. J.*, 63:1268–1272, 1970.
36. Olken, S. M., Bladroe, R., and Newark, H. The ultrasonic diagnosis of primary carcinoma of the gallbladder. *Radiology*, 129:481–482, 1978.
37. Paruleker, S. G. Ligaments and fissures of the liver: Sonographic anatomy. *Radiology*, 130:409–411, 1979.
38. Purdom, R. C., Thomas, S. R., Kereiakes, J. G., Spitz, H. B., Goldenberg, N. J., and Krugh, K. B. Ultrasonic properties of biliary calculi. *Radiology*, 136:729–732, 1980.
39. Reptopoulos, V. D. Ultrasonic pseudocalculus effect in postcholecystectomy patients. *Am. J. Roentgenol.*, 134:145–148, 1980.
40. Raptopoulos, V., D'Orsi, C., Smith, E., Renter, K., Moss, L., and Kleinman, P. Dynamic cholecystosonography of the contracted gallbladder: The double-arc-shadow-sign. *Am. J. Roentgenol.*, 138:275–278, 1982.
41. Rice, J., Sauerbrei, E. E., Semogas, P., Cooperberg, P. G., and Burhenne, H. J. Sonographic appearance of adenomyomatosis of the gallbladder. *J. Clin. Ultrasound*, 9:336–337, 1981.
42. Robbins, S. L. *Pathology*, 3rd ed., pp. 946–962, 1967. W. B. Saunders, Philadelphia.
43. Sanders, R. C. The significance of sonographic gallbladder wall thickening. *J. Clin. Ultrasound*, 8:143–146, 1980.
44. Sanders, R. C., and Zerhouni, E. The significance of ultrasonic gallbladder wall thickness. *Presented at 23rd Annual Meeting of the American Institute of Ultrasound in Medicine, San Diego, Oct. 19–23, 1978.*
45. Sarnaik, S., Slovis, T. L., Corbett, D. P., Emanii, A., and Whitten, C. F. Incidence of cholelithiasis in sickle cell anemia using the ultrasonic gray-scale technique. *J. Pediatr.*, 96:1005–1008, 1980.
46. Sherman, M., Ralls, P. W., Quinn, M., Halls, J., and Keats, J. B. Intravenous cholangiography and sonography in acute cholecystitis: Prospective evaluation. *Am. J. Roentgenol.*, 135:311–313, 1980.
47. Shlaer, W. J., Leopold, G. R., and Scheible, F. W. Sonography of the thickened gallbladder wall, a non-specific finding. *Am. J. Roentgenol.*, 136:337–339, 1981.
48. Simeone, J. F., Mueller, P. R., Ferrucci, J. T., Harbin, W. P., and Wittenberg, J. Significance of non-shadowing focal opacities at cholecystosonography. *Radiology*, 137:181–185, 1980.
49. Slovis, T. L., Hight, D. W., Philippant, A. I., and Dubois, R. S. Sonography in the diagnosis and management of hydrops of the gallbladder in children with mucocutaneous lymph node syndrome. *Pediatrics*, 65:789–794, 1980.
50. Sommer, F. G., and Taylor, K. J. W. Differentiation of acoustic shadowing due to calculi and gas collections. *Radiology*, 135:399–403, 1980.

51. Taylor, K. J. W., Jacobson, P., and Jaffe, C. C. Lack of an acoustic shadow on scans of gallstones: A possible artifact. *Radiology*, 131:463–464, 1979.
52. Yeh, H. C., Ultrasonography and computed tomography of carcinoma of the gallbladder. *Radiology*, 133:167–173, 1979.
53. Yeh, H. C., and Rabinowitz, J. G. Ultrasonography and computed tomography of the liver. *Radiol. Clin. North Am.*, 18:321–338, 1980.
54. Yum, H. Y., and Fink, A. H. Sonographic findings in primary cancer of the gallbladder. *Radiology*, 134:693–696, 1980.

Ultrasound Annual 1982,
edited by Roger C. Sanders.
Raven Press, New York © 1982.

How Relevant is the Small Parts Scanner in the Investigation of the Carotid Artery?

Edward Robert Lipsit

*Department of Radiology, The George Washington University Medical Center; and
Department of Radiology, Columbia Hospital for Women, Washington, D.C. 20037*

This chapter critically examines a relatively new method for detecting atherosclerotic disease in the extracranial carotid artery. Noninvasive ultrasonic imaging of the carotid artery is of much importance, since stroke is a leading cause of morbidity and mortality in our society, and the association of carotid artery disease and cerebrovascular accidents is well known (21).

This new technique utilizes the "duplex" scanner—high-resolution real-time ultrasonic imaging with simultaneous Doppler signal analysis. An in-depth discussion of all noninvasive techniques is beyond the scope of this chapter. Nonetheless, an overview of other techniques is presented to illustrate the accuracy and ultimate role of the duplex scanner in detecting atherosclerotic disease in the carotid artery.

INSTRUMENTATION AND BASIC PHYSICS

The duplex scanner performs two separate functions that combine to provide useful diagnostic information. Although high-resolution real-time B-scan ultrasound (small parts scanner) is an excellent method for visualizing carotid anatomy (55), combination with Doppler signal analysis provides even greater accuracy (18,44,45,60). Our discussion of instrumentation covers high-resolution real-time scanners, continuous wave and pulsed Doppler systems, and the combined duplex scanning system.

High-Resolution Real-Time Systems

At present there are a number of commercially available systems known as "small parts" scanners. This term characterizes their application. These scanners are used to image small superficial structures. High resolution is achieved by utilizing high-frequency transducers with a frequency of 7.5–10 MHz. Physical limitations result in a small field of view. Most systems image an area approximately 3 to 5 cm in length and 4 to 8 cm in depth. The "trade-off" in the limited field of view is offset by superior resolution; a 7.5-MHz system will achieve axial and lateral resolution approximating 0.25 mm and

0.8 mm, respectively. The frame rate for each system will vary, and most units offer digital scan converters capable of pre- and postprocessing data manipulation. A digital caliper that enables a rapid measurement of distance, circumference, and area is available on some units. Aside from software considerations, a major difference in the available real-time systems concerns the geometry and design of the scan probe. The following systems illustrate this fact.

The Microview system (Picker Corp.) utilizes a 10-MHz transducer within a flexible self-contained water bath. The real-time image is achieved by reciprocating linear motion. The fluid tension within the water bath can be readily adjusted by means of a valve. The probe is mounted on a large counterbalanced pivoting scan arm with a broad and relatively immobile base. The transducer housing rotates 350 degrees to achieve proper alignment. A square image is created.

The Mark 500 System (Advanced Technology Laboratories, Inc.) employs a rotating transducer assembly with a variable frequency of 3.0, 5.0, or 7.5 MHz selected by the operator. This probe, which is quite small, has a hard surface; it is freely positioned, since its only connection to the scan converter is by cable. A pulsed Doppler system with graphic display is included in this portable unit. A pie-shaped image is created.

The BioSound (Bio-Dynamics, Inc.) scan probe consists of an 8-MHz stationary transducer within a fluid-filled lucite block. The real-time image is produced by an oscillating acoustic mirror. The probe is hand-held, and the entire system can be transported. Although the effective scan surface of the lucite housing is small, the probe housing is inflexible. Probe heads with surfaces measuring 4 cm and 8 cm as well as a duplex probe providing an audible Doppler signal are available.

Diasonics, Inc. offers a duplex module that is compatible with its standard real-time systems. The probe is hand-held and the scan surface is a flexible membrane. A 7.5-MHz imaging transducer is housed within a fluid chamber and is moved mechanically in a reciprocating fashion to generate a sector image. A separate 3-MHz Doppler transducer is contained within the probe housing, and the beam axis is always aligned in the plane of the real-time image.

Although these representative systems vary in features and physical characteristics, they all produce a high-resolution real-time image at the expense of depth penetration and field of view. These units are, therefore, suitable only for superficial imaging and are appropriate for the study of the carotid vasculature.

Doppler

The Doppler technique is based on the following principle: When a sound wave interacts with a moving structure, such as red blood cells within a vessel, the moving structure will cause the reflected or back-scattered signal frequency to be shifted by an amount proportional to the interface velocity along the sound beam axis. This phenomenon is described by the equation:

$$\Delta f = \frac{2\bar{V}f_0 \cos \theta}{C}$$

Δf = Doppler frequency shift in hertz
$\bar{V}$ = vector velocity of interface (cm/sec), e.g., red blood cells
f_0 = frequency of impinging sound beam (hertz)
C = velocity of sound in soft tissue (1,540 m/sec)
θ = angle in degrees between the sound beam axis and the vector velocity.

When a Doppler sound beam passes through a blood vessel, a number of events takes place. Small amounts of energy are absorbed and radiated by each red blood cell. If the red blood cells are moving within the vessel, a frequency shift will be detected. The magnitude and direction of this shift will be proportional to the velocity of the red blood cells. If the sound beam fills the entire vessel lumen, the back-scattered signal will consist of all Doppler shifts caused by the red blood cells passing through the sound beam. The velocity of the red blood cells will vary depending upon their location within the lumen. Peak velocity will be at the center of the vessel. Variations in the "frequency spectrum" of the Doppler signal will be related to blood flow disturbances produced by a stenosis or vessel wall irregularity due to atheromatous plaque (20). A narrowed lumen will be a region of high flow velocity and will be represented as a high-pitched, audible Doppler signal. Poststenotic turbulent flow will create a harsh sound as high kinetic energy stored in the red blood cells is released (2).

Continuous Wave Doppler

A discussion of Doppler instrumentation centers around two basic types: continuous-wave and pulsed. As the name implies, continuous-wave Doppler systems utilize a continuous sound beam. The beam is characterized only by frequency and width. Sound beams are usually in the 3 to 10 MHz range, with a very narrow beam (less than 1 mm) possible at 10 MHz. The lateral width of the beam is the only specification for resolution. A variety of simple hand-held units are available for the evaluation of superficial arterial and venous flow. These devices produce an audible signal, and practitioners must be thoroughly familiar with its interpretation.

Pulsed Doppler

This type of Doppler system is used in today's duplex scanners and is more complex than continuous-wave instruments. Doppler shifts are detected at selected points along the path of the sound beam. This "pulsed" source emits regular bursts of sound into the tissues, and the depth of the returning signal is determined by the time it takes for the signal to strike an interface (i.e., red blood cells) and return to the transducer. This technique provides high axial resolution (approximately 1 mm), and the "range gate" allows evaluation of flow at a specified point below the skin surface. The combination of real-time imaging and pulsed Doppler enables the examiner to pinpoint the Doppler analysis within the vessel lumen.

As in other areas of diagnostic ultrasound, various physical parameters influence the nature of the pulsed Doppler signal (Table 1). The transducer frequency and the pulse repetition rate will determine the overall depth limitation of the system. The most useful Doppler frequency will depend on the clinical application, and in most situations, a transducer frequency of 3 to 5 MHz and an appropriate pulse repetition rate will provide the optimal combination of depth penetration and sample volume for evaluating the carotid artery.

In practice, the point at which blood flow velocity is measured is called the sample volume. Its location depends on a time delay and is manually set by a depth control. In theory, the sample volume for a commercially available 3-MHz system is 2 mm in diameter and 3 mm in length. The audible or recorded shift in the original Doppler signal frequency will reflect the sum of all red blood cell movement within the sample volume.

Doppler Imaging

Doppler imaging or Doppler arteriography is an interesting approach in which the examiner "maps out" the vasculature on the basis of the Doppler signal (49). A continuous-wave Doppler transducer is fixed to a mechanical arm and an image is produced on a storage oscilloscope that corresponds to each point that flow velocity exceeds a predetermined threshold. This type of examination has a number of potential pitfalls. There is a lack of depth resolution, and two vessels "in line" may not be detected as separate entities unless flow is in opposite directions and a direction-sensitive instrument is used (51). Also, images are limited to the longitudinal plane.

In addition to continuous-wave imaging systems, units are now available that employ pulsed Doppler signals and allow the operator to select a range

TABLE 1. *Doppler "trade offs"*

↑ Frequency	↓ Length of sample volume
↑ Sensitivity	Longer sound bursts
	↑ Length of sample volume
↓ Width of signal	↓ Transducer diameter
	↑ Transducer focusing

gate beneath the skin surface (8,35). The depth of penetration is usually divided into six discrete tissue compartments. Spectral frequency analysis, which is discussed later in this chapter, is also included in some units and may increase diagnostic accuracy. Despite these additional modifications, all of these systems are dependent upon blood flow for imaging, and a totally occluded lumen will result in nonvisualization of the vessel.

Duplex Scanners

The combination of the small parts scanner and pulsed Doppler instrumentation is a most logical step, since the evaluation of small sample volumes requires imaging for precise localization. This type of system provides both morphologic and hemodynamic information; the ability to generate such data truly distinguishes this method from other noninvasive tests. At this point a discussion of "duplex data" is in order.

Volumetric blood flow, expressed in liters or milliliters per minute, is generally considered to represent a measure of tissue perfusion or an index of possible tissue viability (2). When measured concomitantly with blood pressure, we can estimate vascular blood flow resistance. Unfortunately, the noninvasive measurement of volumetric blood flow is no simple task. We must initially determine the area of the vessel lumen (cm^2) and then the average flow velocity (cm/sec) over that area. The calculation of average flow velocity requires a knowledge of the precise angle between the sound beam and the axis of blood flow. The calculation of volumetric flow rate is described by the equation: $Q = VA$, where

Q = volumetric flow rate (liters/min or cm^3/sec)
V = average velocity (cm/sec) along vessel axis at 90 degrees to the plane of the cross-sectional area
A = area of blood vessel lumen (cm^2).

Since measurement of volumetric flow is technically difficult, other parameters related to velocity have evolved. Commercially available systems compute information that may include:

(a) Average velocity over vessel lumen presented as an analog waveform.
(b) Velocity at a point within the vessel presented as an analog waveform.
(c) Velocity distribution summed across the lumen of a blood vessel and displayed as a frequency spectrum.
(d) Velocity distribution within a small sample volume or point located within the vessel and displayed as a frequency spectrum.

The clinical significance and value of the duplex information are discussed following a brief description of high-resolution real-time and duplex scanning technique.

SMALL PARTS SCANNING AND PULSED DOPPLER TECHNIQUE

The examination begins with the patient supine and the neck extended. The patient's head is turned away from the examiner. With a high-resolution real-time system, the carotid artery and jugular vein are located in a transverse plane with the probe positioned just above the clavicle. The probe is slowly moved in a cephalad direction until the carotid bulb and bifurcation are identified (Fig. 1). As the examiner performs this transverse scan the presence and location of a stenotic region are observed and recorded. This transverse view is ideal for locating and assessing the site of maximum luminal compromise (Fig. 2). The degree of stenosis is estimated by measuring the diameter and cross-sectional area of the vessel lumen in a region of stenosis and comparing these with a normal-appearing segment of the vessel.

A longitudinal scan of the carotid vasculature is then performed. The probe is turned 90 degrees and the common carotid artery is located. Ideally, one can image the carotid artery from its origin to a point several centimeters above the bifurcation (Fig. 3). This longitudinal or sagittal view allows the examiner to determine the length of a lesion. Atheromatous plaques and stenotic areas are detected and recorded with the bifurcation used as a reference point (Fig. 4). This longitudinal approach occasionally allows simultaneous visualization of both the internal and the external carotid branches. Difficulty may be encountered as a result of the anteromedial location of the external branch in the majority of patients. When simultaneous imaging is achieved, it is usually accomplished in younger patients, who are more able to

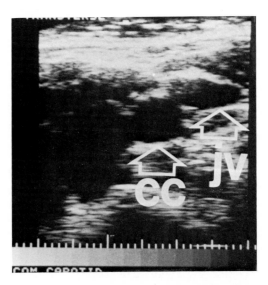

FIG. 1. Transverse scan of the left common carotid artery (CC) and the more lateral jugular vein (JV).

extend their necks fully and who usually possess less tortuous vasculature (Fig. 5). In addition, a posterior longitudinal approach just below the patient's ear may allow an image of both branches when the anterior approach is unsuccessful. In most cases the experienced examiner will be able to differentiate the two separately imaged branches by gently rocking the transducer in a medial-lateral direction.

If a duplex system is used, the Doppler portion of the examination is then performed. In our laboratory we use a Diasonics duplex module as an addition to our general-purpose real-time system. A longitudinal view of the carotid vessel is obtained and the Doppler cursor is activated. This appears as a white line superimposed upon the real-time image. This cursor line shows the Doppler beam angle. A cross mark is positioned along the cursor line and indicates the range gate or sample volume location (Fig. 6). A Doppler signal is then initiated, and the examiner listens for the characteristic pulsating sound indicating the best placement of the probe, range gate, and angle selection. The beam positioning takes place with simultaneous real-time imaging. Physical limitations require that the frame rate be decreased from 20 to 4 frames per second. At this point the examiner measures several flow parameters. Average flow velocity, which varies with the cardiac cycle, is displayed as a waveform on a portion of the video monitor. Spectral analysis of the Doppler frequencies is then performed. This analysis is performed at peak systole and the information computed by the duplex system is graphically displayed in a corner of the monitor (Fig. 7). Quantitative flow of red blood cells within the vessel lumen is the last parameter to be measured. When the examiner is satisfied with the information being obtained, the B-scan image is frozen, as well as the graphic displays of the velocity waveform and frequency spectrum. A cursor is then rotated by a joy stick to align it with the direction of blood flow. Once aligned, the spectrum plot can be

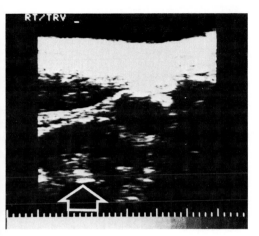

FIG. 2. Echogenic material partially occludes the lumen of the right internal carotid artery (*arrow*).

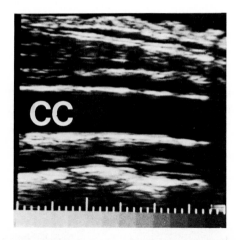

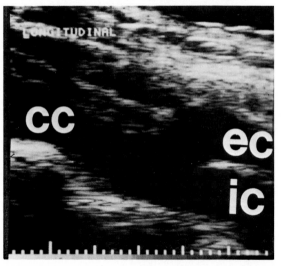

FIG. 3. Top: Longitudinal view of the common carotid artery (cc). **Bottom:** The common carotid artery (cc) bifurcates to form the external (ec) and the more posterior internal carotid artery (ic).

converted to a display of velocity (Fig. 8). Velocity measurements are made in the common carotid in a site free of plaque, at the site of any visualized stenosis, and at points distal to luminal narrowing. This information has practical value; we divide the average velocity at peak systole, measured at the site of maximum narrowing in the internal carotid artery, by that measured at a normal site in the common carotid. Narrowing results in increased velocity that is reflected in ratios greater than 1.0. Ratios exceeding 1.5 are nearly always associated with high-grade lesions (11). The clinical value of this calculation and other Doppler data are discussed in the next section.

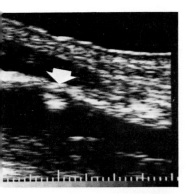
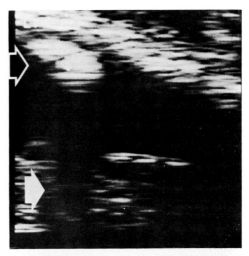
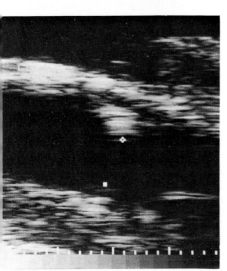
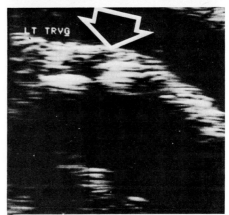

FIG. 4. Top left: Longitudinal scan (10 MHz) Echogenic plaque is identified at the origin of the internal carotid artery (*arrow*). **Top right:** Magnified view shows echogenic plaque (*open arrow*) casting an acoustic shadow (*closed arrow*). **Bottom left:** Digital calipers measure the sonolucent lumen. **Bottom right:** Transverse view confirms the presence of plaque along the anterior vessel wall (*open arrow*).

CLINICAL VALUE OF CAROTID SONOGRAPHY

In the preceding sections we discussed equipment, technique, and the theory behind carotid ultrasonography. It is only natural for the reader now to raise the following questions: Is carotid sonography an accurate method of examination and will it prove helpful to the clinician? Will it ultimately benefit the patient? These questions cannot be easily answered. We must critically evaluate this new technique, both as an isolated modality and also against

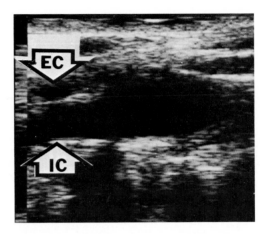

FIG. 5. Longitudinal scan demonstrating simultaneous imaging of both the internal (ic) and external (ec) carotid branches.

other currently used noninvasive techniques. The limitations of other non-invasive techniques, both alone and in combination, are significant and will be presented. Arteriography, the "gold standard" for carotid evaluation, is an invasive technique that carries a low but definite risk to the patient (36). In the following paragraphs, data will be presented both from personal experience and from the efforts of numerous leading investigators. Although conclusions will be presented, the reader is invited to analyze and interpret the data and form an individual opinion concerning this new modality. It should be noted that investigations are ongoing, and future evaluations will reflect changes in technology and increased experience in using this relatively new technique.

A discussion of small parts scanning, both as an isolated technique and combined with pulsed Doppler instrumentation, should begin with a brief review of alternative noninvasive tests. Ackerman (1) divides these examinations into direct and indirect testing methods. Indirect examinations include oculoplethysmography (OPG), supraorbital photoplethysmography (PPG), and Doppler ultrasonic flow detection. These tests depend upon significant changes in collateral cerebral blood flow, pressure, or pulse-wave transit time (35). Such examinations have proven most useful in the detection of internal carotid artery stenosis of 50% or greater (10,13,29). Unfortunately, cerebral emboli may arise from lesions creating a stenosis of less than 50% (42). Direct tests include carotid phonoangiography (CPA) and Doppler arteriography (Doppler imaging). Although these direct techniques may prove clinically useful, they also have important limitations that will now be discussed.

Carotid phonoangiography (CPA) allows analysis of a vascular bruit. Blood flow associated with an internal carotid artery stenosis greater than 80 to 90% may be inadequate to cause a bruit (25). Further, 33% of severe carotid lesions (residual lumen equal to or less than 2 mm) may lack a detectable bruit (1). For

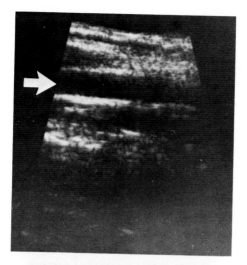

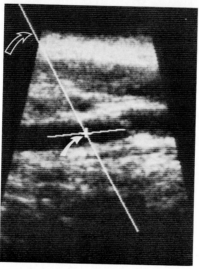

FIG. 6. Top: Longitudinal high-resolution image (7.5 MHz) shows lumen of the common carotid artery (*arrow*). **Bottom:** Longitudinal duplex image. A cursor line indicates the Doppler beam angle (*open arrow*) while a cross mark (*closed arrow*) identifies the location of the sample volume.

these reasons, a negative exam is not conclusive. CPA as an isolated method of examination also results in unacceptably high false-positive rates, ranging from 10 to 61% (10,38). A large bruit may be present with minimal arterial stenosis (39).

Doppler arteriography can be performed with continuous-wave or pulsed Doppler signals. Some investigators believe that pulsed systems are more sensitive (6,54). It would appear that these systems can provide highly accu-

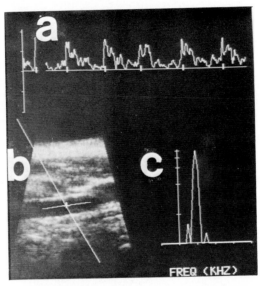

FIG. 7. Duplex video monitor shows velocity wave form (a), high-resolution image with Doppler beam angle and location (b), and spectral analysis graphic display (c).

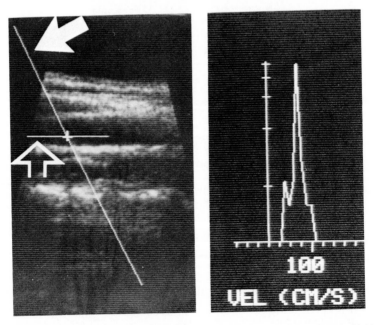

FIG. 8. Calculation of average flow velocity. **Left:** A cursor (*open arrow*) is aligned with blood flow. The axis of the Doppler beam (*closed arrow*) is shown and angle θ can therefore be determined. **Right:** Graphic display of velocity.

rate information in the appropriate clinical setting (1,28,35). Here again, limitations are encountered, since this method relies solely on blood flow for imaging and cannot resolve very slow velocities. Therefore, a totally occluded vessel will not be imaged. Several studies comparing continuous-wave Doppler imaging with conventional arteriography have concluded that this technique is of limited value for predicting stenoses less than 50% (3,57). Berry et al. (8) investigated the accuracy of pulsed Doppler imaging as an isolated technique. Two hundred and twenty-six extracranial carotid arteries were examined in 118 patients. Selective arteriograms were available for correlation. The pulsed Doppler instrumentation used in this study permitted evaluation over a depth of 0.5 to 5.0 cm, and this distance was divided into six "time gates." Overall accuracy was 85%. Further analysis of this statistic reveals that accuracy for a normal arteriogram was 75% and 87% for total occlusion. Lesser degrees of disease, however, were difficult to detect and resulted in an accuracy of only 38%.

The indirect techniques identify hemodynamic changes in distal vascular beds such as the orbital and cerebral circulations. These tests include peri-orbital directional ultrasound, OPG, and supraorbital PPG. As previously noted, these methods are most accurate when stenosis is severe; small ulcerated lesions without significant stenosis will often result in a false-negative examination (5). These small atheromatous ulcers, however, may be the source of platelet aggregates and cholesterol emboli resulting in transient ischemic attacks (37). When a severe lesion is present, these indirect tests cannot reliably differentiate between internal carotid occlusion, which is usually not amenable to surgical reconstruction, and a severe stenosis, which can be surgically corrected (9,12,22,23). If bilateral, symmetrical hemo-dynamic lesions are present, this could result in a false-negative indirect examination (30).

The Doppler ultrasonic flow detector is used to identify reversal of flow in the ophthalmic collateral circulation. This is the basis for a positive examination. The flow reversal occurs in the presence of severe internal carotid stenosis. The phenomenon, however, is not necessarily an index of clinical significance (34). Barnes et al. (4) investigated the combined use of a directionally sensitive Doppler ultrasonic velocity detector and supraorbital PPG. The overall accuracy for the Doppler technique was 97% and for the PPG 95% in detecting or excluding a 50% or greater stenosis. The examinations were in agreement in 89% of the vessels studied, and in these cases the diagnostic accuracy was 99%. Although these figures are impressive, the techniques were not sensitive to less than 50% stenosis. Further, the authors of this study emphasize that more than half the patients with symptoms of cerebral isch-emia on the basis of extracranial atherosclerotic disease suffer emboli from plaques that do not significantly obstruct the lumen.

Oculoplethysmography assesses the timing of the ocular arterial pulse and is helpful in evaluating the functional significance of a stenotic lesion and in

TABLE 2. *Vascular disease rating*

0 = Normal
1 = Less than 25% narrowing
2 = 25 to 75%
3 = More than 75%
4 = Complete occlusion

monitoring the patency of the carotid artery following endarterectomy (32). O'Donnell et al. (43) compared CPA, OPG, and Doppler arteriography both as isolated tests and in combination. As expected, this study demonstrated that a combined approach was most effective, although still associated with a false-negative rate of 23%. Kartchner et al. (33) reported that OPG and CPA combined have an accuracy of 89%. This degree of accuracy was obtained only when a stenosis was greater than 40% of the diameter of the vessel lumen.

In summary, these tests provide useful information only if the clinician is aware of the limitations inherent in each method. The lack of a single suitable substitute for carotid arteriography has led to the investigation of small parts scanners and duplex systems as methods for detecting atherosclerotic disease in the extracranial carotid artery.

During a 12-month period beginning in the fall of 1979, two commercially available small parts scanners were evaluated at The George Washington University Medical Center. These systems were compared with conventional carotid arteriography. The units reviewed included a BioSound and a Micro-view high-resolution real-time scanner. These scanners have been discussed in the preceding section on instrumentation. Patients undergoing arteriography with demonstration of the carotid artery bifurcation in at least two projections were examined with either of the two systems. These scans were performed immediately before or following the arteriogram. The videotaped real-time sonograms and the corresponding arteriograms were discussed and interpreted by three participating physicians as a group and impressions were then recorded.

Vascular disease was graded on a scale of 1 through 4 (Table 2). A gated pulsed Doppler system proved necessary in detecting complete occlusions. Unfortunately, this device was available only toward the conclusion of the study and its use was not reflected in our data. Our examinations were performed as outlined in the previous discussion of real-time technique.

We studied 31 patients, and of these 55 carotid arteries were imaged with sufficient detail to allow evaluation. Thus 89% of our examinations were successful in visualizing the carotid artery bifurcation (Fig. 9). Other investi-

---

FIG. 9. 8-MHz real-time images. **A:** Longitudinal scan of the common carotid artery (cc). **B:** The carotid bifurcation is well demonstrated. cc, Common carotid; ec, external carotid; ic, internal carotid. **C:** Transverse scan shows the common carotid lumen (cc) between thyroid (th) and jugular vein (jv).

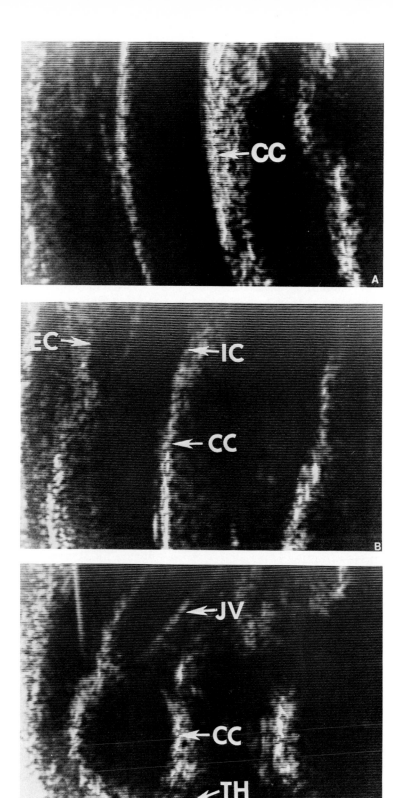

gators have reported even higher success rates (17). In 37 of 55, or 67%, carotids, there was perfect correlation of both sonography and arteriography, i.e., identical grades were independently assigned (Fig. 10). This included 14 normals and 23 abnormals. The abnormal cases were assigned to categories 1 through 3. The two cases of total occlusion of the internal carotid artery were missed. In 12 carotid arteries, atherosclerotic disease was identified; however, the degree of disease was incorrectly estimated. Five diseased carotid vessels were overcalled, i.e., we thought that there was moderate to severe (Category 2 or 3) disease when, in fact, there was only minimal disease found at arteriography. Six carotid arteries were undercalled, i.e., we underestimated the amount of disease present (Fig. 11). In two of these six arteries it was thought that there was a patent although stenotic vessel when, in fact, total occlusion of the internal carotid artery was found at arteriography. In one case, atherosclerotic disease was identified, but we were unsure whether the involved vessel was the internal or external carotid artery. In all, there were five false-negative cases and one false-positive case on the basis of the arteriographic findings.

We found that the specificity of the sonogram for atherosclerotic plaque was 93%. In our single false-positive case, we interpreted minimal plaque at the carotid bulb when a normal carotid bulb was seen at arteriography. If we combine the cases in which we correctly identified the extent of disease with those in which we identified plaque, but misinterpreted the quantification of disease, the overall sensitivity of ultrasound in detecting atheromatous plaque was 87%. Thus in 49 of 55 cases, or 89%, we were able to state accurately whether the vessel was diseased. The predictive value of a positive test was 35/36, or 97%. The predictive value of a negative test was 14/19, or 74%.

An investigation of our five false-negative cases was particularly revealing. Of these cases, two were clearly outside the limit of ultrasound detection. In both of these cases, there was significant stenosis of the internal carotid artery with disease starting 2 to 3 cm above the origin of the vessel. The physical dimensions of both real-time transducers limited us to the most proximal portion of the internal carotid artery. A retrospective analysis of these cases failed to demonstrate the known vascular compromise. Fortunately, most lesions will be within the field of view and lie within 2 cm of the bifurcation (24).

Although most investigators believe that high-resolution imaging is not capable of detecting small plaques and ulcerated plaque with an accuracy equal to or greater than arteriography, this point remains controversial (1,17,30). Personal experience indicates that the accurate detection of ulceration is difficult, if not impossible, with equipment available at present. Further, acute thrombi and very soft plaque may be very difficult to detect. The acoustic impedance of such lesions may be similar to that of blood and result in an apparently "clear" lumen (Fig. 12). Plaques can usually be identified as

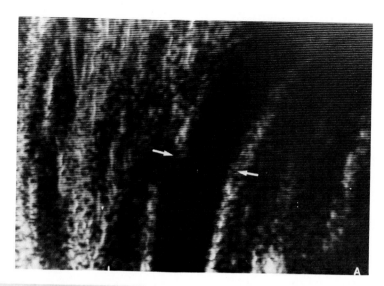

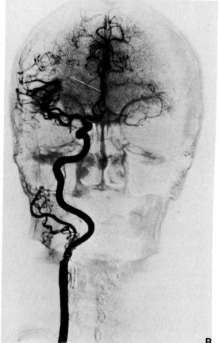

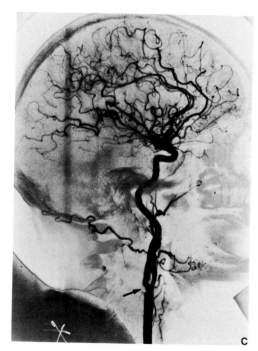

FIG. 10. Sonogram/arteriogram correlation. **A:** Longitudinal sonogram shows mild narrowing at the origin of the internal carotid (*arrows*). **B and C:** Frontal and lateral arteriograms confirm a stenotic lesion (*arrow*).

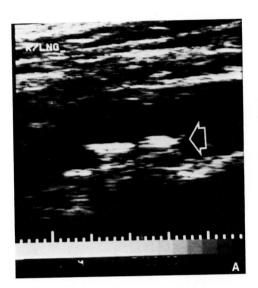 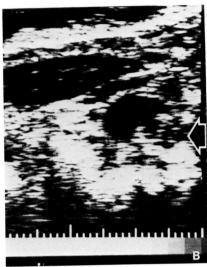

FIG. 11. Error in quantitation. **A and B:** Longitudinal and transverse scans (10 MHz) show echogenic plaque along the posterior wall of the vessel (*open arrow*). Luminal compromise appears to be less than 50%.

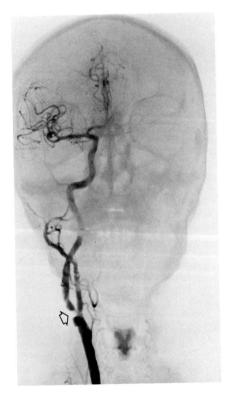

FIG. 11. continued. **C:** Frontal arteriogram reveals high-grade stenosis (*open arrow*).

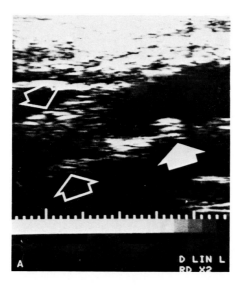

FIG. 12. Sonographic appearance of complete occlusion. **A:** Longitudinal scan shows low-level echoes within the lumen of the internal carotid artery (*open arrows*). Highly echogenic plaque is seen more proximally (*closed arrow*).

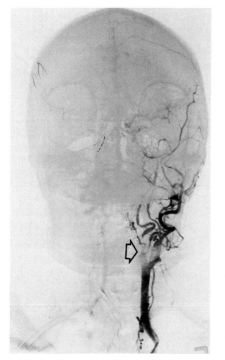

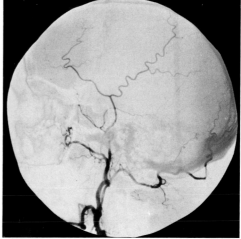

FIG. 12. continued. **Left and right:** Frontal and lateral arteriograms reveal total occlusion of the internal carotid (*open arrow*).

echogenic intraluminal material, and acoustic shadowing can be seen as a result of the presence of dense calcific deposits (Fig. 13). When small, plaque may be imaged; however, it may not produce the hemodynamic changes necessary for Doppler detection (Fig. 14). In such cases the real-time examination will identify disease not detected with other noninvasive techniques. A report by Cooperberg et al. (17) correlates arteriographic findings with high-resolution real-time ultrasound in 26 patients evaluated for transient ischemic attacks. This series reveals a high degree of technical success and diagnostic accuracy. There were, however, seven cases of ultrasound/ arteriogram "mismatch." In six of these cases plaque categorized as "small" was at the center of the diagnostic controversy. In the cases in which plaque was identified, there was an inability to detect ulceration reliably and estimate the degree of stenosis; similar conclusions were drawn from the study performed at The George Washington University Hospital.

Perhaps the most ambitious recent study of carotid sonography was carried out by Comerota et al. (16), who studied more than 1,000 patients with a small parts scanner. In addition, OPG and CPA and periorbital Doppler ultrasonography were performed. Of these patients, 150 underwent bilateral carotid arteriography and comparisons were then made. Stenotic lesions were graded according to the diameter reduction of the arterial lumen. The real-time B-scan sensitivity for a 40 to 69% stenosis was 75%, whereas a 70% or greater stenosis resulted in a sensitivity of only 44%. This disappointing figure is not surprising; only 6 of 16 (38%) totally occluded vessels were correctly identified, and this underscores the need for a method of hemodynamic evaluation. This inability of B-scan imaging to demonstrate total occlusion reliably is well documented (28,30,46). Obviously, the addition of the pulsed Doppler instrument fills this need for a physiologic test. A comparison of small parts scanning with OPG and CPA revealed that the scan was more reliable for less severe stenoses (less than 70%), whereas the OPG/CPA was superior in detecting high-grade lesions. Such lesions produce significant hemodynamic abnormalities. A most interesting finding concerned four of 12 scan/arteriogram "mismatch" patients who underwent carotid endarterectomy. According to the authors, in each instance the surgeon thought that the sonogram more accurately represented the extent of disease. This may be highly significant, since most studies have used the arteriogram as the ultimate basis for comparison. Some investigators have concluded that contrast angiography underestimates the disease process and that closer correlation with surgical findings occurs when lesions are demonstrated by ultrasound (19,25,26,40).

Humber et al. (30) compared small parts scanning with OPG in 81 carotid arteries. Sensitivity for high-resolution real-time imaging was 76%; however, specificity was only 55%. Accuracy for this technique was 60%, as compared with 77% for OPG. A combined approach resulted in the highest accuracy— 82%. Shortcomings for the ultrasound technique included an inability to

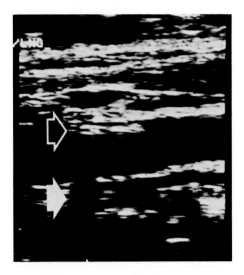

FIG. 13. Longitudinal scan (10 MHz) demonstrating calcified plaque (*open arrow*) casting an acoustic shadow (*closed arrow*).

detect fresh thrombus and difficulty in quantitating the degree of vascular narrowing.

The potential value of duplex scanning will now be discussed. This method combines a real-time image of the carotid bifurcation with flow velocity data generated by a simultaneous gated pulsed Doppler signal. By using the real-time image to measure the Doppler angle and spectral analysis to measure the frequency content of the Doppler signal, instantaneous flow velocity can be calculated. This capability of instantly calculating the mean flow velocity at peak systole was investigated by Blackshear et al. (11). A retrospective analysis of 68 common carotid arteries and internal carotid arteries in 39 patients who had undergone arteriography was combined with a prospective study of 30 arteries in 15 normal controls. This study produced an interesting application for mean velocity information. The authors found that a ratio of the mean peak internal carotid artery velocity to mean peak common carotid artery velocity was less than 0.8 in normals and greater than 1.5 in all high-grade stenoses (60% or greater decrease in diameter). A false-negative rate of 10% was encountered. They conclude that instantaneous velocity information can be of value in estimating the degree of stenosis, since the velocity of flow through a stenosis is related to the degree of narrowing. Since atheromatous plaque progressively narrows the lumen of a vessel, the velocity of flow at that site must increase if constant flow is to be maintained. Absolute velocity measurements will be of limited value because such a measurement will be affected by several variables: vessel size, vessel elasticity, myocardial contractility, and cardiac output. Normal absolute velocities exist as ranges

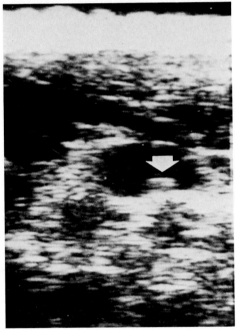

FIG. 14. Top and bottom: Longitudinal and transverse scans show a small intraluminal deposit that does not significantly compromise the vessel lumen (*arrow*).

within the carotid system. The velocity ratio better differentiated normal, low-grade and high-grade stenoses than the absolute velocity measurement data.

In addition to the measurement of absolute and relative velocity, other useful types of information are generated by duplex systems. Using the audible Doppler signal, the operator can detect high-frequency shifts and harsh tones due to hemodynamically significant lesions. Velocity waveforms are identified on a graphic display and differ for the internal and external carotid circulations (31). Analysis of the Doppler signal for spectral content allows detection of flow disturbances related to vascular narrowing (7,11,50,52). This spectral broadening may prove particularly useful; stenotic lesions producing diameter reductions of less than 50% can result in spectral broadening without evaluation of peak frequency (9). These parameters are used in combination with the real-time image to detect and describe carotid lesions accurately.

Because commercially available duplex systems are a recent innovation, they have yet to be fully evaluated in clinical practice. Blackshear et al. (9) investigated duplex scanning and ultrasonic arteriography as compared with conventional contrast arteriography. When a high-grade stenosis or total occlusion was present, audible analysis of the Doppler signal resulted in a correct diagnosis in 88% of cases. A 92% detection rate was achieved with duplex scanning. Further, the spectral analysis of the Doppler signals obtained with the duplex scanner detected all of the high-grade stenoses.

Experience with high-resolution real-time systems indicates that the accuracy of the duplex scanner should be greater than that of the real-time system alone, since real-time B-scan accuracy has been limited by the lack of a simultaneous physiologic test (14). Complete occlusions often escape detection, and this is just the type of lesion that would be readily diagnosed by pulsed Doppler systems. In the George Washington University study, it is reasonable to assume that two missed cases of total occlusion would have been correctly identified. In addition, two distal internal carotid lesions, which could not be imaged because of their location, might very well have shown proximal hemodynamic abnormalities that would have led to further investigation.

At this point a word of caution is appropriate. This method of examination requires meticulous technique as well as experience in both real-time scanning and Doppler data interpretation. It is technically difficult. Even in skilled hands, numerous potential sources of error exist (59). A knowledge of the patients vascular anatomy is essential, since an anatomic variant for the position of the external carotid artery at the bifurcation is reported to be present in 5 to 17% of patients studied (48,56). The Doppler sample volume must be within the center of the vessel lumen. Circumferential high-grade vascular narrowing over a long length may not demonstrate a frequency shift or spectral broadening. Artifacts in the real-time image can simulate disease. Obviously, a thorough knowledge of these and other pitfalls is necessary.

Finally, our review of testing methods should include a new modality that may revolutionize the field of arteriography. Intravenous "video" or "digital" angiography (IVA) is a means of obtaining high-quality arterial or venous images following a peripheral contrast injection. This is accomplished by computer enhancement of fluoroscopic images (41). This technique can demonstrate stenosis and occlusion of the carotid artery and will often show plaque and arterial ulceration (47,53). Diagnostic accuracy has been reported to be as high as 97% (15,58). A recent paper by Zweibel et al. (60) compares continuous-wave Doppler imaging with frequency spectral analysis, small parts scanning, and IVA. Selective biplane magnification arteriography was used as the standard of comparison. While this series is small and represents an ongoing investigation, the initial findings support the concept that small parts scanning combined with Doppler information is more accurate than small parts scanning or IVA alone.

CONCLUSIONS AND RECOMMENDATIONS

At present, noninvasive methods for studying the carotid bifurcation are not a substitute for conventional carotid arteriography. As previously described, all noninvasive tests have limitations. Although contrast arteriography may not always accurately demonstrate the disease process, it is nonetheless the "gold standard" with which other tests are compared. In most noninvasive vascular laboratories, the challenge is to approach arteriographic accuracy while utilizing techniques that present no known morbidity to the patient. The duplex system provides both direct and indirect information that, in combination, allows a high degree of accuracy in detecting atheromatous plaque, stenotic lesions, and total vascular occlusion. Unfortunately, small ulcerations within plaque cannot be identified with confidence. The duplex method combines the best features of other noninvasive methods; abnormalities are diagnosed because of abnormal hemodynamics as recorded by pulsed Doppler signal analysis or by structural abnormalities directly imaged with a high-resolution real-time display. While a combination of noninvasive tests will usually prove most accurate, the duplex scanner should become a routinely used method for noninvasive testing.

How should the duplex scanner be used? As already stated, this new approach is not a replacement for contrast arteriography. There are, however, several clinical situations that appear most appropriate for duplex application:

Evaluation of a carotid bruit in the asymptomatic patient.

Patients with borderline cardiovascular symptomatology not warranting immediate arteriography.

Individuals with suggestive symptoms, but who are at increased risk for an invasive contrast procedure.

Patients who are candidates for surgical procedures that may result in a hypotensive state.

Patients who require multiple carotid studies, i.e., monitoring patency following carotid endarterectomy.

Identification of total vascular occlusion. Such a diagnosis will obviate the need for surgery in most cases (12,27).

In this author's opinion, the greatest impact on carotid imaging may be IVA; an "invasive" procedure that probably carries no more risk than the intravenous contrast injection performed during an intravenous urogram. This new technique of IVA will probably undergo changes reminiscent of early CT. With technologic improvement and increasing user experience, this method should result in a diagnostic accuracy approaching that of conventional film/screen arteriography. The duplex scanner may have its ultimate role as a complementary technique and result in a combined approach accurate enough to limit the application of conventional arteriography to only the most difficult diagnostic cases. Each patient examination will produce a "conventional appearing" arteriogram, a profile of pulsed Doppler flow data and a high-resolution B-scan image detailing transverse as well as longitudinal vascular anatomy. Such comprehensive data should result in an accurate diagnosis and an appropriate therapeutic plan.

In summary, the duplex scanner is a new technologic development that is a natural culmination of recent refinements in both high-resolution real-time ultrasound and pulsed Doppler signal analysis. As with all new technology, continuing improvement in instrumentation resulting in greater diagnostic accuracy is to be expected. The preliminary experience of this author and leading investigators is most promising. Although not a substitute for conventional arteriography, this method should become the cornerstone of noninvasive evaluation of the carotid bifurcation.

ACKNOWLEDGMENTS

I wish to thank a number of individuals for their assistance in the preparation of this chapter. Michael Kornreich, M.D. and William Studdard, M.D. contributed much time and effort toward the investigation of small parts scanning at The George Washington University Hospital. David O. Davis, M.D., Chairman, Department of Radiology, The George Washington University Medical Center, supported the clinical research that was presented in this chapter. Washington Ultrasound Associates (W. J. Cochrane, M.D., W. R. Forster, M.D., L. M. Glassman, M.D., E. J. McDonald, M.D., D. H. Rosenfeld, M.D., M. A. Thomas, M.D.) provided encouragement, support, and a true dedication to evaluating new ultrasound technology. Finally, thanks to Toni Whittenburg for typing the manuscript.

REFERENCES

1. Ackerman, R. H. A perspective on noninvasive diagnosis of carotid diseases. *Neurology*, 29:615–622, 1979.
2. Baker, D. W. Applications of pulsed Doppler techniques. *Radiol. Clin. North Am.*, 18:79–103, 1980.
3. Barnes, R. W., Bone, G. E., Reinertson, J. E., Slaymaker, E. E., Hokanson, D. E., and Strandness, D. E. Noninvasive ultrasonic carotid angiography: Prospective validation by contrast arteriography. *Surgery*, 80:328–335, 1976.

4. Barnes, R. W., Garrett, W. V., Slaymaker, E. E., and Reinertson, J. R. Doppler ultrasound and supraorbital photoplethysmography for noninvasive screening of carotid occlusive disease. *Am. J. Surg.*, 134:183–186, 1977.
5. Barnes, R. W., Russell, H. E., Bone, G. E., and Slaymaker, E. E. Doppler cerebrovascular examination: Improved results with refinement in technique. *Stroke*, 8:468–471, 1977.
6. Barnes, R. W. Doppler ultrasonic arteriography and flow velocity analysis in carotid artery disease. In: *Noninvasive Diagnostic Techniques in Vascular Disease*, edited by E. F. Bernstein, pp. 40–48, 1978. C. V. Mosby, St. Louis.
7. Barnes, R. W. Noninvasive diagnostic techniques in peripheral vascular disease. *Am. Heart J.*, 97:241–258, 1979.
8. Berry, S. M., O'Donnell, J. A., and Hobson, R. W. Capabilities and limitations of pulsed Doppler sonography in carotid imaging. *J. Clin. Ultrasound*, 8:405–412, 1980.
9. Blackshear, W. M., Phillips, D. J., Thiele, B. L., Hirsch, J. H., Chikos, P. M., Marinelli, M. R., Ward, K. J., and Strandness, D. E. Detection of carotid occlusive disease by ultrasonic imaging and pulsed Doppler spectrum analysis. *Surgery*, 86:698–706, 1979.
10. Blackshear, W. M., Thiele, B. L., Harley, J. D., Chikos, P. M., and Strandness, D. E. A prospective evaluation of oculoplethysmography and carotid phonoangiography. *Surg. Gynecol. Obstet.*, 148:201–205, 1979.
11. Blackshear, W. M., Phillips, D. J., Chikos, P. M., Harley, J. D., Thiele, B. L., and Strandness, D. E. Carotid artery velocity patterns in normal and stenotic vessels. *Stroke*, 11:67–71, 1980.
12. Blaisdell, W. F., Clauss, R. M., Galbraith, J. G., Imparato, A. M., and Wylie, E. J. Joint study of extracranial arterial occlusions. IV. A review of surgical considerations. *JAMA*, 208:1889–1895, 1969.
13. Bone, G. E., and Barnes, R. W. Limitations of the Doppler cerebrovascular examination in hemisphere ischemia. *Surgery*, 79:577–580, 1976.
14. Carlsen, E. N. Instrumentation considerations in establishing a clinical ultrasound facility. *Clin. Diagn. Ultrasound*, 5:1–19, 1980.
15. Chilcote, W. A., Modic, M. T., Pavlicek, M. S., Little, J. R., Furlan, A. J., Duchesneau, P. M., and Weinstein, M. D. Digital subtraction angiography of the carotid arteries: A comparative study in 100 patients. *Radiology*, 139:287–295, 1981.
16. Comerota, A. J., Cranley, J. J., and Cook, S. E. Real time B-mode carotid imaging in diagnosis of cerebrovascular disease. *Surgery*, 6:718–729, 1981.
17. Cooperberg, P. L., Robertson, W. D., Fry, P., and Sweeney, V. High resolution real time ultrasound of the carotid bifurcation. *J. Clin. Ultrasound*, 7:13–17, 1979.
18. Cooperberg, P. L., Li, D. K., and Sauerbrei, E. E. Abdominal and peripheral applications of real time ultrasound. *Radiol. Clin. North Am.*, 18:59–77, 1980.
19. Croft, R. J., Ellam, L. D., and Harrison, M. J. Accuracy of carotid angiography in the assessment of atheroma of the internal carotid artery. *Lancet*, i:997–1000, 1980.
20. Felix, W. R., Sigel, B., Gibson, R. J., Williams, J., Popky, G. L., Adelstein, A. L., and Justin, J. R. Ultrasound detection of flow disturbances in arteriosclerosis. *J. Clin. Ultrasound*, 4:275–282, 1977.
21. Fields, W. S., North, R. R., Hass, W. K., Galbraith, G., Wylie, E. J., Ritnov, G., Burns, M. H., MacDonald, M. C., and Myers J. S. Joint study on extracranial arterial occlusion as a cause of stroke. I. Organization of study and survey of patient population. *JAMA*, 203:955–960, 1968.
22. Fields, W. S. Selection of stroke patients for arterial reconstructive surgery. *Am. J. Surg.*, 125:527–529, 1973.
23. Fields, W. S., and Lemak, N. A. Joint study of extracranial arterial occlusion as a cause of stroke. X. Internal carotid artery occlusion. *JAMA*, 235:2734–2738, 1976.
24. Fisher, C. M., Gore, I., Okabe, N., and White, P. D. Atherosclerosis of the carotid and vertebral arteries—extracranial and intracranial. *J. Neuropathol. Exp. Neurol.*, 24:466–476, 1965.
25. Gompels, B. M. High definition imaging of carotid arteries using a standard commercial ultrasound "B" scanner. A preliminary report. *Br. J. Radiol.*, 52:608–619, 1979.
26. Green, P. S. Real-time, high-resolution ultrasonic carotid arteriography system. In: *Non-Invasive Diagnostic Techniques in Vascular Disease*, edited by E. F. Bernstein, pp. 29–39, 1978. C. V. Mosby, St. Louis.

27. Hafner, C. D., and Tew, J. M. Surgical management of the totally occluded internal carotid artery: A ten-year study. *Surgery*, 89:710–716, 1981.
28. Hobson, R. W., Berry, S. M., Katocs, A. S., O'Donnell, J. A., Jamil, Z., and Savitsky, J. P. Comparison of pulsed Doppler and real-time B-mode echo arteriography for noninvasive imaging of the extracranial carotid arteries. *Surgery*, 87:286–293, 1980.
29. House, S. L., Mahalingam, K., Hyland, L. J., Ferris, E. B., Comerota, A. J., and Cranley, J. J. Noninvasive flow techniques in the diagnosis of cerebrovascular disease. *Surgery*, 87:696–700, 1980.
30. Humber, P. R., Leopold, G. R., Wickbom, I. G., and Bernstein, E. F. Ultrasonic imaging of the carotid arterial system. *Am. J. Surg.*, 140:199–202, 1980.
31. Johnston, K. W., deMorais, D., Kassam, M., and Brown, P. M. Cerebrovascular assessment using a Doppler carotid scanner and real time frequency analysis. *J. Clin. Ultrasound*, 9:443–449, 1981.
32. Kartchner, M. M.. McRae, L. P., and Morrison, F. D. Noninvasive detection and evaluation of carotid occlusive disease. *Arch. Surg.*, 106:528–535, 1973.
33. Kartchner, M. M., and McRae, L. P. Noninvasive evaluation and management of the "asymptomatic" carotid bruit. *Surgery*, 82:840–847, 1977.
34. LoGerfo, F. W., and Mason, G. R. Directional Doppler studies of supraorbital artery flow in internal carotid stenosis and occlusion. *Surgery*, 76:723–728, 1974.
35. Lusby, R. J., Woodcock, J. P., Skidmore, R., Jeans, W. D., Hope, D. T., and Baird, R. N. Carotid artery disease: A prospective evaluation of pulsed Doppler imaging. *Ultrasound Med. Biol.*, 7:365–370, 1981.
36. Mani, R. L., and Eisenberg, R. L. Complications of catheter cerebral arteriography: Analysis of 5000 procedures. II. Relation of complication rates to clinical and arteriographic diagnoses. *AJR*, 131:867–869, 1978.
37. McBrien, D. J., Bradley, R. D., and Ashton, N. The nature of retinal emboli in stenosis of the internal carotid artery. *Lancet*, i:697–699, 1963.
38. McDonald, P. T., Rich, N. M., Collins, G. J., Andersen, C. A., and Kazloff, L. Doppler cerebrovascular examination, oculoplethysmography, and ocular pneumoplethysmography use in detection of carotid disease. A prospective clinical study. *Arch. Surg.*, 113: 1341–1349, 1978.
39. McRae, L. P., and Kartchner, M. M. Oculoplethysmography: Timed comparison of ocular pulses and carotid phonoangiography. In: *Non-Invasive Diagnostic Techniques in Vascular Disease*, edited by E. F. Bernstein, pp. 91–111, 1978. C. V. Mosby, St. Louis.
40. Mercier, L. A., Greenleaf, J. F., Evans, T. C., Sandok, B. A., and Hattery, R. R. High-resolution ultrasound arteriography: A comparison with carotid angiography. In: *Non-Invasive Diagnostic Techniques in Vascular Disease*, edited by E. F. Bernstein, pp. 231–244, 1978. C. V. Mosby, St. Louis.
41. Mistretta, C. A., Crummy, A. B., and Strother, C. M. Digital angiography: A perspective. *Radiology*, 139:273–276, 1981.
42. Moore, W. S., and Hall, A. D. Importance of emboli from carotid bifurcation in pathogenesis of cerebral ischemic attacks. *Arch. Surg.*, 101:708–715, 1970.
43. O'Donnell, T. F., Pauker, S. G., Callow, A. D., Kelly, J. J., McBride, J. K., and Korwin, S. The relative value of carotid noninvasive testing as determined by receiver operator characteristic curves. *Surgery*, 87:9–19, 1980.
44. Olinger, C. P. Ultrasonic carotid echoarteriography. *AJR*, 106:282–295, 1969.
45. Perry, S., and Renault, P. *B-Scan of Peripheral Vessels. National Center for Health Care Technology Assessment Report Series, Vol. 1, No. 12*, 1981. U.S. Department of Health and Human Services, Washington, D.C.
46. Phillips, D. J., Powers, J. E., Eyer, M. K., Blackshear, W. M., Bodily, K. C., Strandness, D. E., and Baker, D. W. Detection of peripheral vascular disease using the Duplex Scanner III. *Ultrasound Med. Biol.*, 6:205–218, 1980.
47. Pond, G. D., Smith, J. R., Hillman, B. J., Ovitt, T. W., and Capp, M. P. Current clinical applications of digital subtraction angiography. *Appl. Radiol.*, 10:71–79, 1981.
48. Prendes, J. L., McKinney, W. M., Buananno, F. S., and Jones, A. M. Anatomic variations of the carotid bifurcation affecting Doppler scan interpretation. *J. Clin. Ultrasound*, 8:147–150, 1980.

49. Reid, J. M., and Spencer, M. P. Ultrasonic Doppler technique for imaging blood vessels. *Science*, 176:1235–1236, 1972.
50. Reneman, R. S., and Spencer, M. P. Local Doppler audio spectra in normal and stenosed carotid artery in man. *Ultrasound Med. Biol.*, 5:1–11, 1979.
51. Spencer, M. P., Reid, J. M., and Davis, D. L. Cervical carotid imaging with a continuous wave Doppler flowmeter. *Stroke*, 5:145–154, 1974.
52. Spencer, M. P., and Reid, J. M. Quantitation of carotid stenosis with continuous-wave (C-W) Doppler ultrasound. *Stroke*, 10:326–330, 1979.
53. Stieghort, M. F., Strother, C. M., Mistretta, C. A., Crummy, A. B., Sackett, J. F., Lieberman, R., and Turski, P. Digital subtraction angiography: A clinical overview. *Appl. Radiol.*, 10:43–49, 1981.
54. Sumner, D. S., Russell, J. B., Ramsey, D. E., Hajjar, W. M., and Miles, R. D. Noninvasive diagnosis of extracranial carotid arterial disease: A prospective evaluation of pulsed-Doppler imaging and oculoplethysmography. *Arch. Surg.*, 114:1222–1229, 1979.
55. Taylor, K. J. Small field imaging. *Clin. Diagn. Ultrasound*, 5:106–121, 1980.
56. Teal, J. S., Rumbaugh, C. L., Bergeron, R. T., and Segall, H. D. Lateral position of the E.C.A.: A rare anomaly? *Radiology*, 108:77–81, 1973.
57. Weaver, R. G., Howard, G. H., McKinney, W. M., Ball, M. R., Jones, A. M., and Toole, J. F. Comparison of Doppler ultrasonography with arteriography of the carotid artery bifurcation. *Stroke*, 11:402–404, 1980.
58. Weinstein, M. A., Modic, M. T., Buonocore, E., and Meaney, T. F. Digital subtraction angiography: Clinical experience at the Cleveland Clinic Foundation. *Appl. Radiol.*, 10:53–66, 1981.
59. Zweibel, W. J., and Crummy, A. B. Sources of error in Doppler diagnosis of carotid occlusive disease. *AJR*, 137:1–12, 1981.
60. Zweibel, W. J., Sackett, J. F., Strother, C. M., and Crummy, A. B. Comparison of ultrasonography and intravenous video angiography for carotid diagnosis. *Paper 1701. Presented at 26th Annual AIUM Scientific Session, Aug. 17–21, 1981, San Francisco.*

Ultrasound Annual 1982,
edited by Roger C. Sanders.
Raven Press, New York © 1982.

Operative Real-Time B-Mode Ultrasound Scanning

Bernard Sigel, Julio C. U. Coelho, and Junji Machi

The Department of Surgery, Abraham Lincoln School of Medicine
University of Illinois College of Medicine, Chicago, Illinois 60680

Obtaining information at surgical operations is accomplished by exploration that requires tissue dissection and manipulation. This may include operative maneuvers that may lengthen the procedure and increase risk to the patient. Such risk is often unavoidable and constitutes the necessary price for more complete knowledge of the pathology prior to the making of operative decisions. Operative exploration morbidity and time may be significantly reduced by the use of imaging methodology during some surgical procedures. Such imaging may provide information that could not be obtained safely by dissection and, consequently, was not previously obtained during operations.

Plain and contrast radiography have to date been preeminent in the field of operative imaging. Soon after X-ray procedures were introduced into medicine, radiography was applied to many areas of surgery. Major developments in orthopedic surgery would not have been possible without the use of fluoroscopy and radiography in the operating room. Contrast radiography with air and radiodense media significantly facilitates the work of surgeons during neurosurgical, biliary, and vascular operations.

The example set by radiography was a precedent for the operative use of ultrasonic imaging after this technology came into common use during the last 2 decades. The early use of ultrasound scanning during surgical procedures was mostly with A-mode imaging. In 1961, Schelegel and associates reported on the use of A-mode ultrasound to locate renal calculi during operation (16). Hayashi (9), Knight (14), Eiseman (7), and their respective associates employed A-mode scanning to detect calculi in common bile ducts at operation. These workers reported good results but these and other attempts with A-mode did not lead to widespread use of operative ultrasound (1,6).

The reason for a lack of general acceptance of operative ultrasound may relate to the differences in images between ultrasound and radiography. Ultrasonic images are less similar to our usual concepts of anatomic structure than radiographic images. Transmission radiographic images portray rather faithfully a gray-scale that conforms to our expectations of tissue density. Reflective ultrasonic images, on the other hand, are based on mismatches in acoustic impedance between tissue structures and are more difficult to associate with our ideas about anatomic detail. A-mode presents the greater problems, particularly when used by practitioners who are less familiar with ultrasound imaging (e.g., surgeons).

The introduction of B-mode scanning provided two-dimensional imaging that presented structural detail more acceptably than A-mode. However, the early compound B-mode scanners were relatively bulky and had additional limiting features such as restrictive motion of transducers attached to articulated arms, which made their use in an operating room setting difficult. This situation changed with the development of real-time B-mode scanners. Recent advances in technology have made available relatively compact and portable real-time B-mode scanners with small transducers operating at frequencies of 7 to 10 MHz. These "small parts" instruments provide high resolution but shallow penetration. This is an ideal trade-off for operative ultrasonography in which tissues and organs may be scanned by direct contact and there is no need to traverse covering layers of skin, fat, fascia, and muscle.

Operative ultrasound using high-frequency B-mode real-time scanning has been applied by different groups in a number of surgical fields. In this chapter we consider in some detail the areas in which we have had the greatest experience. These have been in biliary (10–12,17,18,20,25–27), pancreatic, (19,20,22,25,26,28), renal (4,5,13,23,25,26), and vascular surgery (2,3,12, 24–26,29). We also briefly survey the use of operative ultrasound in areas in which our personal experience has been less extensive or is a summary of the work of others.

METHODOLOGY

Operative ultrasonography has been performed with high-frequency small parts scanners. B-mode scanners that are used for eye examinations and that may be sterilized may be employed at operation. We have employed High Stoy real-time, B-mode instruments.[1] Initially, we used the Bronson-Turner model with limited dynamic range and later the SP-100-B unit with a greater gray-scale display. Both instruments employ a 6-mm diameter transducer that is mechanically driven at 30 sweeps per second to produce a sector angle of 18 degrees. The transducer has a focal distance of 2 cm and operates at 7.5 MHz. The transducer is housed in a cylindrical, stainless steel probe that is tapered at the transducer end. A mylar window is situated at the transducer end to permit passage of the sound beam. A video scan converter is used to simultaneously display the images on a video monitor and to record real-time images on a ¾-inch video tape recorder (Fig. 1).

---

[1]High Stoy Technological Corporation, 2 Nevada Drive, Lake Success, New York 11042.

FIG. 1. Photo of a High Stoy SP-100-B ultrasound system. (Reproduced by permission of High Stoy Technological Corporation, Lake Success, NY.)

The probe or scan head is approximately 4 cm in diameter and is sterilized by cold ethylene oxide. The probe and coaxial cable are placed directly in the operative field.

Operative scanning is performed after the opening incisions are completed and the tissues and organs have been exposed. It may be employed during several phases of the operation. Ultrasound examination may be performed early, during the course of the operation, or at the end of the procedure. The surgeon first performs a standard exploration that consists of inspection and palpation. This occurs prior to any extensive dissection or commitment to a plan of management. If ultrasound might be useful at this stage to help detect or localize abnormal structures, operative ultrasonography is performed next. It is used during operation to detect further lesions or localize structures, or at the completion of operation just prior to closure. Utilization at the end of an operation is to provide assurance that correctable abnormalities have not inadvertently been left behind.

The technique of operative ultrasonography is basically the same for all applications. The probe is first positioned over the tissue surface to be scanned. Usually the probe end is maintained about 5 to 10 mm away from the tissue surface. Sterile saline is then added to provide acoustic coupling. The

5 to 10 mm withholding of the probe provides an acoustic window, permits visualization of the surface, and avoids distortion of tissue by compression with the probe.

Scanning maneuvers are performed either by manually moving the probe along the line of the transducer sector oscillation (longitudinal scanning) or at right angles to the transducer motion (transverse scanning). Alternately, if space constraints prevent probe motion along a scan path, the probe may be angulated to widen the image field. Most structures of interest at operation are encountered at 4 cm or less of tissue depth.

All scanning images are viewed on the video monitor. Images are either recorded on videotape or are photographed with a still camera, employing film that can be immediately developed.

ULTRASOUND SCANNING DURING BILIARY OPERATIONS

There are three principle indications for imaging ultrasonography during biliary operation: (a) detection of calculi in the gallbladder, (b) locating the common bile duct, and (c) detection of calculi in the common bile duct.

Diagnosis of gallbladder calculi is not a common indication for operative ultrasonography because preoperative studies or the initial surgeon's manual exploration have usually detected calculi in the gallbladder if they are present. Preoperative studies have usually included ultrasound scanning or oral cholecystography or both. At operation the surgeon can usually palpate stones unless the gallbladder wall is thickened or the gallbladder is tensely distended. Operative ultrasonography is useful if preoperative studies have not been performed or are inconclusive, or if palpation is not fully informative. Incomplete palpatory findings include not only failure to detect gallbladder calculi but also difficulty in recognizing the presence of small calculi associated with larger more readily palpable gallstones. Knowledge of the presence of small gallbladder calculi may be important in a decision regarding opening of the common bile duct.

Locating the common bile duct during biliary operation is usually not a problem. However, on occasion, as a result of previous operation or inflammation, the site of this structure is obscured. Tedious, time-consuming dissection is often required to locate the common bile duct safely under such circumstances. In these situations, operative ultrasonography may be useful in locating the common bile duct. Discerning the location of the hepatic artery in relation to the common duct may also be accomplished with ultrasonography.

Detecting calculi in the common bile duct is the most important application of imaging ultrasound during biliary tract operations. At operations for biliary calculi, the surgeon usually removes the gallbladder that contains stones or that appears to be grossly abnormal. At these operations, the surgeon must also be as certain as possible that calculi have not been left behind in the common bile duct and its tributaries. Stones within the common bile duct are often not palpable. Surgically opening the duct to explore for calculi significantly increases morbidity and mortality and is generally avoided as a routine procedure because common duct calculi are found at only 10 to 15% of gallbladder operations. Exploration of the common duct is usually performed if factors indicative of common bile duct stones are present or if an operative cholangiogram is positive. Another method to evaluate the common bile duct is with ultrasound. Operative ultrasonography of the common bile duct has been performed to determine its utility as a screening procedure for detecting stones in the common bile duct.

The technique of common bile duct ultrasonography is illustrated in Fig. 2. The entire duct is scanned from the biliary radicals to its entry into the common duct. The retro-duodenal portion of the duct is examined either through the duodenum or from its posterior lateral aspect after mobilization of the duodenum. We usually scan the terminal portion of the common bile duct using both approaches.

Results

Detection of gallbladder calculi by operative ultrasound has not been necessary in most cases. However, it was useful in 18% of biliary operations in which preoperative or operative information was inconclusive. In these circumstances, operative ultrasound was consistantly correct in identifying the presence or absence of gallbladder calculi (18).

At several operations, ultrasound was used to help locate a common bile duct whose position was obscured by inflammation and dense adhesions. While our experience is limited, the ultrasonic discovery of bile ducts in these

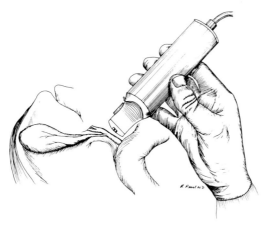

FIG. 2. Illustration of longitudinal scanning over the supraduodenal portion of the common bile duct. (From ref. 25.)

few cases has appreciably shortened search time of what would otherwise have been difficult and lengthy dissections.

The principle use of operative ultrasonography has been to detect calculi in the common bile duct. Figure 3 illustrates calculi in the common bile duct. Our experience with common bile duct scanning in the first 150 biliary tract operations at which ultrasound has been employed (27) has recently been analyzed. At most of these operations, operative cholangiography was also performed. The validity of the ultrasonography and radiography was determined by opening the common bile duct for exploration in cases in which this was clinically indicated and by following the course of patients after operation.

In the series of 150 patients who underwent biliary operation and operative ultrasonography, choledochotomy was performed in 45 patients to examine the common bile ducts. Thirty-six of these explorations were for suspected calculi and the remainder were for tumor. Stones were found in 81% of the ducts explored for this diagnosis. The validity indices[2] to measure operative ultrasonography and operative cholangiography are sensitivity, specificity, efficiency (accuracy), and predictability of negative and positive tests. These were as follows (data for ultrasound are listed first): sensitivity, 89% and 91%; specificity, 97% and 93%; efficiency, 95% and 92%; predictability of a negative test, 98% and 98%; predictability of a positive test, 86% and 74%. The differences between the ultrasound and the radiographic results were not statistically significant. On the basis of these data, we concluded that operative ultrasonography was comparable in accuracy to operative cholangiography. The use of both procedures during operation tended to reduce the number of negative common bile duct explorations that would have been performed if operative cholangiography were the only test employed. We believe that, with sufficient experience in performance and interpretation, ultrasound may become the initial operative screening procedure for detection of stones in the common bile duct.

---

[2]Definition of validity indices:

$$\text{Sensitivity} = \frac{\text{True positives} \times 100}{\text{True positives} + \text{false negatives}}$$

$$\text{Specificity} = \frac{\text{True negatives} \times 100}{\text{True negatives} + \text{false positives}}$$

$$\text{Efficiency} = \frac{(\text{True positives} + \text{true negatives}) \times 100}{\text{All examinations}}$$

$$\text{Predictability of a negative test} = \frac{\text{True negatives} \times 100}{\text{True negatives} + \text{false negatives}}$$

$$\text{Predictability of a positive test} = \frac{\text{True positive} \times 100}{\text{True positives} + \text{false positives}}$$

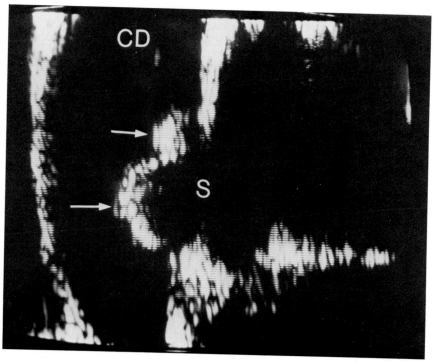

FIG. 3. Sonogram of a longitudinal section through a dilated common bile duct (CD) shows two calculi (*arrows*). Acoustic shadow (S) is also seen. (From ref. 25.)

ULTRASOUND SCANNING DURING PANCREATIC OPERATION

Surgery of the pancreas is performed most commonly either for inflammatory disease or for tumors. Imaging ultrasonography has been employed during both types of operation.

During operation for pancreatic pseudocyst, abscess, or chronic pancreatitis, ultrasonography has been helpful to establish diagnoses, localize structures, and exclude certain abnormalities (20,22,28). Operative ultrasound may establish a diagnosis of pancreatic pseudocyst, abscess, or dilated pancreatic ducts that was not made on preoperative testing. In instances in which preoperative studies have identified a pseudocyst, abscess, or dilated pancreatic ducts, imaging ultrasound at operation has been used to facilitate localization. Conversely, ultrasound has been employed to rule out the presence of pseudocysts, abscesses, and dilated pancreatic ducts in cases in which there has been preoperative or operative suspicion of their presence.

At operation for pancreatic tumors, ultrasonography has been used to help establish the diagnosis, evaluate vascular invasion, and provide guidance for needle biopsy of the pancreas. Reflective ultrasound, in our experience, cannot, at present, distinguish between tumors and pancreatitis in terms of the acoustic characteristics of the two lesions. Both pancreatic tumors and chronic inflammation tend to produce increased echogenicity. However, there are other features besides tissue signature that aid the surgeon during operations for pancreatic tumors.

The technique of ultrasound scanning of the pancreas is illustrated in Fig. 4. Scanning is performed best after the pancreas has been exposed by transection of the gastrocolic or, less frequently, the gastrohepatic ligament. The scan path is usually along the long axis of the gland. It is important to maintain the transducer probe a few millimeters away from the gland surface to image the entire thickness of the pancreas without the distortion produced by the weight of the transducer-probe.

Results

We have reported our results with ultrasonography at operations in which six glands were normal and 35 revealed evidence of pancreatitis or its complications (28).

Figure 5 shows a sonogram obtained at operation for a normal pancreas. The pancreatic ducts appear as a lacy network of sonolucent spaces amid the more echogenic parenchymal tissue and stroma. Major blood vessels in close proximity to the gland are usually detected with ease.

Figure 6 demonstrates a problem of localizing a pancreatic pseudocyst at operation. Preoperative studies detected a pseudocyst. However, at operation, the entire pancreas was enlarged and it was not possible on the basis of initial palpation to determine the exact site of the pseudocyst. Ultrasonography revealed that an area of pancreatic enlargement anterior to the superior mesenteric vessels represented inflammatory swelling while another area at

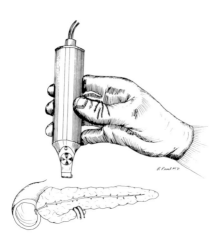

FIG. 4. Illustration showing transverse scanning of the pancreas in the region of the superior mesenteric vessels. (From ref. 25.)

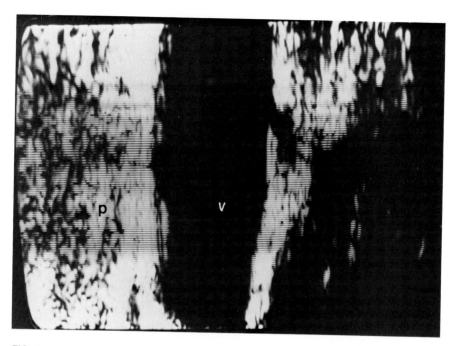

FIG. 5. Sonogram showing a transverse section through a normal pancreas (P). Deep to the pancreatic tissue is the superior mesenteric vein (V) receiving a tributary on its deep surface. (From ref. 25.)

the junction of the body of the tail of the gland contained the pseudocyst. Needle aspiration confirmed both findings.

The objective of most operations for pancreatic pseudocyst is to provide an internal drainage system between the cyst and some portion of the gastrointestinal tract such as the stomach, duodenum, or jejunum. In addition to accurate detection and localization of the pseudocyst, other factors are also important in selection of the most appropriate site of entry into the cyst. One is the thickness of the pseudocyst wall at various locations. Cyst wall thickness may vary three- to fourfold in different regions of the same pseudocyst. Ultrasound may accurately determine the various wall thicknesses and aid in selection of the most optimal thickness for suture placement.

Another factor in drainage site selection is avoidance of inadvertent injury to surrounding structures whose location may not be readily apparent as a result of distortions or contractions in anatomy produced by the pseudocyst or its associated inflammation. Figure 7 shows an example of this problem; the pseudocyst is adherent to the duodenum, and this relationship was not readily apparent from the external appearance of the tissues. In other situations, the pseudocyst may be in close proximity to other structures such as the splenic artery or common bile duct, whose integrity must be safeguarded

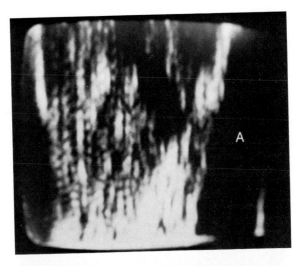

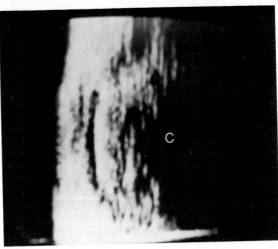

FIG. 6. A composite of two sonograms taken at the same operation. The **upper** sonogram shows internal echoes resulting from inflammatory swelling on the gland immediately superficial to the superior mesenteric artery (**A**) but no cystic cavity. The **lower** sonogram reveals the pseudo-cyst between the body and tail of the gland with a distinct cyst wall and sonolucent cavity (C). (From ref. 25.)

when the pseudocyst is opened. Ultrasound has been helpful in recognizing the presence of neighboring structures. In particular, the real-time feature of ultrasonography has been useful in identifying adjacent arteries because of their pulsation.

Ultrasonography may be helpful in detection and localization of a pancreatic abscess at operation. Such scanning may assist the surgeon in lending certainty that all major extensions and loculations of the abscess have been discovered and adequately drained. Figure 8 illustrates a pancreatic abscess.

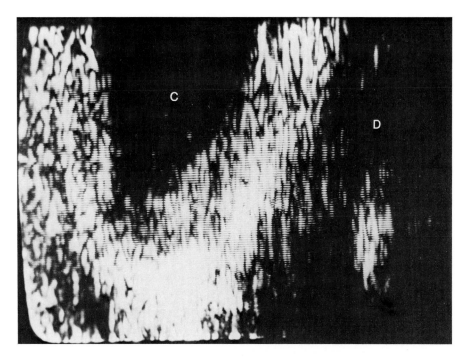

FIG. 7. Sonogram through a pseudocyst (C) in the head of the pancreas in close proximity to the duodenum (D). (From ref. 25.)

Ultrasound scanning has been useful in demonstrating dilated pancreatic ducts at operation for chronic pancreatitis. To palliate the disabling pain often exacerbated by eating, operation may be indicated. The objective of these operations is either to drain an obstructed and dilated pancreatic ductal system internally or to remove the pancreas. Our preference has been to perform a drainage procedure (Puestow operation) whenever possible and to extirpate the gland only if such a drainage procedure were not feasible. Figure 9 shows a dilated pancreatic duct that was not readily found by the surgeon and that was accurately localized by operative ultrasound.

In the 35 operations for inflammatory disease of the pancreas, ultrasound provided additional assistance at 21 procedures. This consisted of making an intitial diagnosis of pseudocyst, helping to select drainage sites in palpable and nonpalpable pseudocysts, localizing dilated pancreatic ducts, and excluding the presence of suspected but absent pseudocysts, abscesses, and dilated pancreatic ducts.

Our reported experience with operative ultrasonography for pancreatic tumors has been with 14 patients (22). Twelve of these had carcinoma of the head or periampullary region, which produced jaundice. One patient had a carcinoma of the body of the pancreas and another patient had a gastrin-producing tumor in the body of the gland (Zollinger-Ellison tumor).

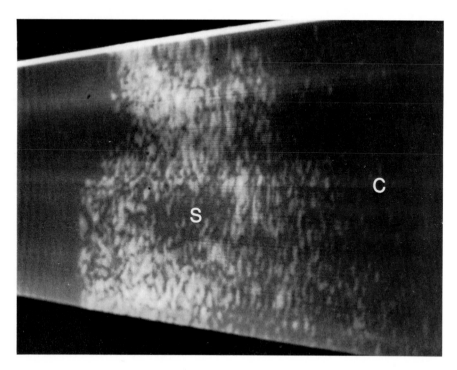

FIG. 8. Sonogram of a pancreatic abscess taken through the transverse mesocolon. The distinction between the abscess wall and cavity (C) is not clear. A small satellite abscess (S) is located within the thickened mesocolon. (From ref. 25.)

As mentioned previously, present real-time, B-mode ultrasonography cannot distinguish pancreatic carcinoma from chronic pancreatitis on the basis of tissue characterization. Operative ultrasonic imaging may, nevertheless, provide information that may be useful during tumor surgery. This relates to a determination of the cause of common duct obstruction, information about involvement of the superior mesenteric vein, and assistance with performance of pancreatic needle biopsy.

Often patients with carcinoma of the pancreas are subjected to operation because of obstructive jaundice. This diagnosis can readily be made on the basis of preoperative diagnostic procedures. The cause of bile duct obstruction, however, may not be apparent from such preoperative studies or from inspection and palpation at operation. Operative ultrasound is a useful means of determining the cause of obstruction simply and relatively early during the operation. Figure 10 demonstrates the distal common bile duct in a patient who underwent operation for obstructive jaundice. Operative ultrasound, in this patient, identified a tumor shelf within the duct. This, plus ultrasonic evidence of no obstructing biliary calculi, provided strong inferential evi-

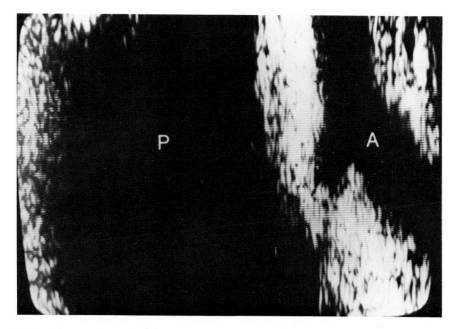

FIG. 9. Sonogram through a dilated pancreatic duct (P). An artery (A) is depicted deep to the pancreatic duct. (From ref. 25.)

dence of pancreatic tumor which was subsequently confirmed by a tissue diagnosis. Thus, common bile duct obstruction by tumor may produce a triad of findings on ultrasound scanning. These are confirmation of common duct dilatation, the absence of impacted stones in the duct, and a characteristic abnormal termination of the common duct. The abnormal termination is either external compression or internal infiltration by tumor.

Figure 11 shows pancreatic tumor invasion of a superior mesenteric vein. In this patient, attempted dissection of the tumor away from the vein was not successful.

Pancreatic biopsy during cancer surgery is a controversial issue. Often the tumor itself is missed and areas of surrounding pancreatitis are submitted for tissue examination. Further, complications of pancreatic biopsy are a common and serious hazard in the postoperative period. This results from persistant leakage of pancreatic juice from an opened pancreatic duct, particularly if the ductal system is obstructed and the obstruction is not relieved. If the surgeon elects to perform a needle biopsy, ultrasonic guidance at operation may help in two respects: first, to locate the tumor more precisely; second, to avoid puncture of dilated pancreatic ducts. A Zollinger-Ellison tumor was discovered by direct inspection. Ultrasound revealed this to be relatively echo-free when compared with surrounding pancreatic tissue.

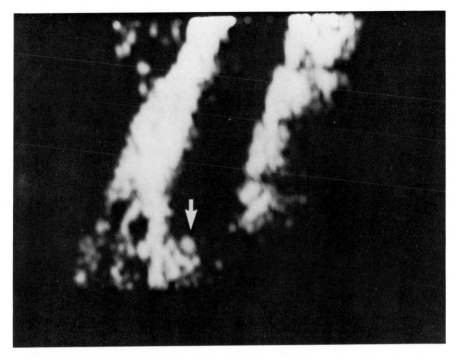

FIG. 10. Sonogram showing a longitudinal section through a common duct with infiltrating tumor producing an obstructing shelf (*arrow*). (From ref. 25.)

This finding is consistent with our other experience that endocrine tumors tend to be more sonolucent than the surrounding tissues. On this basis, operative ultrasonography may be a useful diagnostic adjunct in locating pancreatic endocrine tumors that are not readily seen or palpated by the surgeon.

ULTRASONIC SCANNING DURING RENAL OPERATIONS

Cook and Lytton were the first to employ real-time, B-mode ultrasound scanning during operations for renal lithiasis (4). Their pioneering work constitutes the first application of real-time, B-mode ultrasonography during any type of operation.

There are four indications for employing imaging ultrasound during operation for renal calculi. The first relates to accurate localization of the calculi.

The diagnosis of renal calculi is made preoperatively on the basis of pain and contrast radiography. Between the time of the last X-ray and operative exposure of the kidney, calculi may have changed position. Operative ultrasound may be used to localize calculi precisely and determine if a change in position has occurred.

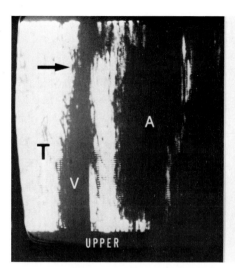 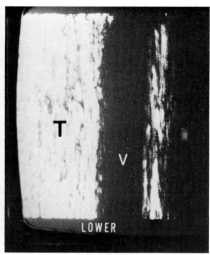

FIG. 11. Composite of two sonograms showing pancreatic tumor (T) and the superior mesenteric vein (V). The sonogram on the **left** is of a higher segment of the vein showing a narrowing of the vein by tumor (T) most marked at the *arrow*. The superior mesenteric artery (A) is also seen with tissue of increased echogenicity encroaching on both vessels. The sonogram on the **right** shows a lower segment of the vein and tumor (T) that is not compressing the vein (V). (From ref. 25.)

A second indication for operative ultrasonography is to guide an exploratory needle to the intrarenal stone. Calculi that are fixed within a minor calyx often cannot be palpated and the surgeon must resort either to the use of extended incisions through renal parenchyma or to use of operative radiography, nephroscopy, or plasma coagulum to adhere to the calculus. In these

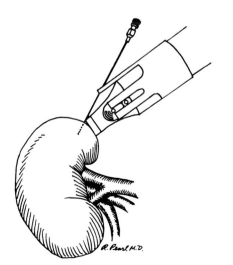

FIG. 12. Illustration of ultrasonic guidance of an exploratory needle to the site of a stone using a needle-guide adapter attached to the transducer-probe. (From ref. 25.)

circumstances, ultrasound may be used to place a needle at the location of the stone (Fig. 12). A relatively small incision is made along the needle tract to extract the calculus. This permits stone removal with less tissue dissection and injury, and significantly shortens the search process.

A third indication for imaging ultrasonography is aid in detecting calculi that are nonopaque on X-ray. Such calculi can be detected only by radiography using contrast material that may be difficult to utilize in the operative setting after the kidney has been opened. Ultrasonography at renal lithotomy surgery has been useful to survey the kidney after stone extraction to make certain that all calculi and fragments have been removed.

Not all intrarenal masses are visible at the time of surgery if they do not distort the renal capsule. Sonography can be helpful in showing the sites of small intranephric masses that are not neoplastic (e.g., angiomyelipoma), allowing local resections rather than nephrotomy.

Results

We have reported on the use of real-time, B-mode ultrasound scanning at 16 operations to remove calculi from the kidney (23). The stones varied in size from 3 to 20 mm. Figure 13 shows a small calculus detected at operation.

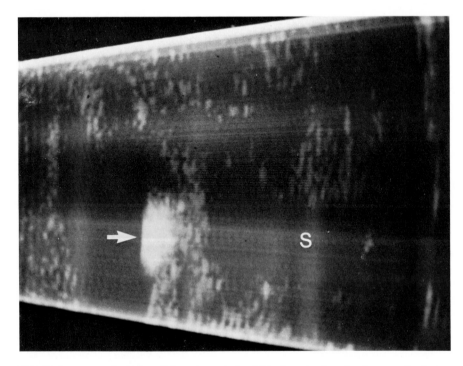

FIG. 13. Sonogram of a calyceal stone (*arrow*) about 3 to 4 mm in diameter. An acoustic shadow (S) is also seen. (From ref. 25.)

Operative ultrasonography was considered helpful in 15 of the 16 operations. In one of our early experiences, ultrasound failed to detect two stone fragments after extraction of a staghorn calculus that were discovered by operative radiography. In cases in which ultrasonography proved useful, the following benefits were noted. In 11 patients, calculi were accurately localized by the use of ultrasound. In one of these patients, preoperative studies identified three radioopaque calculi. Operative ultrasonography localized these but also detected three additional calculi that were nonopaque in the preoperative X-rays. In six patients, ultrasound helped to locate retained fragments or indicated complete clearance after extraction of large calculi.

ULTRASONIC SCANNING DURING VASCULAR OPERATIONS

There are two main indications for the use of imaging ultrasound during vascular reconstruction operations. These are site selection for arteriotomy and detection of vascular defects immediately after restoration of blood flow but before closure of the operative incision. The need to use ultrasound to select an arteriotomy site seldom occurs because of the accuracy of preoperative arteriography. Occasionally, operative ultrasonography may supplement preoperative arteriography and the clinical findings following exposure of blood vessels. By far, however, the most important application of imaging ultrasound during vascular surgery is to detect vascular defects at the end of the procedure.

The defects for which the examination is performed are strictures, thrombi, and intimal flaps that may have been inadvertently left behind following endarterectomy or anastomosis. Such defects may be initially hemodynamically insignificant and produce no changes that may be identified visually, by palpation, or by Doppler ultrasound examination. By ultrasonically imaging the site of endarterectomy or anastomosis, the surgeon is provided information about the possible presence of defects that could lead to later complications. This information is made available during surgery, which permits the surgeon to take remedial action at the same operation if a significant defect is discovered.

To evaluate the feasibility of employing real-time ultrasound to detect vascular defects at operation, we performed two animal studies (2,3). These experimental studies were intended to determine the sensitivity of ultrasonography and to compare its accuracy with that of arteriography.

In the first study, defects simulating those encountered at operation were created to determine the size limits that could be detected by high-resolution imaging ultrasound (2). These experiments revealed that 68% of intimal flaps undermined 1 mm and 100% of intimal flaps 2 or more millimeters in size were demonstrable by operative scanning of the exposed blood vessels. In the second study, the accuracy of ultrasound was compared with that of single-exposure arteriography and serial biplane arteriography in detecting vascular defects (3). This study showed that all three diagnostic procedures were highly specific in detecting all lesions. All three tests were highly sensitive in recognizing strictures. However, ultrasound was more sensitive, at a statistically significant level, in detecting small thrombi and intimal flaps than either single exposure or serial biplane arteriography.

On the basis of these preliminary findings, we applied imaging ultrasonography at vascular reconstruction operations. The technique of scanning a bypass anastomosis is illustrated in Fig. 14.

Results

We have reported our initial clinical experience in 55 patients in whom real-time, B-mode ultrasonography was performed during vascular operations (24). Vascular defects were found in about 20% of the patients examined. Lane has reported a higher occurrence of vascular defects discovered by operative ultrasonography (12). Most of the defects were intimal flaps that occurred at an endarterectomy or anastomotic site or at the placement location for vascular clamps. The majority of the defects were small or not in a strategic location and did not warrant the risk of reentry. However, in instances in which the defects were significant, the vessels were reentered and sizable defects were corrected with no complications in the early postoperative period. Figure 15 illustrates a significant intimal flap that was corrected after discovery by ultrasonography.

ULTRASONOGRAPHY DURING ENDOCRINE SURGERY

We have had limited experience with ultrasonic imaging of endocrine tumors during operation. Most of the tumors were readily located by the surgeon and ultrasound did not significantly aid in their discovery at operation.

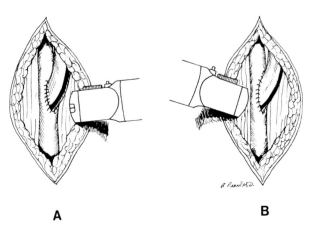

| A | B |

FIG. 14. Illustration of lateral scanning of an end-to-side anastomosis. The anastomosis is usually scanned from both sides. (From ref. 25.)

All endocrine tumors that we have examined at operation [parathyroid adenomas (21) and adrenal and pancreatic tumors (25)] were relatively sonolucent when compared with surrounding tissue (Fig. 16). This suggests that the ultrasonographic appearance of endocrine tumors may be distinctive and may serve to distinguish them in situations in which their presence may be difficult to detect on preoperative testing at exploration (e.g., tumors within the substance of the pancreas).

ULTRASONOGRAPHY DURING BRAIN SURGERY

Our personal experience has been limited in this field. Small series have been reported by others (8,15,30). Operative ultrasonography may be useful in delineating the edges of brain tumors and in identifying cystic structures associated with tumors. A recent experimental study favorably compared operative ultrasound with computerized tomography in locating areas of hemorrhage and bone fragments following cranial trauma (8).

Accurate placement of shunts within the anterior horn of the lateral ventricles is possible under ultrasonic guidance in infants (31). A transducer is placed on the anterior fontanelle out of the sterile field and the shunt is inserted from a lateral approach. The shunt tube can be seen throughout its interventricular course.

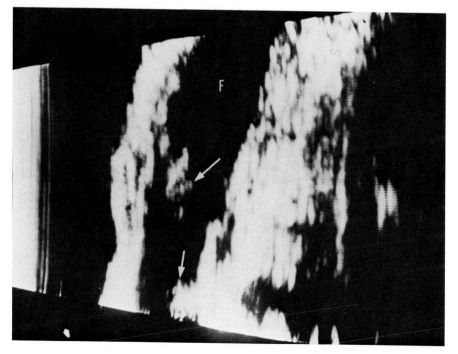

FIG. 15. Sonogram of a superficial femoral artery (F) 2 cm distal to a femoral-graft anastomosis illustrating two intimal flaps, approximately 3 and 4 mm in size (*arrows*). The location of the intimal flaps corresponds to the placement site for an occlusive clamp. (From ref. 25.)

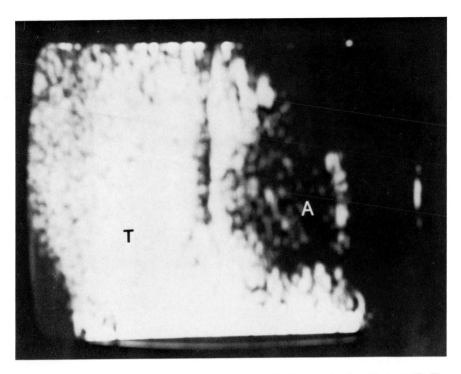

FIG. 16. Sonogram showing a parathyroid adenoma (A) deep to the thyroid gland (T). The adenoma is relatively sonolucent. The capsule of the adenoma shows increased echogenicity. (From ref. 25.)

CONCLUSIONS

Operative ultrasonography appears to be a promising adjunct to diagnosis, localization, and surveillance in a number of surgical fields. Although first used successfully at surgical operations more than 20 years ago, imaging ultrasound has not come into common usage because of problems with equipment utilization in the operating room setting, and the necessity for experience in the performance and interpretation of ultrasonic images. Recent advances in technology have made available compact high-resolution B-mode ultrasound scanners, which are relatively simple to use and produce images that are easier to interpret. These developments have greatly improved the prospects for operative imaging.

A number of studies cited in this review attest to the efficacy of operative ultrasound imaging. Greater utilization of this technology may facilitate surgical operations by reducing tissue dissection and the need for contrast radiography with its well-known disadvantages. We believe that there should now be a wider application of operative ultrasound in the various fields of surgery.

More extensive application, however, will depend upon the fulfillment of certain conditions that, we believe, are essential to the future development of operative ultrasonography.

The conditions are a willingness on the part of the surgeon to perform ultrasonic scanning and to participate in the interpretation of images and a good working relationship between surgeons and ultrasonographers. Only with surgeons experienced in ultrasonography and collaborating with their colleagues in ultrasonography can this new technology be established. Such experience and collaboration has precedent. The development of operative radiography would not have been possible without good working relationships between surgeons and radiologists. The same should be true for operative ultrasonography, resulting in advancement of both disciplines and, ultimately, better management of patients undergoing surgical procedures.

REFERENCES

1. Andaloro, V. A., Schor, M, and Marangola, J. P. Intraoperative localization of a renal calculus using ultrasound. *J. Urol.*, 116:92–93, 1976.
2. Coelho, J. C. U., Sigel, B., Flanigan, D. P., Schuler, J. J., Spigos, D. G., and Nyhus, L. M. Detection of arterial defects by real-time ultrasound scanning during vascular surgery: An experimental study. *J. Surg. Res.*, 30:535–543, 1981.
3. Coelho, J. C. U., Sigel, B., Flanigan, D. P., Schuler, J. J., Spigos, D. G., Tan, W. S., and Justin, J. An experimental evaluation of arteriography and imaging ultrasonography in detecting arterial defects at operation. *J. Surg. Res.*, 32:130–137, 1982.
4. Cook, J. H., and Lytton, B. Intraoperative localization of renal calculi during nephrolithotomy by ultrasound scanning. *J. Urol.*, 117:543–546, 1977.
5. Cook, J. H., and Lytton, B. The practical use of ultrasound as an adjunct to renal calculous surgery. *Urol. Clin. North Am.*, 8:319–329, 1981.
6. Dempsey, P. J., Shoaf, M. D., Weinerth, J. L., and Paulson, D. F. Experimental basis for intraoperative calculus localization by ultrasound. *Invest. Urol.*, 14:380, 1977.
7. Eiseman, B., Greenlaw, R. H., and Gallagher, J. Q. Localization of common duct stones by ultrasound. *Arch. Surg.*, 91:195–199, 1965.
8. Enzmann, D. R., Britt, R. H., Lyons, B., Buxton, T. L., and Wilson, D. A. Experimental study of high-resolution ultrasound imaging of hemorrhage, bone fragments, and foreign bodies in head trauma. *J. Neurosurg.*, 54:304–309, 1981.
9. Hayashi, S., Wagai, T., Miyazawa, R., Ito, K., Ishikawa, S., Uematsu, K., Kikuchi, Y., and Uchida, R. Ultrasonic diagnosis of breast tumor and cholelithiasis. *West. J. Surg.*, 70:34–40, 1962.
10. Lane, R. J., and Crocker, E. F. Operative ultrasonic bile duct scanning. *Aust. NZ J. Surg.*, 49:454–458, 1979.
11. Lane, R. J., and Glazer, G. Intraoperative B-mode ultrasound scanning of the extra-hepatic biliary system and pancreas. *Lancet*, ii:334–337, 1980.
12. Lane, R. J. Intraoperative B-mode Scanning. *J. Clin. Ultrasound*, 8:427–434, 1980.
13. Lytton, B., and Cook, J. H. Intraoperative ultrasound. In: *Ultrasound in Urology*, edited by M. I. Resnick and R. C. Saunders, p. 340, 1979. Williams & Wilkins, Baltimore.
14. Knight, R. P., and Newell, J. A. Operative use of ultrasonics in cholelithiasis. *Lancet*, i:1023–1025, 1963.
15. Rubin, J. M., Mirfakhraee, M., Duda, E. E., Dohrmann, G. J., and Brown, F. Intraoperative ultrasound examination of the brain. *Radiology*, 137:831–832, 1980.
16. Schlegel, J. U., Diggdon, P., and Cuellar, J. The use of ultrasound for localizing renal calculi. *J. Urol.*, 86:367–369, 1961.
17. Sigel, B., Spigos, D. G., Donahue, P. E., Pearl, R., Popky, G. L., and Nyhus, L. M. Intraoperative ultrasonic visualization of biliary calculi. *Curr. Surg.*, 36:158–159, 1979.

18. Sigel, B., Coelho, J. C. U., Spigos, D. G., Donahue, P. E., Renigers, S. A., Capek, V., Nyhus, L. M., and Popky, G. L. Real-time ultrasonography during biliary surgery. *Radiology*, 137:531–533, 1980.
19. Sigel, B., Coelho, J. C. U., Spigos, D. G., and Nyhus, L. M. Ultrasound and the pancreas. *Lancet*, ii:1310–1311, 1980.
20. Sigel, B., Coelho, J. C. U., Spigos, D. G., Donahue, P. E., Wood, D. K., and Nyhus, L. M. Ultrasonic imaging during biliary and pancreatic surgery. *Am. J. Surg.*, 141:84–89, 1981.
21. Sigel, B., Kraft, A. R., Nyhus, L. M., Coelho, J. C. U., Gavin, M. P., and Spigos, D. G. Identification of a parathyroid adenoma by operative ultrasonography. *Arch. Surg.*, 116:234–235, 1981.
22. Sigel, B., Coelho, J. C. U., Nyhus, L. M., Velasco, J. M., Donahue, P. E., Wood, D. K., and Spigos, D. G. Detection of pancreatic tumors by ultrasound during surgery. *Arch. Surg. (In press.)*
23. Sigel, B., Coelho, J. C. U., Sharifi, R., Waters, W. B., and Spigos, D. G. Ultrasonic scanning during operation for renal calculi. *J. Urol.*, 127:421–424, 1982.
24. Sigel, B., Coelho, J. C. U., Flanigan, D. P., Schuler, J. J., and Spigos, D. G. Ultrasonic imaging during vascular surgery. *Arch. Surg.*, 117:764–767, 1982.
25. Sigel, B. (ed.). *Operative Ultrasonography*. Lea & Febiger, Philadelphia, *(In press)*.
26. Sigel, B., Coelho, J. C. U., Justin, J., and Machi, J. Principles and applications of operative ultrasound. *J. Clin. Surg. (In press)*.
27. Sigel, B., Coelho, J. C. U., Nyhus, L. M., Donahue, P. E.. Velasco, J. M., and Spigos, D. G. Comparison of cholangiography and ultrasonography in the operative screening of the common bile duct. *World J. Surg. (In press)*.
28. Sigel, B., Coelho, J. C. U., Donahue, P. E., Nyhus, L. M., Spigos, D. G., Baker, R. J., and Machi, J. Ultrasonic assistance during surgery for pancreatic inflammatory disease. *Arch. Surg.*, 117:712–716, 1982.
29. Sigel, B., Coelho, J. C. U., Flanigan, D. P., Schuler, J. J., Spigos, D. G., and Machi, J. Comparison of B-mode real-time ultrasound scanning and arteriography in detecting vascular defects during operation. *Radiology (In press)*.
30. Voorhies, R. M., and Patterson, R. H. Preliminary experience with intraoperative ultrasonographic localization of brain tumors. *Radiol./Nucl. Med.*, 10:8–9, 1980.
31. Skolnik, A., and McLane, D. G. Intraoperative real-time ultrasonic guidance of ventricular shunt placement in infants. *Radiology*, 141:515–517, 1981.

Ultrasound Annual 1982,
edited by Roger C. Sanders.
Raven Press, New York © 1982.

Critical Review of Measurement Techniques in Obstetrics

*,**D. Graham and **R. C. Sanders

*Departments of Gynecology and Obstetrics and **The Russell H. Morgan Department of Diagnostic Radiology; The Johns Hopkins Medical Institutions, Baltimore, Maryland 21205

As the technology available has improved, ultrasound has progressively become an increasingly invaluable aid in the management of normal and abnormal pregnancies. Although initially, by means of bistable techniques, the only obstetrical measurement that could be obtained with reasonable frequency was the biparietal diameter, improved equipment resolution combined with further knowledge of sonographic patterns of fetal growth have led to a proliferation in the number of physical measurements that may now be used in fetal assessment. In this chapter the use of different physical measurements in the assessment of gestational age and fetal growth will be discussed.

CROWN–RUMP LENGTH

Sonographic Measurement

Measurement of the crown–rump length (CRL) for assessment of gestational age was first proposed by Robinson (55). This measurement, which is the longest length of the embryo excluding the limbs, is not necessarily identical to the embryonic CRL, since no account is made of the variations in the degree of flexion of the fetal body. This measurement is of value from 6 to 14 weeks after the last menstrual period; after that time, the degree of flexion of the body will significantly influence the measurement obtained.

Advantages of CRL

At present, CRL appears to be the single most accurate measurement for assessment of gestational age (GA). Although the measurement is easy to obtain, both with static and real-time scanners, the rapidity of obtaining satisfactory pictures of a moving fetus with real-time makes it the more satisfactory method. No significant differences have been found in the measurements obtained with each technique (2,3). The reproducibility of the technique is also high: a study to determine the reproducibility showed that repeated measurements may be made within ± 1.2 mm (55).

Several factors account for the accuracy of the technique:

(a) The rapid incremental increase in CRL in the first trimester, the mean CRL increasing from 10 mm at 7 weeks to 83 mm at 14 weeks;
(b) There is minimal biologic variation on a weekly basis at this stage in pregnancy;
(c) The dimension is easy to obtain and measure using real-time equipment.

The major factors leading to errors in the measurement of CRL are:

(a) Underestimating the GA, by not obtaining the longest length of the embryo;
(b) Overestimating the GA by including the yolk sac, which may be positioned adjacent to the rump, in the CRL measurement;
(c) Differences in the length of the preovulatory phase of the cycle.

Measurement of the CRL allows accurate estimation of the GA and the estimated date of confinement. Robinson (55), in a series of 35 patients with "certain" dates who were examined blindly, found that the ultrasonic estimate of age was within 3 days of the menstrual age in all but one patient. In a statistical survey (56), the same author found that a prediction could be made to within ±4.7 days in 95% of cases if a single measurement was made and to within even narrower limits given three independent measurements. Similar results have been obtained by Drumm (18) and by Kurjak et al. (39).

In a longitudinal study Deter et al. (15) found somewhat less impressive results. In this study of 20 patients, with known dates of conception [based on basal body temperature (BBT) measurements and intercourse records] he found that although most (70.5%) of the CRL measurements were within the normal range given by Robinson and Fleming (56), 29.5% were >2 standard deviations from the mean (27.9% below and 1.6% above). Closer examination of the results did, however, show reasonable explanations for 78% of the abnormal measurements, principally genetically reduced growth potential or a prolonged follicular phase. Nelson (49) similarly showed that a significant number of CRL measurements obtained by dynamic image scanning fell outside the 95% confidence limits given by Robinson (56).

BIPARIETAL DIAMETER

The biparietal diameter (BPD) is the most commonly obtained measurement obtained in pregnancy. Although it is used as a measure of gestational age, it actually accurately reflects only fetal size, gestational age being extrapolated from this fetal size by assuming that the size is at the 50th percentile. The BPD may be obtained from 10 to 11 weeks until delivery and is a measurement that is relatively easy to obtain by means of both static and especially real-time systems. The manner in which BPD measurements have been obtained has changed since the technique was introduced. Originally BPD was measured using A-mode techniques, a technically difficult maneuver. Bistable equipment allowed easier and more accurate determination of the BPD, and when combined with A-mode technique gave very accurate results. A range of error of 1.5 to 2.4 mm was achieved between 20 and 30 weeks, and outside this range, the error was only slightly greater (12,46). Introduction of gray scale has led to even greater improvement in the technique of BPD measurement, since one may now visualize a significant amount of intracranial anatomy, allowing choice of a more exact and reproducible level. At this time the level most widely utilized is that which shows the thalamus, cavum septum pellucidum, and interhemispheric fissure. Although it appears that this is not invariably the widest transverse diameter (10), this is the level that is most universally used and should therefore be retained. Most workers measure the diameter from the outer border of the skull to the inner border of the echoes from the opposite side of the skull. This measurement will be smaller than that measured postnatally with calipers, since the latter measurement includes the thickness of the scalp and also the thickness of the cranial wall not measured sonographically.

Sources of Error in BPD Measurement

Different Techniques

Use of a standard level allows more accurate comparison of results from different centers and also more reliably allows assessment of BPD growth from sequential studies in any one individual. One might expect that different BPD results might be obtained from measurements obtained with bistable, B-scan, and real-time but such measurements have been shown to have no statistical differences (11,17,41). It appears that real-time systems allow more rapid visualization of the correct BPD level.

Observer Error

Lawson et al. (41) has shown that interobserver error in 14% of patients was statistically different and was in the magnitude of ± 1 mm, as compared with the overall mean. Two further studies (12,17) showed that the standard deviations taken at 70% confidence limit were 1.74 mm and 2.64 mm, respectively. The effect of this error of magnitude (± 2 SD = 5.28 mm) would be dependent on the gestational age. In the second trimester, when the fetal head is growing relatively rapidly, this degree of error would not affect GA assessment by more than 1 to 2 weeks. Near term, however, with significant slowing of the fetal head growth, the discrepancy would be much greater, approaching as much as 4 to 5 weeks.

Osinusi (52) similarly obtained an overall standard deviation of 1.1 mm. In 80% of cases (in 80 patients), the individual standard deviations of the results were less than or equal to the estimated standard deviation of the group as a whole. The variability of measurement related to BPD was least in the range 59 to 83 mm.

Docher and Seltatree (17) showed that a real-time scanner allowed a relatively inexperienced operator to obtain reasonably accurate results. The mean standard deviations of the two operators in this study were 1.09 and 1.18 mm.

Different Populations

Kurz et al. (40), in a study of 17 different BPD charts, considered differences in populations as a possible factor in the difference seen between some of these charts. However, this was thought not to be a significant factor, since no individual study was shown to have consistently higher or lower BPD measurements. Similar findings were obtained by Sabbagha et al. (57). A subsequent statistical comparison of four B-scan studies showed that composite BPDs derived from all four groups did not differ significantly from the BPDs in each single group (58).

It has been suggested that there may be differences in BPD between Whites and Blacks. However, Hohler et al. (32), in a study of 338 examinations done on 209 pregnant patients, found no significant statistical difference in the relationships between BPD and gestational age for White and Black patients. Sabbagha et al. similarly have shown that between 20 and 40 weeks' gestation, the mean BPDs of White and Black populations are almost identical (57).

Use of Different Tissue Velocities

An assumed velocity of 1,540 m/sec is used as the reference in the United States. However, in a number of the studies derived in Europe, a velocity of 1,600 m/sec was used. Use of this different velocity results in an error of 4%, an error that assumes greater significance in the last weeks of pregnancy.

Changes in Head Shape with Premature Rupture of Membranes and with Prematurity

With premature rupture of membranes (PROM), especially with prematurity, there may be lateral flattening of the head resulting in a somewhat dolichocephalic shape. The degree of dolichocephaly (or brachycephaly) of the skull may be measured by calculation of the cephalic index (CI) using the formula:

$$CI = \frac{BPD}{OFD} \times 100$$

where OFD is the occipitofrontal diameter.

Hadlock (29) showed that variations in the shape of the fetal skull might adversely affect the accuracy of the BPD measurement in estimating fetal age. On the basis of established postnatal citeria, in each case the CI was in the dolichocephalic or bradycephalic range. It was shown that a cephalic index >1 SD from the mean (<74, >83) may be associated with a significant alteration in BPD measurement expected for a given gestational age. Cephalic index does not appear to show any significant change from 14 to 40 weeks. Prenatal values of CI have been found to agree with those obtained postnatally by Haas (56), who found a mean value of 81.7, and Jordaan (36), who found a mean value of 80.6 in 50 neonates delivered by Ceasarean section.

Where the CI shows that the head is dolichocephalic or bradycephalic, the head circumference (HC), which is not significantly altered by minor changes in head shape, should be used (5,31). Accuracy of the HC in predicting GA in the third trimester (± 2 to 3 weeks) is comparable to the accuracy of the BPD during this period (5,31,62). In Hadlock's study, the HC measurements were within ± 1 week of the true gestational age (19,62).

Positional Problems

In approximately 5% of examinations, the fetus is in a position that makes measurement of the BPD very difficult or impossible, i.e., face up, face down, or deeply engaged. In such instances, alternative measurements for GA may be used, e.g., femoral length (q.v.).

The importance of fetal lie has been studied by Poll (53), who found that the BPD measurements of an infant in breech position were frequently below the accepted norms, even with a normally growing fetus. This has not, however, been a finding reported by other authors.

Intrinsic BPD Differences

Intrinsic differences in biparietal diameters at any given menstrual gestational age may be a result of genetic variations in head size in fetuses of the same conceptual age or of differences in actual GA attributable to differences in the length of the preovulatory phase of the cycle.

Use of Inappropriate Nomogram

Although most nomograms show the same general curve, there are often significant differences in values assigned to any particular GA. For example, the mean GA assigned to 8.0 cm in the Hobbin chart is 33 weeks (30) but only 31 weeks in the Weiner chart (65). When the variation of ± 2 to 3 weeks is added to this, significant errors in the assignment of GA may result.

Use of Single Measurements

Overreliance on a single random measurement of BPD rather than serial studies, especially during the last trimester of pregnancy, may result in inaccuracies in estimations of gestational age. Because of the wide variation in head size a single BPD obtained for GA in the last weeks of pregnancy is of extremely limited value in assigning GA and in fact, may be quite misleading, especially in the patient in whom there is symmetrical intrauterine growth retardation (IUGR) with a smaller than expected head size. In such instances, the GA may be significantly underestimated, leading to problems in clinical management. Where single measurements are used for GA assessment, it is considered that the closest prediction is obtained if the BPD is obtained prior to 26 weeks of pregnancy (58). Measurement prior to this time will reduce errors caused by normal variation or by IUGR, which most often occurs after this time.

Whether or not a second BPD estimation later in pregnancy is of any value in allowing more accurate assignment of gestational age has been a subject of some disagreement. O'Brien (50) evaluated the efficacy of a second BPD between 18 and 30 weeks of gestation in the prediction of the onset of labor, the accuracy of prediction being determined by analysis of the difference between the sonographically predicted EDC and the date of onset of spontaneous labor. Although it was found that the combined result of both examinations did not add to the accuracy of EDC prediction, the sample size of 50 patients was small.

To avoid the problems inherent in extrapolation of GA from single measurements, Sabbagha et al. (59) proposed the Growth Adjusted Sonographic Age (GASA). This method is based upon the observation that when studied sequentially fetuses may be classified into below average, average, or above average groups on the basis of their BPD growth. Two separate BPDs are obtained in the usual fashion, the first at approximately 20 to 26 weeks of pregnancy and the second at least 9 weeks later. The incremental growth obtained is then compared with that expected in each of the three groups and the fetus assigned to one of those groups. The GA assigned at the first BPD estimation is retained if the fetus shows average growth. If growth is above or below average, the original GA is recalculated. With this method, an accuracy of predictions of GA of ± 1 to 3 days (95% confidence limits) is claimed (59). Not only is the GASA thought to define better the duration of pregnancy, but also it allows placement of the fetus in a specific percentile growth bracket. By assignment of a specific growth percentile, it is hoped that fetuses at high risk for IUGR might be more easily detected.

When IUGR is suspected, serial estimation of BPD will aid in diagnosis since the observed increase in BPD over a period of time will be less than expected. In study of 1,200 normal women, Campbell (6) showed that the

BPD normally grows at 3.5 mm per week between 16 and 22 weeks and then gradually decreases, so that in the last weeks of pregnancy it may be as little as 1 mm per week.

Application of Normal Values to Abnormal Pregnancies

To define more precisely an abnormal value, biparietal nomograms should be derived from values obtained from normal patients. The most important conditions that one might expect to lead to alterations in BPD are macrosomia and IUGR. Aantaa (1) measured BPDs in 24,560 cases using B-scan techniques and compared the increase in BPD in normal pregnancies, in those complicated by diabetes or toxemia and in fetuses thought to be small-for-dates. Only the last of these groups demonstrated a statistically significant difference. Similarly, Ogata (51) performed serial measurements of the BPD in a group of diabetic patients. These measurements in all fetuses conformed to the growth patterns for fetuses of nondiabetic mothers. Wladimiroff (67) also found that in large-for-gestational age (LGA) infants, the BPD proved to be of very little value, since it was >90th to 95th percentile of the normal curve in only 7% of cases.

From the above, the following conclusions may be assumed on BPD measurement:

(a) The BPD is a relatively accurate, easily performed measurement that may be applied from 10 to 11 weeks of gestation up to delivery.

(b) Accuracy of the measurement decreases as gestational age increases from an accuracy of ±1 to 1.5 weeks in the first trimester (2,55) down to ±3 to 4 weeks (2 SD) in late third trimester (2,3,55).

(c) The nomogram that best fits measurements obtained in the local population should be used.

(d) A consistent level for measurement should be used: one which includes the thalami and the cavum septum pellucidum.

(e) Serial measurements of BPD may increase the accuracy of gestational age prediction.

(f) Although serial BPD measurements will detect a number of IUGR fetuses, this may not occur until the insult is severe. In a significant percentage of cases, BPD growth will be normal in IUGR (6,32,66).

ABDOMINAL CIRCUMFERENCE

The fetal abdomen is most conveniently and reproducibly measured through the liver, using a level that includes the horizontal portion of the portal sinus. This level is at the same level as the stomach and slightly cephalad to the kidneys. Where the fetus is prone and shadowing from the spine prevents visualization of the portal sinus, a section at the level of the stomach is satisfactory but less accurate. The abdominal circumference (AC), which may be measured from early in the second trimester, indirectly measures liver size. This is decreased in IUGR because of decreased glycogen deposition and conversely in macrosomia, increased liver size together with increased subcutaneous tissue will result in an enlarged AC.

Measurement of the AC can be combined with measurement of the head circumference, measured at the level of the plane of the BPD to estimate:

(a) The HC/AC ratio: Normally AC is less than HC until 32 to 36 weeks, when the ratio approaches unity (7). After 32 to 36 weeks, AC becomes greater than HC. Measurement of the HC/AC ratio may be used in screening for IUGR, macrosomia, and fetal anomalies. The following possibilities exist:

Normal H/A:	(i) Normal fetus
	(ii) Symmetrically growth-retarded infant
H/A ↑ :	(i) Asymmetric IUGR
	(ii) ↑ Head size, e.g., in hydrocephalus
H/A ↓ :	(i) Macrosomia
	(ii) Microcephaly

(b) The AC may be combined with the BPD in a refined technique to estimate fetal weight (64). Warsoff used these two parameters to produce a computer-derived nomogram for fetal weight for which an accuracy of ± 30 g/kg (1 SD) was claimed. Other investigators (15) using this method, however, claim that the method systematically underestimates weight. This has also been found by the original investigators who have now revised their original formula (60).

The AC has also been used in the investigation and screening of macrosomia. Ogata (51) used serial estimates of fetal BPD and AC to determine intrauterine growth and found that although the BPD in all fetuses conformed to growth patterns for fetuses of nondiabetic mothers, there were found to be two patterns of AC growth:

(a) normal growth
(b) accelerated growth.

In the group that showed accelerated growth, there was subsequently found to be more immunoreactive insulin in amniotic fluid, the infants weighed more at birth, and had more subcutaneous fat. A significantly enlarged abdominal circumference was found to be a more sensitive screening criterion for gestational diabetes than the conventional clinical criteria used for performance of a glucose tolerance test (GTT).

Wladimiroff (67) in 42 normal infants and 30 LGA infants found that in the detection of LGA the ultrasound measurement of the fetal chest area (a less reproducible level than AC) appeared to be superior to BPD measurements, since the latter was >90% of the normal curve in only 7% of cases of macrosomia whereas measurement of the fetal chest area resulted in a detection rate of 80%.

Abdominal circumference may be used in several ways to estimate fetal weight indirectly (8,45).

WEIGHT ESTIMATION

A reliable sonographic method to estimate fetal weight would be of immense value in clinical obstetrics, since many clinical decisions must be based on expected fetal weight, e.g., premature rupture of membranes and premature labor, macrosomia, IUGR, third-trimester bleeding, and preeclampsia. Several methods are at present available.

Clinical Methods

Clinical methods of weight estimation depend on abdominal palpation of the fetus, a weight estimation being based on the clinician's experience. Such methods are notoriously unreliable, since they tend to place most fetuses near the average and both overestimate the SGA fetus and underestimate the LGA fetus. Thus, Loeffler (45) showed that although the prediction of birth weight by palpation was satisfactory when the fetus was of average size, its accuracy decreased at each extreme of the birth weight range. The error was such that about 60% of predictions for babies weighing <2,300 g at birth had been inaccurate by over 450 g and in each case the predicted birth weight has been underestimated. Lind (44) showed that guess work alone, based on the normal distribution of birth weight, would give a similar error (1 SD = ~440 g).

Biparietal Diameter

Weight estimation from a single BPD is as inaccurate as the clinical method described. The standard error derived by Morrison (48) was ±494 g at term.

The method of weight assessment from BPD was refined by Jordaan (37), who used a relationship for estimating brain weight from the BPD and the cross-sectional area of the BPD plane (NCSA). Subtracting the brain weight from the birth weight gave the somatic weight, which was then related empirically to the AC. A weight estimate could therefore be obtained from the measurements of the BPD, NCSA, and AC. This method appears to be too cumbersome for routine clinical use.

Abdominal Circumference

Several charts have been derived showing a relationship between abdominal circumference and estimated fetal weight. Such methods include those of Campbell and Wilkin (8) and Poll (53). The former method, which is based on

AC alone, was found by Deter (14) to overestimate systematically the fetal weight (mean deviation 5.3%) with a variability of 13.9% (1 SD).

Poll (53) used a method that estimated AC from a single measurement. This AC was converted into a weight percentile for the maturity at the time of measurement so that the weight at delivery could be predicted, the tables of Thompson et al. (61) being used. This method assumes, however, that where there is an interval between measurement and actual delivery, the rate of fetal growth in the interval remains unchanged.

Kearney et al. (38) found that calculations of fetal weight from abdominal circumference were within 15% of the true fetal weight in the small fetus of under 2.5 kg. The relative inaccuracy of the method with larger fetuses was considered to be of little practical importance.

BPD and AC

Warsoff (63) and Warsoff et al. (64) combined measurements of the BPD and AC at the standard levels to provide a more accurate method of weight estimation. A standard deviation of ± 100 g/kg was stated originally but with experience, a standard deviation of ± 30 g/kg (1 SD) is now claimed.

In comparing these two methods in his study population, Deter (14) found that Warsoff method systematically overestimated the weight (mean duration 1.6%) whereas that of Warsoff et al. gave systematic underestimates (mean duration, 3.2%). The variability between the two Warsoff methods was similar (SDs, 8.8 and 8.4%) and significantly lower than the Campbell-Wilkin method described above (SD, 13.9%). When separate weight subclasses were examined, both the mean deviations and their variability were significantly lower with the two Warsoff methods. Deter found that of the three methods examined, that of Warsoff showed the most consistency among the weight subclasses and gave the best estimates of fetal weight. This original formula has now been revised to correct for systematic underestimates (60).

Fetal Volume

It is considered that of all methods, one that could accurately predict fetal volume would be expected to give the best estimate of fetal weight, since mass = volume $\times$ density (provided that density remains fairly constant). Changes in individual fetal tissue density have been studied by Morrison (44), who found that individual tissue component densities tended toward unity, and although the greatest change in overall density would be expected in the last weeks of pregnancy with the accumulation of body fat, such density shift would be only marginal and be less than $\pm 0.01\%$.

While fetal volume measurements would be expected to give the most accurate weight predictions, such measurements do present a major technical difficulty, since ultrasound scanners provide only a two-dimensional cross-sectional representation of the body part being examined. Morrison (44) measured fetal volume by serial echograms of cross-sectional area taken in parallel planes along the length of the fetus at set intervals with subsequent addition of the cross-sectional areas. He found that the correlation between birth weight and the ultrasonic measurement of fetal volume was highly significant ($p > 10^{-6}$), and that the best standard error achieved was such that in a normal term infant it would be $\pm 3\%$ of the birth weight. The major disadvantage of this method is that the estimation of fetal cross-sectional area is much more liable to error from operator inexperience than is BPD measurement, since any error made in the cross-sectional measurements is magnified 36 times in calculating fetal volume.

A three-dimensional computer-based system of displaying fetus, placenta and uterus has been devised by Brinkley et al. (4). This method lends itself to estimation of fetal volume and fetal weight but is not yet available for routine use.

LIMB LENGTHS

Recently a great deal of attention has focused on the value of limb length measurement in the assessment of gestational age and in disorders of fetal growth and in the diagnosis of skeletal malformations (20,21,35,54). The femur length has been the long bone most often measured and is the easiest to visualize completely, since its range of movements are the most limited and because of its shape, it is easy to recognize. Although this is a measurement that is easy to obtain with real-time sonography, there is unfortunately no consistency to the end points that are used. The most accurate results are obtained when two or three images are obtained with real time and the longest image used. Measurement may be from the most proximal visualized portion of the femoral neck or from the area of the greater trochanter with the distal end point in each instance being the visualized femoral condyles. The two different measurements may be 1 to 2 mm different. The former measurement appears to be the more consistently obtainable.

Femoral length measurements may be obtained in almost all second- and third-trimester pregnancies, more often than a BPD may be obtained. The femoral length appears to grow in a linear fashion to term at a rate of 2 to 3 mm per week. Several studies have examined the growth rate and the predictive accuracy of the femoral length measurement. Hohler found a liner growth rate to term and a predictive accuracy very similar to that of the BPD. An almost constant relationship beween BPD and femoral length of 79 $\pm$ 6% was found in 90% of cases after 22 weeks of gestation. Farrant (20) measured limb lengths in patients with certain dates and with biparietal diameter measurements in agreement with the expected gestation. A number of abortuses also had limb lengths measured by sonography and radiography which showed a mean error of 0.54 mm. The X-ray study showed that the ultrasound scans visualized only the ossified portion of the diaphysis. It appeared in this study that limb bone lengths correlated more closely with BPD than with gestational age calculated from the patients' LMPs. (In a separate study, it had been found that the gestation calculated from ultrasound BPD measurement disagreed with that from LMP by more than 7 days in over 14% of patients.) The subsequent obstetric courses of these patients had suggested that the ultrasound assessment of gestational age was more accurate than that from LMP. This is presumably due to the fact that gestational age calculated from LMP may contain errors due to mistaken dates, length of cycle and early or late ovulation.

Measurement of limb lengths will undoubtedly increase in value in the evaluation of growth disorders and in the diagnosis of fetal dwarfism syndromes. The growth pattern of long bones in growth retarded fetuses is not yet fully evaluated, but it appears that limb length is less severely affected than trunk circumference or body weight. A method that can estimate weight per unit length would undoubtedly increase the sensitivity of IUGR screening.

In the diagnosis of dwarfism syndromes, lengths of all the long bones of the limbs are relatively easy to obtain even in early midtrimester. In most of the short limb dwarfism syndromes, examination of the fetus prior to 20 weeks will show abnormality. However, normal limb lengths at this stage of pregnancy have been reported in a patient with heterozygous achondroplasia (21).

TOTAL INTRAUTERINE VOLUME

A method of estimation of total intrauterine volume (TIUV) was proposed by Gohari et al. (24). This value is derived by measurement of the length of the uterus (from uterine fundus to the urinary bladder at the level of the cervix) and the maximum width and depth of the uterus along this axis. The three measurements are multiplied together and then by 0.5233 to give the TIUV in milliliters; this is the formula for a prolate ellipse. This measurement measures the total contribution of the fetal volume, the placental mass, and the volume of amniotic fluid, all of which are decreased in IUGR.

In the original method, Gohari classified the resultant volume into one of three groups:

  (a) Normal: in this group, there were no examples of IUGR;
  (b) Gray zone: 1.0 to 1.5 SD below the mean with a 30% incidence of IUGR;
  (c) Abnormal zone: >1.5 SD below the mean with a 60% incidence of IUGR.

At a threshold of 1.5 SD below the mean, there was a 75% sensitivity, 100% specificity, and 100% accuracy of a positive prediction and a 91% accuracy of a negative prediction. Using a 1.0 SD below the mean, they reported a 100% sensitivity, 78% specificity, 65% accuracy of a positive prediction, and 100% accuracy of a negative prediction.

Subsequent work by Chinn et al. (9) and Levine (43) has, however, found a quite different utility. In their studies, which looked at similar groups of patients with a similar prevalence of disease and virtually identical method of estimation of TIUV, quite different results were obtained. These workers found that the accuracy of a negative test was 89%. Of major significance was the fact that a number of profoundly growth-retarded fetuses were missed by this screening test. It was considered that the difference in the results of the different groups was attributable to differences in the shape and tolerance limits of the respective normal curves.

It is considered that the TIUV is of limited clinical value for several reasons:

(a) It appears that the sensitivity for detection of IUGR is poor and that the accuracy of a positive prediction is even more unsatisfactory;

(b) There are significant errors that may be introduced in the actual measurement. Measurement errors are compounded when the three separate dimensions are multiplied together. Small errors in measurement lead to large differences in the final volume obtained (27). The method also suffers from the fact that it is assumed that all uteri have the shape of a prolate ellipse.

(c) The method is subject to even greater error when the BPD has to be used as an estimator of GA as it has to be in significant percentages of patients. Where there is symmetrical IUGR, there will be a smaller than average BPD, and hence an underestimate of GA. An abnormal TIUV may then falsely appear to be normal.

Pitfalls

Grossman (27) also has pointed out some pitfalls of TIUV measurement.

(a) The original Gohari article, although referring to TIUV, has illustrations that suggest that they include the thickness of the uterine wall. Also, Levine et al. (43) indicate that their measurements are made to the endometrial surface when possible but they include the uterine wall when the inner margin is not clearly defined.

(b) This inclusion of uterine wall will cause a significant increase in TIUV. Therefore, Grossman suggests using outer measurement of uterine wall, which is usually better defined.

(c) The bladder may have effect on uterine dimensions (26): measurements with bladder full and empty showed wide and unpredictable changes in TIUV. Since it is impossible to define a full bladder, measurements must be done with the bladder empty to standardize the technique for reproducible results.

(d) Uterine contractions may alter the shape.

(e) There may be problems with selection of longest dimensions.

(f) Fleisher et al. (23) showed that using the standard formula for a prolate ellipse was inaccurate even in renal transplants and one would therefore expect uterine measurements to be less accurate. It is suggested that the problem may be eliminated by calculating volume based on planimeter measurements of the area of successive 1-cm planes. This, however, is quite time-consuming and adds the potential for significant error inherent in making a large number of difficult measurements.

INTERORBITAL DISTANCE

The use of the femoral length as a parameter in assessment of gestational age has already been discussed. An alternative method has been proposed by Jeanty et al. (34), who measured interorbital distance and binocular distances. These measurements are especially easy to obtain when the fetus is face-up, a position that makes obtaining the BPD more difficult. The values obtained may be used not only for assessing gestational age but also for diagnosis of malformations associated with hyper- or hypotelorism.

VENTRICULAR SIZE

Prior to the development of high resolution, gray-scale equipment the diagnosis of hydrocephalus was usually made late in the course of the disease by a markedly enlarged BPD or by a significant discrepancy between head circumference and trunk circumference. Present equipment, however, allows visualization of much intracranial anatomy including the third and lateral ventricles. The ventricular size may therefore be measured directly and the ratio of the ventricular size to the BPD may be correlated with gestational age (13,22,33). The lateral ventricles are relatively large prior to 17 weeks of gestation; however, the ratio of ventricular size to head diameter falls as gestational age increases. While measurement of the frontal horns or lateral walls of the body of the ventricles are more reliable, hydrocephalus may initially affect principally the posterior horns, which are less consistently measurable and have greater normal variation.

RENAL SIZE

From early in the second trimester, the fetal kidneys and bladder may be readily visualized. The relative size of the fetal kidney and fetal abdomen has been quantitated by Grannum et al. (25), who produced a nomogram of mean kidney circumference against abdominal circumference. This measurement may be useful in the diagnosis of fetuses at risk for conditions that lead to altered renal size such as infantile polycystic disease and multicystic kidney, although in the latter condition, the cystic components are usually readily visible.

A similar chart, correlating renal length and diameter with gestational age, has been produced by Lawson (42).

UMBILICAL VEIN DIAMETER

Measurement of the umbilical vein diameter as an indicator of severity of rhesus hemolytic anemia has been reported by DeVore et al. (16) and by Mayden (47). The diameter of umbilical vein is measured both in the umbilical cord and as it passes through the liver. The former diameter is usually found to be the larger. It is suggested that the umbilical vein may dilate within the liver and in the amniotic fluid in response to severe Rh disease, and that the umbilical vein diameter in the umbilical fluid dilatation may occur prior to the rise in ΔOD. It appears, however, that this measurement is only of predictive value when increased and that normal values may be obtained in affected infants.

SUMMARY

Increasing resolution of current ultrasound equipment has improved considerably the ability to visualize normal intrauterine anatomy, allowing measurement of a number of fetal parameters. This in time has led to increased knowledge of normal and abnormal fetal growth patterns, allowing prediction of such abnormalities early enough to intervene and lower fetal morbidity and mortality rates.

REFERENCES

1. Aantaa, K., and Forss, M. Growth of the fetal biparietal diameter in different types of pregnancies. *Radiology*, 137:167–169, 1980.
2. Adam, A. H., Robinson, H. P., and Dunlap, C. A comparison of crown rump length measurements using a real-time scanner in an antenatal clinic and a conventional B-scanner. *Br. J. Obstet. Gynecol.*, 86:521, 1979.
3. Bovicelli, L., Orseni, L. F., Rizzon, N., et al. Estimation of gestational age during the first trimester using real-time measurements of fetal crown rump and BPD. *J. Clin. Ultrasound*, 9:71, 1981.
4. Brinkley, J. F., McCallum, W O., and Muramatsu, S. In: *Proceedings of the Annual Meeting AIUM, New Orleans, 1980.*
5. Campbell, S. Fetal head circumference against gestational age. In: *The Principles and Practice of Ultrasonography in Obstetrics and Gynecology*, 2nd ed., edited by R. Sanders and A. E. James, p. 454, 1980. Appleton-Century-Crofts, New York.
6. Campbell, S., and Dewhurst, C. J. Diagnosis of the small-for-dates fetus by serial ultrasonic cephalometry. *Lancet*, ii:1002–1006, 1971.
7. Campbell, S., and Thomas, A. Ultrasound measurement of the fetal head to abdomen circumference ratio in the assessment of growth retardation. *Br. J. Obstet. Gynecol.*, 84:165–174, 1977.
8. Campbell, S., and Wilkin, D. Ultrasonic measurement of fetal abdominal circumference in the estimation of fetal weight. *Br. J. Obstet. Gynecol.*, 82:689, 1975.
9. Chinn, D. H., Filly, R. A., and Callen, P. W. Prediction of IUGR by sonographic estimation of TIUV. *J. Clin. Ultrasound*, 9:175–179, 1981.
10. Christie, A. D. Critical assessment of fetal biometry. In: *Handbook of Clinical Ultrasound*, edited by M. deVlieger, J. H. Holmes, E. Kazner, et al., pp. 153–159. John Wiley, New York.
11. Cooperberg, P. L., Chow, T., Kite, V., et al. Biparietal diameter: A comparison of real-time and conventional B-scan techniques. *J. Clin. Ultrasound*, 421–423, 1976.

12. Davison, J. M., Lind, T., Farr, V., et al. The limitation of ultrasonic fetal cephalometry. *J. Obstet. Gynecol. Br. Commnw.*, 80:769–775, 1973.
13. Denkhaus, H., and Winsberg, F. Ultrasonic measurement of the fetal ventricular system. *Radiology*, 131:781, 1979.
14. Deter, R. L., Hadlock, F. P., Marrist, R. B., et al. Evaluation of three methods for obtaining fetal weight estimates using dynamic image ultrasound. *J. Clin. Ultrasound*, 9:421–425, 1981.
15. Deter, R. L., Harrist, R. B., Madlock, F. P., et al. The use of ultrasound in the assessment of normal fetal growth: A review. *J. Clin. Ultrasound*, 9:481–493, 1981.
16. DeVore, G. R., Mayden, K., Tortora, M., et al. Dilatation of the fetal umbilical vein in rhesus hemolytic anemia: A predictor of severe disease. *Am. J. Obstet. Gynecol.*, 141:464, 1981.
17. Docher, M. F., and Seltatree, R. S. Comparison between linear array real-time ultrasonic scanning and conventional compound scanning in the measurement of the fetal biparietal diameter. *Br. J. Obstet. Gynecol.*, 84:924, 1977.
18. Drumm, J. E. The prediction of delivery date by ultrasonic measurement of fetal crown rump length. *Br. J. Obstet. Gynecol.*, 84:1–5, 1977.
19. Dubowitz, L. M., Dubowitz, V., and Goldberg, C. Clinical assessment of gestational age in the newborn infant. *J. Pediatr.*, 77:1–10, 1970.
20. Farrant, P., and Meire, H. B. Ultrasound measurement of fetal limb lengths. *Br. J. Radiol.*, 51:660, 1981.
21. Filly, R. A., Golbus, M. S., Carey, J. C., et al. Short limbed dwarfism: Ultrasonographic diagnosis by mensuration of fetal femoral length. *Radiology*, 138:653, 1981.
22. Fiske, C. E., Filly, R. A., and Callen, P. W. Sonographic measurement of lateral ventricular width in early ventricular dilatation. *J. Clin. Ultrasound*, 9:303, 1981.
23. Fleisher, A., Bressler, E., Dodson, M., et al. Predictive value of volume and parenchymal changes for the sonographic detection of acute renal allograft rejection. *Presented at Annual Meeting of Radiologic Society of North America, Dallas, Texas, Nov. 1980.*
24. Gohari, P., Berkowitz, R., and Hobbins, J. C. Prediction of IUGR by determination of TIUV. *Am. J. Obstet. Gynecol.*, 127:255, 1977.
25. Grannum, P., Bracken, M., Silverman, R., et al. Assessment of fetal kidney size in normal gestation by comparison of kidney circumference/abdominal circumference ratio. *Am. J. Obstet. Gynecol.*, 136:249, 1980.
26. Grossman, M. Significance of bladder distension in uterine volume determination. *Am. J. Obstet. Gynecol.* (In press, 1982).
27. Grossman, M., Flynn, J. J., Aufrichty, D., et al. Pitfalls in ultrasonic determination of TIUV. *J. Clin. Ultrasound*, 10:17–20, 1982.
28. Haas, L. L. Roentgenological skull measurements and their diagnostic applications. *AJR*, 67:197–209, 1952.
29. Hadlock, F. P., Deter, R. L., Carpenter, R. J., et al. Estimating fetal age: Effect of head shape on BPD. *AJR*, 137:83–85, 1981.
30. Hobbins, J. C. Yale Nomogram, Siemans Corporation, Electromedical Division, 1979.
31. Hoffbauer, H., Arabin, P. B., and Baumann, M. C. Control of fetal development with multiple ultrasonic body measures. *Contrib. Gynecol. Obstet.*, 6:147–156, 1979.
32. Hohler, C. W., Lea, J., and Collins, H. Screening for IUGR using the ultrasound BPD. *J. Clin. Ultrasound*, 4:187–191, 1977.
33. Jeanty, P., Dramaix-Wilmet, M., and Delbeky, D., et al. Ultrasonic evaluation of fetal ventricular growth. *Neuroradiology*, 21:127, 1981.
34. Jeanty, P., Dramaix-Wilmet, M., Van Gansbeke, D., et al. Fetal ocular biometry. *Presented at Annual Meeting, AIUM, San Francisco, 1981.*
35. Jeanty, P., Kirkpatrick, C., Dramaix-Wilmet, M., et al. Ultrasonic evaluation of fetal limb growth. *Radiology*, 140:165, 1981.
36. Jordaan, H. V. F. The differential enlargement of the neurocranium in the full term fetus. *S. Afr. Med. J.*, 50:1978–1981, 1976.
37. Jordann, H. V. T., and Clark, W. B. Prenatal determination of fetal brain and somatic weight by ultrasound. *Am. J. Obstet. Gynecol.*, 136:54, 1980.
38. Kearney, K., Vigneron, N., Frischman, P., et al. Fetal weight estimation by ultrasonic measurement of abdominal circumference. *Obstet. Gynecol.*, 51:156–162, 1978.
39. Kurjak, A., Cecuk, S., and Breyer, B. Prediction of maturity in the first trimester by ultrasonic measurement of the fetal crown rump length. *J. Clin. Ultrasound*, 4:83–84, 1976.

40. Kurtz, A. B., Wapner, R. J., Kurtz, R. J., et al. Analysis of biparietal diameter as an accurate indicator of gestational age. *J. Clin. Ultrasound*, 8:319–326, 1980.
41. Lawson, T. L., Albacelli, J. N., Greenhouse, S. W., et al. Grey scale measurement of the biparietal diameter. *J. Clin. Ultrasound*, 5:17–20, 1975.
42. Lawson, T. L., Foley, W. D., Berland, L. L., et al. Ultrasonic evaluation of fetal kidneys. *Radiology*, 138:153, 1981.
43. Levine, S. C., Filly, R. A., and Creasy, R. K. Identification of fetal growth retardation by ultrasonographic estimation of TIUV. *J. Clin. Ultrasound*, 7:21, 1979.
44. Lind, T. The estimation of fetal growth and development. *Br. J. Hosp. Med.*, 3:501, 1970.
45. Loeffler, F. E. Clinical foetal weight prediction. *Br. J. Obstet. Gynecol.*, 74:675–679, 1967.
46. Lunt, R. M., and Chard, T. Reproducibility of measurement of fetal BPD by ultrasonic cephalometry. *J. Obstet. Gynecol. Br. Commnw.*, 81:682–685, 1974.
47. Mayden, K. L. The umbilical vein diameter in Rh isoimmunization. *Med. Ultrasound*, 4:119, 1980.
48. Morrison, J., and McLennan, M. J. The theory, feasibility and accuracy of an ultrasonic method of estimating fetal weight. *Br. J. Obstet. Gynecol.*, 83:833, 1976.
49. Nelson, L. H. Comparison of methods for determining crown rump measurements by real-time ultrasound. *J. Clin. Ultrasound*, 9:67, 1981.
50. O'Brien, W. F., Coddington, C. C., and Cefalo, R. C. Serial ultrasonographic BPDs for prediction of EDC. *Am. J. Obstet. Gynecol.*, 138:467, 1980.
51. Ogata, E. S., Sabbagha, R., Metzger, B. E., et al. Serial ultrasonography to assess evolving fetal macrosomia. *JAMA*, 243:3405, 1980.
52. Osinusi, B. O., Hall, A. J., Adam, A. H., et al. Reproducibility of BPD measurements obtained with a R-T scanner. *Br. J. Obstet. Gynecol.*, 87:467, 1980.
53. Poll, V., and Kasby, C. B. An improved method of fetal weight estimation using ultrasound measurements of fetal abdominal circumference. *Br. J. Obstet. Gynecol.*, 86:922, 1977.
54. Queenan, J. T., O'Brien, G. O., and Campbell, S. Ultrasound measurement of fetal limb bones. *Am. J. Obstet. Gynecol.*, 138:297, 1980.
55. Robinson, H. P. Sonar measurement of the fetal crown-rump length as a means of assessing maturity in the first trimester of pregnancy. *Br. Med. J.*, 4:28–31, 1973.
56. Robinson, H. P., and Fleming, J. E. E. A critical evaluation of sonar crown-rump length measurements. *Br. J. Obstet. Gynecol.*, 82:702–710, 1975.
57. Sabbagha, R. E., Barton, F. B., and Barton, B. A. Sonar biparietal diameter. I. Analysis of percentile growth differences in two normal populations using same methodology. *Am. J. Obstet. Gynecol.*, 126:479–484, 1976.
58. Sabbagha, R. E., and Hughey, M. Standardization of sonar cephalometry and gestational age. *Obstet. Gynecol.*, 52:402–406, 1978.
59. Sabbagha, R. E., Hughey, M., and Depp, R. Growth adjusted sonographic age: A simplified method. *Obstet. Gynecol.*, 51:383–386, 1978.
60. Shepard, M. J., Richards, V. A., Berkowitz, R. L., et al. An evaluation of two equations for predicting fetal weight by ultrasound. *Am. J. Obstet. Gynecol.*, 142:47, 1982.
61. Thompson, A. M., Bullewicz, W. Z., and HyHen, T. R. The assessment of fetal growth. *J. Obstet. Gynecol. Br. Commonw.*, 75:903, 1968.
62. Usher, R., and McLean, F. Intrauterine growth of liveborn caucasian infants between 25 and 44 weeks of gestation. *J. Pediatr.*, 74:901–910, 1969.
63. Warsoff, S. L. Ultrasonic estimation of fetal weight for the detection of IUGR by computer assisted analysis. Thesis, Yale University School of Medicine, pp. 61–63, 1971.
64. Warsoff, S. L., Gohari, P., Berkowitz, R. L., et al. The estimation of fetal weight by computer-assisted analysis. *Am. J. Obstet. Gynecol.*, 128:881, 1977.
65. Weiner, S. N., Flynn, M. J., Kennedy, A. W., et al. A composite curve of ultrasonic biparietal diameters for estimating gestational age. *Radiology*, 122:781–786, 1977.
66. Whetham, J. C. G., Muggah, H., and Davison, S. Assessment of IUGR by diagnostic ultrasound. *Am. J. Obstet. Gynecol.*, 127:577, 1976.
67. Wladimiroff, J. W., Bloemsa, C. A., and Wallenburg, C. S. Ultrasonic diagnosis of the large-for-dates infant. *Am. J. Obstet. Gynecol.*, 52:285, 1978.

Ultrasound Annual 1982,
edited by Roger C. Sanders.
Raven Press, New York © 1982.

Ultrasonic Puncture Techniques:
A Practical Guide

Richard Chang

The Russell H. Morgan Department of Radiology and Radiological Science,
The Johns Hopkins Medical Institutions, Baltimore, Maryland 21205

Ultrasound imaging has made a major contribution not only in diagnosis but also in biopsy and aspiration or drainage procedures. Many procedures that were once considered difficult and time-consuming are now quickly and simply performed with some form of ultrasonic localization. In this chapter, we summarize the role of ultrasound in the performance of puncture procedures with emphasis on its practical aspects.

PUNCTURE AFTER TARGET LOCALIZATION FROM A STATIC GRAY SCALE SCAN

Although many specially adapted ultrasonic transducers are available at present or can be simply constructed to assist in puncture performance, the great majority of procedures are still most efficiently performed by using simple localization of a suitable skin entry site, with depth determination obtained with conventional B-scan equipment (Fig. 1A). An impression on the skin entry site is made with a blunt object, e.g., an inverted needle top, and the skin is then prepped and a local anesthetic is infiltrated into the site. A needle stop is carefully positioned along the needle barrel, using a sterile ruler for measurement, so that the needle can be quickly thrust to the exact depth indicated by the ultrasound scan (Fig. 1B).

Some simple considerations are important to ensure successful puncture. A small dermatotomy can significantly reduce resistance to puncture offered by the skin. This is particularly important with long thin needles because they are extremely flexible, and are easily deflected by any additional resistance away from the desired path for puncture. Care should be taken to advance the needle at the same angle used by the transducer for target localization and depth determination. Frequently, it is helpful to have an assistant check the visual appearance of the angle of the needle from another perspective just prior to advancing the needle. The patient should be asked to suspend his respiration in the same phase of respiration used in ultrasonic localization. This is particulary important for small lesions in the liver and kidney, since these organs may have considerable respiratory excursion.

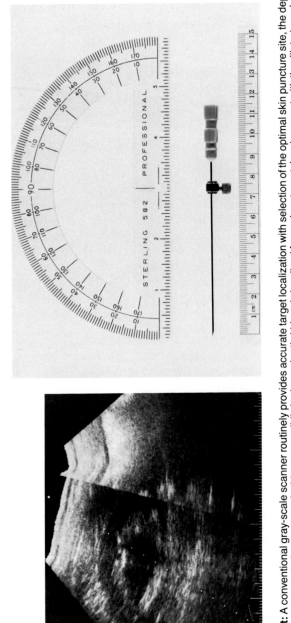

FIG. 1. Left: A conventional gray-scale scanner routinely provides accurate target localization with selection of the optimal skin puncture site, the depth of the target, and the angle for puncture. In this case a mildly hydronephrotic kidney is localized for nephrostomy tube placement with the dilated renal pelvis lying 7 cm deep to the skin. The depth marks are displayed as white dots spaced 1 cm apart. **Right:** Accessories for puncture. Needle with needle stop (Becton-Dickinson Co.); gas-sterilized ruler; gas-sterilized protractor.

One should choose a skin entry site that provides the shortest distance to the target lesion that is compatible with a safe route for puncture. If free ascitic or amniotic fluid is to be aspirated, one should avoid excess pressure on the abdomen because this can displace fluid away from the puncture site. Finally, a quick thrust is recommended because slow advancement of the needle often results in displacement of the target lesion without actual penetration into the lesion.

Although at least 95% of the procedures in an ultrasound department can be performed without any further equipment, occasionally a puncture performed in this manner fails to obtain the expected tissue or fluid. When this occurs, there is no simple way to be certain that the needle tip was accurately positioned in the target lesion. The needle tip could be in an incorrect position because of respiratory or patient motion, or fetal motion in the case of amniocentesis. On the other hand, certain lesions can appear to contain fluid on ultrasound imaging, but can be too viscous to aspirate even with perfect needle placement, as frequently occurs with hematomas. When a procedure seems unsuccessful one should consider some of the other devices designed to improve accuracy in needle placement.

DEDICATED B-MODE ASPIRATION BIOPSY TRANSDUCERS

Special B-mode transducers are available that have a central channel through the active transducer piezoelectric element to permit passage of a needle (Fig. 2).

When mounted to the scan arm, the transducer is offset from the scan arm by a short extension arm, so there is no impediment from the scan arm to insertion of the needle. However, the transducer must be mounted so that the short extension arm must be perpendicular to the sector plane of the scan arm. If mounted improperly, the image can be severely distorted (10).

In actual use, the patient is usually scanned with a conventional transducer. After visualization of the target lesion, the area is prepped, the aspiration transducer is mounted, and a sterile stocking is slipped over the scan arm to ensure a sterile procedure. The target is again scanned, using sterile mineral oil or gel as the contact medium. The depth of the target is then measured, and a needle stop is positioned along the needle at a length equal to the sum of the depth to the lesion and the height of the aspiration transducer. With the patient suspending respiration, the target area is rescanned; as the transducer passes just over the target, the needle is quickly thrust to the depth determined by the needle stop.

There are certain disadvantages of these types of transducers. The presence of the needle channel may degrade the image to a varying extent. Generally, either more gain or output energy is required than with the standard transducer. The size of needle that can be used is limited to small needles, because further enlargement of the needle channel would result in

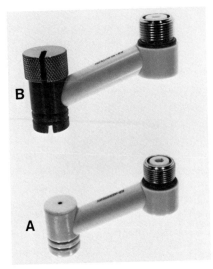

FIG. 2. Specialized puncture transducers for gray-scale scanners have transducers with a central channel for insertion of the puncture needle and a short horizontal extension arm to attach the transducer to the scan arm. Transducer A has central channel only; transducer cannot be freed from needle during a puncture procedure. Slotted transducer (transducer B): The transducer can be separated from the needle during the procedure by moving the transducer away from the needle along the side slot. Nevertheless, these transducers are difficult to use and offer no advantage over methods now available.

further degradation of image quality. The procedure itself can be quite cumbersome to perform. Generally, an assistant should be present so that he or she can operate the hand and foot controls of the ultrasound unit, while the puncturer operates the scan arm with one hand and is ready to advance the needle with the other hand. With some transducers, it is not possible to separate the transducer and needle after puncture of the patient. Even with transducers equipped with side slots, it can be difficult to move the scan arm exactly parallel with the side slot to separate the needle without accidentally jarring or dislodging the needle. When the needle cannot be disengaged, there is a theoretical risk of causing tissue laceration while the needle is fixed by the transducer and the tissues and target organs move with respiration.

As with any transducer guide where the needle channel is an integral part of the transducer, gas sterilization is the preferred method because of the temperature sensitivity of the piezoelectric element. This generally means that the transducer can be used for only one procedure every other day.

More importantly, if a puncture still fails to obtain the expected fluid or tissue, one is still uncertain whether the needle tip actually penetrated the target, or whether it was deflected, or merely displaced the target as the needle was advanced. If the B-scanner has a continuous A-mode display, it can be helpful to check the A-mode display when one is trying to puncture a difficult fluid collection. But in this event, use of an A-mode transducer is far simpler. For these reasons, special B-scan aspiration transducers have not enjoyed widespread use or popularity.

A-MODE TRANSDUCERS

Because of its simplicity, the A-mode transducer (Fig. 3) remains a useful device for puncture of fluid collections. Fluid aspirations are usually unsuccessful because of movement of the collection with respiration, or fetal

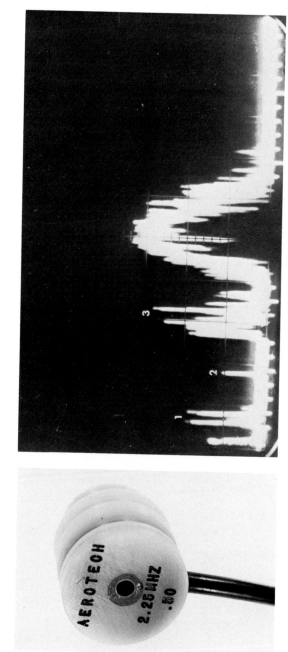

FIG. 3. Left: A-mode transducer. **Right:** Example of A-mode display during puncture of a fluid collection (amniocentesis). (1) Near zone soft tissue; (2) echo from needle tip; (3) echoes from far zone soft tissue. Note that the amplitude of the needle tip echo is small relative to the amplitude of soft tissue echoes in the near or far zone when a small-gauge (20-gauge) needle is used. The needle tip echo is better seen with a 3.5-MHz transducer.

motion in the case of amniocentesis. One first performs a routine scan of the target area with a conventional B-scanner or real-time ultrasound device to localize a good position to place the A-mode transducer on the patient's skin. After the skin has been prepped, the transducer is placed over the site, using sterile mineral oil as contact medium. The A-mode display is observed as the patient breathes and a search is made for an echo-free space with strong anterior wall and posterior wall echoes that represents the cyst or fluid collection. The transducer can be easily tilted or rocked until this display pattern is obtained. The patient is asked to suspend his respiration in that portion of the cycle that optimizes this display pattern when the actual puncture is to be performed. A needle stop can be properly positioned along the needle at a length equal to the depth to the target and height of the A-mode transducer. In amniocentesis, one may have to wait until the fetus moves his extremities and the pocket of amniotic fluid reappears on the A-mode display. It is important to avoid excess pressure on the abdomen, which could displace fluid away from the puncture site.

The A-mode transducer has many advantages. It is small and lightweight, and although it remains attached to the needle throughout the procedure, it causes no significant fixation of the needle at the skin entry site. Because of its small face, it is ideal for those aspirations that are difficult because of narrow rib interspaces. Since it gives a continuous display, one can adjust the angle of puncture to the position of the fluid collection up until the very moment of puncture. Aspiration of a collection can be observed by watching the near and far wall echoes coalesce around the small echo representing the needle tip. Portable displays are available so that the procedure can be performed in the patient's room or in the fluoroscopy room if fluoroscopy is required.

Although the A-mode transducer cannot be used for biopsy of soft tissue lesions, the A-mode transducer is a logical choice for puncture of difficult fluid collections in most cases. The principle exceptions are extremely small fluid collections (less than 1 cm in size) or fluid collections surrounded by adjacent fluid or vascular structures. In each of the cases, it is difficult to recognize the proper "echo-free" space on the A-mode display alone. Fortunately, these situations arise rarely, and because of its simplicity and ease of use, the A-mode transducer remains a valuable addition to the ultrasonographer's armamentarium.

REAL-TIME SCANNING

With real-time ultrasound units (Fig. 4), it is sometimes possible to image both needle and target organs continuously throughout the procedure from a remote site. The transducer is held over one site on the patient's skin, while another skin site is prepped for actual puncture. The angle and depth for puncture are usually determined from a standard B-scanner or previous real-time evaluation. The principle requirement for adequate visualization is that a

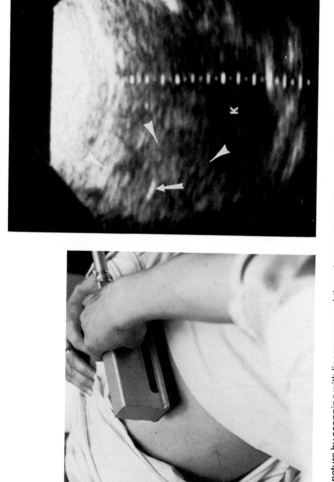

FIG. 4. Left: Monitoring a puncture by scanning with linear array real-time ultrasound at a site remote to the skin puncture site. This method is useful where a large acoustic window is present, such as in puncture of lesions close to or in the liver, or in the large gravid uterus. **Right:** Example of aspiration of a perihepatic hematoma (*arrowheads*). *Arrow* points to needle. K, Right kidney. Transverse scan, using an ATL-real-time sector transducer.

large acoustic window is present to allow the target to be scanned from this remote site as well as from the puncture site where a depth determination was made. In general, this limits use of the technique to either obstetrical procedures, where the large gravid uterus fulfills these requirements, or to lesions in or around the liver, another organ that transmits sound well. A fluid-filled bladder can be used to monitor a culdocentesis. Similarly, a large fluid-filled stomach could potentially be used to perform a puncture procedure on a lesion in the pancreas.

A particularly important application of remote real-time ultrasound monitoring of a puncture procedure is in fetal transfusion for Rhesus (Rh factor) incompatibility. The gravid uterus is an ideal acoustic window. Not only can one monitor actual entry of the needle into the fetal abdomen, but as the transfusion blood enters the abdomen it creates echoes that resemble bubbles on the real-time display. Since the ultrasound is not attached to the needle, one can also monitor the fetal heart so that slowing of the fetal heart rate caused by overtransfusion can be prevented.

In general, however, when only a small acoustic window is present, this method is unsuccessful. Attempts to insert a needle freehand at an oblique axis along the side of a real-time transducer, to image both the lesion and the needle, usually fail. This is because it is very difficult to maintain the needle tract exactly down the scan plane of the ultrasound beam. Thus, the method is limited to use in only a few regions of the body.

REAL-TIME PUNCTURE TRANSDUCERS AND NEEDLE GUIDES

Linear Arrays

Linear array (Fig. 5) aspiration transducers with a central slit are available from a few manufacturers (ADR, Toshiba, Aloka). A crystal element is usually deleted from the linear array to allow for a needle channel. The projected path of the needle is seen on the image as a narrow zone of decreased echoes as a result of absence of the crystal element in the region of the needle channel. The needle length is poorly visualized, but the needle tip will be seen when it is advanced into a fluid collection. Once the needle tip is properly positioned, the needle can be detached from the linear transducer through a slide slot in most versions. The advantage of a linear array is that the depth of the lesion is simply measured as with any linear array.

The disadvantage of the linear array aspiration type of transducer is the difficulty of performing scans or punctures when the target lies deep to the rib interspaces. Sector scanners provide a wider continuous image when scanning through rib interspaces. A second problem is that the angle of puncture is limited to a narrow sector. Oblique punctures may be desired in the placement of nephrostomy tubes or stints. As with any ultrasound transducer, steriliza-

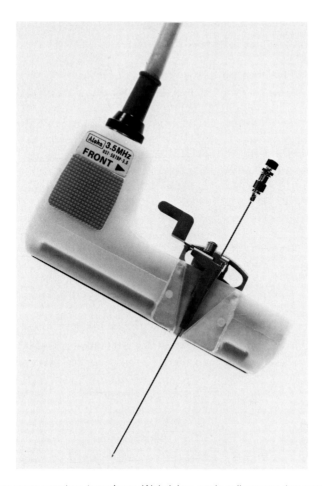

FIG. 5. Linear array puncture transducer (Aloka) has angle adjustment that permits oblique punctures up to 20 degrees from vertical. Oblique punctures are often required because of overlying ribs but in addition often permit better needle visualization than with a vertical puncture. A relative "blind" zone is created by removal of a crystal element for placement of a needle channel, and this "blind" zone is further exacerbated by skin depression and poor acoustic contact when the needle is advanced. It interferes with needle visualization when a vertical puncture is performed.

tion often requires gas sterilization which limits the procedure to a single case every other day.

The size of the channel designed by the manufacturer limits the maximum size of the needle that can be used. Despite the above-mentioned limitations, these puncture transducers can be extremely effective. Ohto et al. have reported percutaneous pancreatic ductography using a linear array in a patient with a dilated pancreatic duct secondary to carcinoma of the head of the pancreas (37).

REAL-TIME SECTOR TRANSDUCERS:
NEEDLE GUIDE ATTACHMENTS

Mechanical real-time sector scan transducers consist of either rotating crystals, oscillating crystals, or acoustic reflectors. Phased array sector scanners consist of multiple-crystal elements, just as in linear arrays. Mechanical sector scanners cannot be easily constructed to contain an integral needle channel, and removal of a crystal element in a phased array for a needle channel would degrade the image even more markedly with a phased array than with a linear array. For these reasons it is more practical to design needle guide attachments for sector real-time transducers than to modify the transducer itself (Fig. 6).

An example of a commercial real-time system with needle guide attachment is shown in Fig. 6A.

Just as in computed tomography (CT), each ultrasound image represents information arising from a certain slice thickness. The slice "thickness," is determined by the width of the ultrasound beam. The purpose of the needle guide is to ensure that the needle remains within the scan thickness when it is advanced toward the target. The scan thickness can be quite thin, and experience shows that it is impossible to advance the needle freehand and still maintain the needle in the scan plane. The first step in the design of a needle guide is, therefore, careful determination of the location of the scan plane. For a simple geometry such as the cylindrical Rohe sector transducer, a notch on the side of the transducer locates the scan plane (Fig. 6B). For this transducer, a simple needle guide attachment can be made from an endplate notched to fit the notch on the side of the transducer, and tapped for screws to serve as needle channels. Each screw can be drilled to accommodate needles of different gauges. Similarly, a more rectangular-shaped transducer such as the Hewlett-Packard (Fig. 6D) lends itself to needle guide attachment. The author has worked with Hewlett-Packard in the design of an effective prototype for real-time punctures (Fig. 6D).

With more complicated transducer configurations such as the ATL unit (Fig. 6C), it is best to make some preliminary physical tests to identify the location of the scan plane. The simplest method is to mount the transducer in a water bath to some reference plate and insert needles in toward the scan plane. As the needle tip enters the scan plane, it will create a bright echo on the real-time image display. This point represents one edge of the scan plane. Similarly, needles can be brought in from the opposite direction to mark the other edge of the scan plane thickness. In this way, the scan plane thickness and location can be determined. The needle guide should be designed so that the needle path lies within the center of the scan thickness. With proper placement the length of the needle is clearly visualized when the needle is placed in the needle guide and scanned in a water bath.

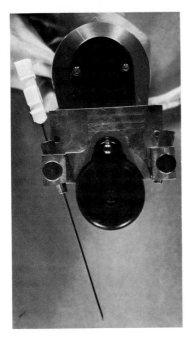

FIG. 6. Real-time sector scanner needle guides. **Top left:** Aloka commercially available, portable mechanical sector scanner with needle guide attachment. **Top right:** Simple needle guide attachment machined by author for Phillips (Rohe) sector scan real-time unit. **Bottom left:** ATL needle guide constructed by author. (A similar commercial device is now available from ATL Co.) **Bottom right:** Needle guide for Hewlett-Packard phased array sector scanner. This guide was developed by Tom Stevens and Dave Miller of Hewlett-Packard, Andover division in collaboration with the author, and has the advantage of continuously adjustable puncture angles.

Not all real-time sector scanners lend themselves to puncture applications or design of needle guides. The ideal instrument is portable, has a small face transducer, and has a high-quality image.

During performance of a puncture, the needle guide attachment must be sterilized, but the transducer itself can be placed within a sterile plastic bag containing transmission gel. The sterilized needle guide is attached over the sterile bag to the transducer and the patient is scanned, using sterile mineral oil as acoustic contact medium, until the target lies along the expected course of the needle. During fluid aspiration, depth determination is not necessary, except to be certain that a needle of sufficient length is used. One will be able to see the needle tip as it enters the fluid collection, so predetermination of exact depth is unnecessary. The thickness or gauge of the needle, however, does strongly affect needle visualization on real-time imaging. The larger the needle, the more distinct is its visualization on real-time imaging. With an 18-gauge (or larger) needle, one can usually visualize the length of the needle down to its tip (Fig. 9B), but with smaller gauge needles (20-gauge or 22-gauge), only the tip can be detected (Fig. 7B). With these smaller-gauge needles, the tip is seen clearly only when it enters a fluid collection. As it passes through soft tissue, the tip is very poorly seen, except in tissues of low echogenicity such as normal liver or kidney. Often the position of the needle is best inferred from real-time observation of the displacement and movement of adjacent tissue as the needle is advanced.

For proper positioning of the transducer for puncture, the operator must have some indication of the expected course of the needle. This can be provided in a number of ways. With fixed-angle commercial needle guide attachments such as the Aloka, software is provided that automatically generates markers for the expected course of the needle. The Hewlett-Packard device has adjustable angles, but the skin entry site is always at a fixed distance from the apex of the sector regardless of angle, and the required angle is predicted from the depth of the target lesion. Alternatively, the needle can be scanned in a sterile water bath just prior to puncture, and measurement cursors can be placed along the needle course for use during actual puncture. In the future, the expected needle course for each angle will be incorporated within the software programming for display on the monitor. For custom-designed needle guides, we scan the needle in a water bath and trace the needle course on a transparent film. This is done for every display scale to generate a mask that is placed back onto the image display whenever we perform a puncture. This mask method is necessary when the instrument does not have measurement cursors. Once the expected course is known, the transducer is moved over the patient until the target lesion lies on the expected course of the needle.

Some examples of real-time needle guide assisted punctures are shown in Figs. 7–9, 24–27. While it is esthetic to visualize both the needle and target lesions simultaneously and continuously during a puncture procedure, it

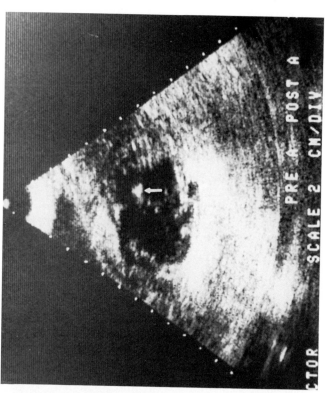

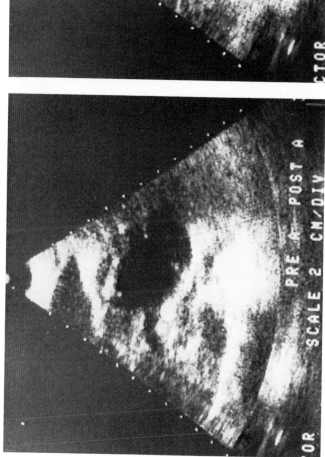

FIG. 7. Puncture of an infected pseudocyst using needle guide for Rohe sector scanner. Transverse scan of pancreas. **Left:** Prior to puncture. **Right:** After puncture, the tip of the 20-gauge needle is seen within the pseudocyst. Pus was aspirated and this collection was surgically drained. Note with thin needles (22-gauge or 20-gauge) only the top (*arrow*) of the needle is usually detected.

takes longer to set up for such a procedure. Most lesions are large and are adequately performed with simple skin site localization and depth determination from a simple static scanner. For this reason, we believe real-time guided procedures are best reserved for specialized indications such as:

(a) Puncture of small fluid collections—small in dimension compared with respiratory excursion, or fetal motion in case of amniocentesis.

(b) Complete aspiration of fluid collections.

(c) When other methods fail to obtain the expected fluid.

(d) Aspiration of a fluid collection adjacent to or surrounded by other similar fluid collections or vascular structures (Fig. 8).

(e) Complex procedures requiring both ultrasound localization and fluoroscopic control, such as nephrostomy tube placement (Fig. 9), abscess drainage (Fig. 27) and biliary drainage (Figs. 24 and 25), fetal drainage procedures, or fetal transfusions.

Although the first three indications could be performed with an A-mode transducer, demonstration of accurate needle tip placement is best demonstrated with a real-time guided puncture. Figure 8 illustrates an example of the fourth indication. This patient had experienced adult polycystic kidney disease, and had developed fever and left flank pain. He was referred for CT, which revealed that one cyst had abnormal attenuation compared with surrounding cysts. The patient was brought to ultrasound for diagnostic aspiration, and puncture was performed with a 20-gauge needle using the Phillips (Rohe) needle guide adaptor. The needle tip could be clearly seen as it advanced into different parts of the cyst, which on ultrasound appeared to have thicker walls. Aspiration, however, resulted in withdrawal of only a tiny blood clot. Culture of this small specimen grew *E. coli*, which also grew on samples of the patient's urine. Following nephrectomy there was dramatic relief of flank pain and fever. The pathologic specimen revealed an infected hemorrhagic cyst. Note that aspiration of this type of lesion could not be performed with an A-mode transducer. One would be confused by the many adjacent cysts. Secondly, if this had been done with any other ultrasound technique, one would not be certain that the needle tip had been accurately placed, since so little material could be obtained. Because the needle tip was clearly seen, we did not pursue further unnecessary and fruitless needle passes. Conversely, if fluid had been obtained after localization and depth determination with a simple gray-scale scanner, one would not be absolutely certain that the fluid came from the correct cyst in this patient with polycystic kidney disease.

SPECIALIZED APPLICATIONS OF ULTRASOUND
Mass Biopsies

Extensive experience has shown that percutaneous fine (22- or 23-gauge) needle biopsy is an effective and safe method of diagnosis of suspected malignancy. An overall success rate for diagnosis in the range of 70 to 88%

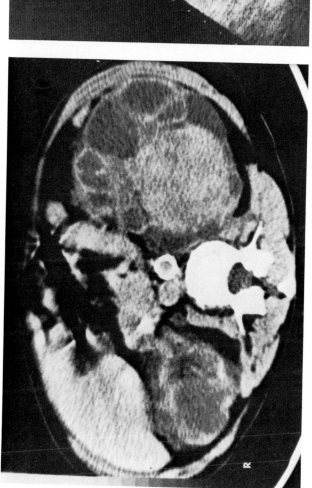
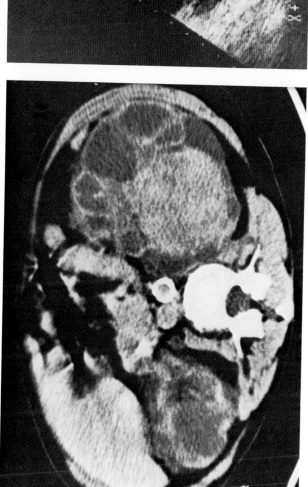

FIG. 8. Cyst aspiration in a patient with polycystic kidney disease and left flank pain and fever. **Left:** CT scan showing a dense cyst surrounded by many smaller cysts in left kidney. **Right:** Transverse scan. Patient lies in the right decubitus position with left side up. The abnormal cyst has a thickened wall (*arrowheads*) on ultrasound.

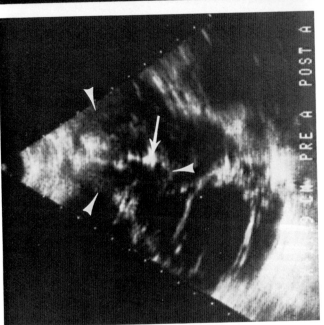

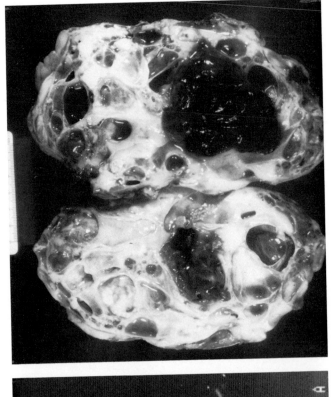

FIG. 8. continued. **Left:** Puncture with a 20-gauge needle into cyst using Rohe needle guide. No free fluid could be obtained; only a small blood clot could be aspirated and was sent for culture. Without real-time monitoring, one would not be certain about needle position, since no fluid was aspirated. Alternatively, without real-time observation, if fluid had been obtained one could not be certain it came from the correct cyst (*arrow*, needle tip). **Right:** Culture of blood clot grew *E. coli* and patient underwent nephrectomy. Examination revealed a hemorrhagic infected renal cyst.

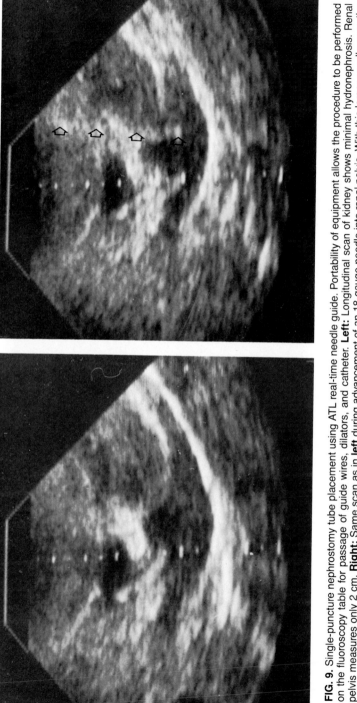

FIG. 9. Single-puncture nephrostomy tube placement using ATL real-time needle guide. Portability of equipment allows the procedure to be performed on the fluoroscopy table for passage of guide wires, dilators, and catheter. **Left:** Longitudinal scan of kidney shows minimal hydronephrosis. Renal pelvis measures only 2 cm. **Right:** Same scan as in **left** during advancement of an 18-gauge needle into renal pelvis. With this larger needle even the length of the needle can be detected (*arrows*). Contrast injection would fill the renal pelvis with echoes, and guide wire advancement can also be monitored with real-time.

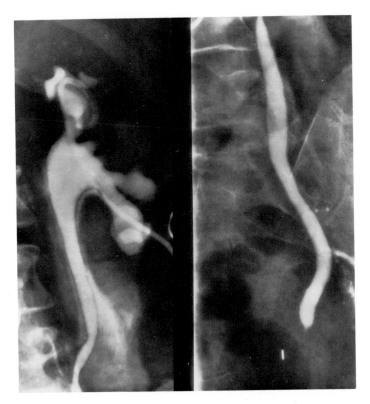

FIG. 9. continued. Placement of nephrostomy catheter. Patient had distal ureteral obstruction secondary to carcinoma of the cervix.

can be expected (9,14,23,39,57). False-negative biopsies occur as a result of inadequate tissue or improper placement of the needle, but no false-positives should occur.

The success rate from fine-needle biopsy of lymphadenopathy is lower for lymphomas (64%) than for metastatic disease (56,57). This is partially because the architecture or morphology of the lymph node is an important criteria in diagnosis of lymphoma. In patients with nodular sclerosing Hodgkin's disease, biopsy aspiration of diagnostic material may be difficult to achieve because the node is replaced with hard, dense fibrous tissue (46). It is rare to be able to define the histologic type of lymphoma from a fine-needle biopsy (56).

Except for isolated case reports, there are only a few known serious complications of fine-needle biopsies. One case was a patient who developed Gram-negative sepsis, after a fine-needle aspiration of the pancreas, presumably as a result of seeding of the needle tract as the needle traversed bowel anterior to the pancreas (14). While there is theoretical risk of seeding of the needle tract with malignant cells following biopsy, only one documented case with use of a fine needle has been reported to our knowledge. This patient had biopsy of a pancreatic carcinoma and developed nodular implants along the course of the tract (13). A case of seeding of renal cell carcinoma along a needle tract has also been reported, but the aspiration biopsy was performed with a larger 18-gauge needle (6).

Occasionally small hematomas occur following fine-needle biopsy, but they are usually insignificant. Lundquist reported a single, intrahepatic hematoma requiring operative management in 2,611 liver biopsies (30). A single case of severe necrotizing pancreatitis has been reported following fine-needle biopsy of the pancreas (11). These isolated case reports represent the only significant complications in the literature.

In general, only minor complications occur in the abdomen following fine-needle biopsy, whereas in the chest there is a 5% incidence of pneumothorax requiring chest tube evacuation (27). Usually the only complication in abdominal or extremity biopsies have been small minor hematomas. Our only unusual complication was a urinoma developing following biopsy of retroperitoneal lymph nodes after lymphangiography using fluoroscopic guidance. Apparently, the thin needle penetrated the right renal pelvis but the small leak healed without intervention.

Two choices of biopsy needles are available (Fig. 10A). The 22-gauge Chiba needles or spinal needles have beveled tips, whereas the Greene needle (Cook 15.0 cm) and Maydayag needles have pencil point-type tip stylets with outer cutting needle. For deeper lesions, the cutting needles with pencil point-like stylets are preferred for two reasons. The cutting needle is more likely to obtain a small core of tissue. Secondly, the tips of beveled needles tend to be displaced in a direction opposite to the bevel face, as a result of increased tissue resistance on the side of the bevel, during advancement of the needle (22). This deflection results in error proportional to the depth to the target lesion. Prior to availability of short 9-cm Greene needle, we used spinal needles for biopsy of superficial lesions because the shorter length needles provided greater rigidity.

After localization of the lesion, the biopsy needle is advanced until the tip lies within the lesion. The stylet is removed, a syringe is attached, and suction is applied continuously while one is making short up-and-down rotary movements of the needle in the lesion. Suction on the syringe is released prior to withdrawal, and a total of three or four passes are made into the lesion to obtain adequate samples. Fewer passes may not yield adequate material for evaluation, and more passes result in greater risk of complications without significantly increasing yield unless one knows that initial passes had been misdirected (14).

A pistol grip handle apparatus (Fig. 10B, Pistol Cameo Syringe-Precision Dynamics Co.) is useful for applying suction, and allows one to perform the biopsy with one hand. It is useful to use a larger syringe size rather than one of a smaller size because the suction is greater if the air in the small dead space of the needle channel can be expanded into the greater volume of a larger syringe. Typically, we use 20-cc syringes.

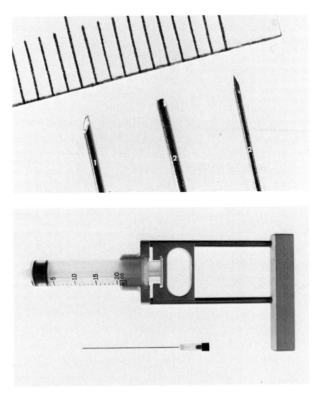

FIG. 10. Top: Fine needles for aspiration biopsy. (1) 15-cm beveled tip Chiba needle and stylet; (2) 22-cm Greene needle with cylindrical cutting edge and a pencil-point stylet. This needle is preferable because it is more likely to produce a core of tissue, and the pencil-point stylet is less likely to cause deviation of the needle during needle advancement than a beveled tip needle. This is an important consideration for deep lesions, although for relatively superficial lesions a short 22-gauge spinal needle may be used. **Bottom:** Pistol-grip aspiration handle permits biopsy to be performed with one hand. The suction is greater with a larger syringe because the air dead space within the needle channel can be expanded to the larger volumes of the larger syringe sizes.

After each pass, the tissue processing method should be determined by the pathologist who will be responsible for examining the tissue. Some authors eject the aspirate onto albuminized glass slides and smear the specimen, as in examination of peripheral blood. The slide is then fixed in Carnoy's solution for appropriate staining (14). At our institution the sample is ejected from the needle and syringe with Hank's solution, and taken to the cytopathology laboratory for concentration and staining.

The choice of imaging for localization depends on the size of the lesion and visibility. Lymph nodes can be abnormal although normal in size, and these are discovered on lymphangiography. Similarly, occasional obstruction of the distal common bile duct by a small mass may require transhepatic cholangiography to pinpoint the mass, and biopsy may be best performed with fluoroscopy. Lesions in the retroperitoneum, such as lymphadenopathy, which are obscured by bowel gas require CT. When the lesion can be imaged with ultrasound, selection of a skin puncture site angle and depth determination (under ultrasound) is usually the simplest and quickest way to perform the biopsy (Fig. 11).

In regard to renal biopsy, Bolton et al. made a comparative study of fluoroscopy with intravenous contrast injection, radionuclide scan, and ultrasound for localization of the lower pole of the kidney, and ultrasound localization was found to be the most precise of these three imaging modalities (53).

Although real-time aspiration transducers and needle guides offer the ultimate imaging modality for aspirations of fluid-containing structures, they are somewhat limited when applied to soft-tissue biopsies. The tip of a 22-gauge needle cannot be seen with most systems in soft-tissue organs. This is easily understood from experience with A-mode transducers (Fig. 4B). The echo amplitude produced by a needle tip in a fluid space represents the maximum echo that can be expected because the incident ultrasound pulse is stronger as it passes through the relatively loss-less fluid medium. The interface between needle tip and surrounding fluid is a strong scattering surface and the return echo also traverses the loss-less fluid medium. Nevertheless, the echo from the tip of a small needle is considerably smaller than the soft tissue echoes in the near and far fields (Fig. 3B). This small echo is detected well only against the echo-free background of a fluid collection. The situation for a needle tip in soft tissue can only be expected to be worse. Consequently, the tip of a 22-gauge needle can be seen only in tissues of relatively low-level echo amplitudes, such as in normal liver or normal renal cortex. In more echogenic organs such as pancreas, or abnormal liver and kidney, real-time devices do not provide improved visualization during a biopsy procedure using fine needles. Larger, for example, 18-gauge, needles can be more easily seen, but do not carry the wide margin of safety associated with the fine-needle biopsy technique. Occasionally the most that can be done is to place an 18-gauge needle so that the tip lies just outside the lesion or organ (provided a safe puncture route is available) and pass a 22-gauge needle through the channel of the 18-gauge needle in a coaxial manner to perform the biopsy. The 18-gauge needle then provides visualization of a portion of the course of the 22-gauge needle, and serves to stabilize the very flexible fine needle.

UROLOGIC APPLICATIONS

There are many indications for needle or catheter intubation of the kidney (Table 1). Prior to ultrasound, this was accomplished by using fluoroscopy after approximate localization of the kidney and renal hilum on plain film and using a thin needle to opacify the collecting system in a manner analagous to percutaneous transhepatic cholangiography. However, this was often time-consuming and frequently required a number of needle passes. Ultrasound has become the procedure of choice for both detection and localization of the nonvisualizing obstructed or hydronephrotic kidney.

Antegrade pyelography and the Whitaker test (31,36,52–54) are helpful in

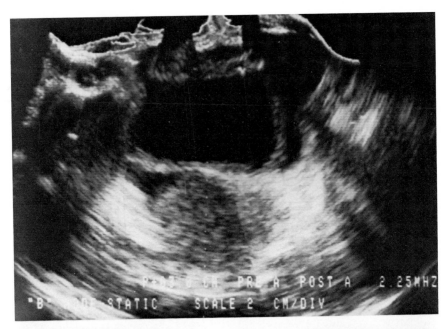

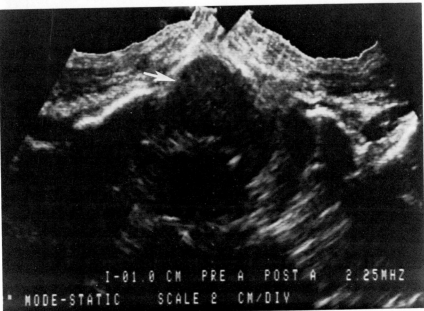

FIG. 11. Mass biopsy in a patient with back pain with history of previous abdominoperineal resection for rectal carcinoma. **Top:** Transverse scan showing presacral mass lying posterior to bladder. Biopsy in this position would require puncturing the urinary bladder. **Bottom:** Patient is turned prone and scan is repeated with mass seen through sciatic notch. Biopsy through sciatic notch (*arrow*) revealed recurrence of rectal carcinoma.

TABLE 1. *Applications of needle or catheter placement in the renal collecting system*

	Refs. (58)
Antegrade pyelography	
Flow-pressure dynamics (Whittaker tests)	31,36,52–54
Percutaneous nephrostomy	
For maligant obstruction	
For benign obstruction of the urinary tract	
For pyonephrosis	
Infusion of stone lysis agents or antibiotics	8,24,45,47,49,51
Percutaneous stone removal	1,35,41
Percutaneous nephroscopy	16,32,55
Biopsies of the collecting system	25
For ureteral fistulas	3,17,28,29,40
Percutaneous ureteral stent placement	2,3
Dilatations of ureteral strictures	43,44

evaluating a dilated collecting system for possible obstruction and for locating the level of obstruction. This is particularly true if the patient has evidence of lower urinary tract infection, in which case a retrograde pyelogram is contraindicated. With a complete obstruction, antegrade pyelogram complements retrograde pyelography in defining the proximal limits of an obstructing lesion (Fig. 12).

Indications for percutaneous nephrostomy tube placement include malignant obstructions to the urinary tract and benign conditions such as infection (Fig. 13A and B), obstructive renal calculi (Fig. 13C and D), nonobstructive renal stones, for stone extraction or infusion stone lysing agents, traumatic or iatrogenic ureteral tears. Even the obstructed transplant kidney may be managed with nephrostomy tube drainage (Fig. 13E–G).

Numerous nephrostomy catheter systems have been devised. A particularly convenient set is the Vance system, which provides needles, guide wire, dilators, and pigtail catheter in one complete set (Fig. 14A and B). Another system that we have used is the Cystocath system, which was originally designed for suprapubic tube placement (Fig. 15). This uses a large trocar and cannula, and a very soft flexible catheter. In this system, the trocar and cannula are thrust into the collecting system, and when urine is obtained the trocar is removed and the smaller flexible catheter is fed through the larger cannula into the renal pelvis before the cannula is removed. The principal advantage of this system is that if the trocar could be successfully placed into the collecting system under ultrasound, fluoroscopy is not required because no guide wires or dilators are used. However, we cannot recommend this system for patients with any clotting abnormalities. One of our patients with lymphoma and thrombocytopenia developed a large perinephric hematoma following placement of a nephrostomy tube using the cystocath system, presumably because the track made by the trocar is larger than the final

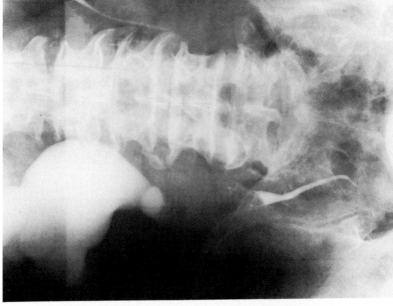

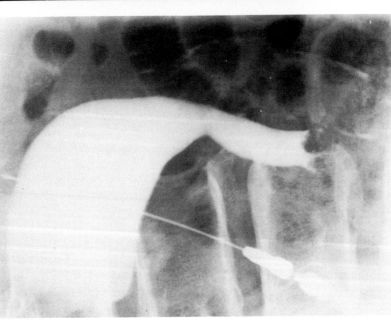

FIG. 12. Left: Antegrade pyelogram in patient with gross hematuria and complete obstruction right ureter. Intraluminal mass seen in right ureter. Following pyelography the collecting system was decompressed and partially replaced with contrast for retrograde pyelogram on the following day. **Right:** Retrograde pyelogram outlines lower edge of lesion and residual contrast outlines upper margin. This lesion was too large for local resection and reanastomosis.

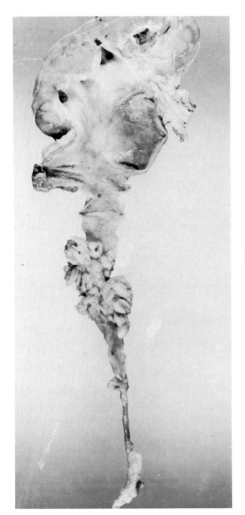

FIG. 12. continued. Nephrectomy specimen. Transitional carcinoma of the right ureter.

nephrostomy tube, and this allowed bleeding along the tube into the peri-nephric space (Fig. 16A and B). A third system is the Malecot catheter system, which also requires larger dilators and a trocar or stylet.

The procedure for placement of a nephrostomy tube is modified for each particular patient, for the catheter system used, and according to the ultra-sound imaging equipment available. If the collecting system is known to visualize after an intravenous contrast injection, ultrasound is not needed, and the nephrostomy tube can be placed under fluoroscopic guidance if an intravenous contrast injection is given in advance at the proper time. In many cases, the collection system does not visualize with intravenous contrast injection, and ultrasound localization is required.

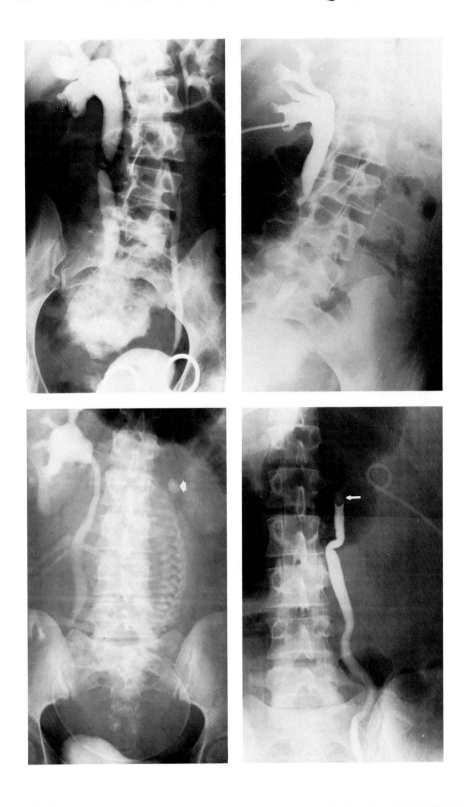

When using a larger trocar system, I prefer puncture under fluoroscopic control after initial opacification of the collecting system using ultrasound via a small 22-gauge needle. Ultrasound localization does not take into account kidney motion during needle advancement unless a real-time needle-guided system is used. Goldberg (16a) noted that when renal biopsies were performed in the prone position, a sandbag under the kidney was helpful in reducing kidney motion during puncture. However, renal biopsy was unsuccessful in one of their obese patients with normal-sized kidneys, presumably because kidney displacement is more difficult to control in the obese patient. In the two-needle pass method, the fine or "skinny" needle can be introduced into the collecting system in either the prone or the semiprone decubitus position, whichever seems more convenient. However, whenever larger needles or catheters are introduced the needle is inserted in a posterior axillary line skin puncture site and into the collecting system through the posterolateral cortex of the kidney. A direct posterior puncture into the collecting system is to be avoided because of danger of injuring the major renal vessels, which are large as they enter the renal hilum. The renal vessels taper as they reach the lateral cortex, so lateral puncture has a much lower risk of vascular injury. In addition, the greater thickness of renal cortex provides better fixation or stabilization of both puncture needle and final catheter, and the renal pelvis depth is greater from this approach, and therefore represents a larger target. When a nephrostomy tube is placed from the posterolateral position, the patient can be comfortable in both supine and decubitus positions. If a ureteral stint is to be placed, there is a more direct course from renal pelvis to ureter for guide wire placement when puncture is from the posterior axillary line. Finally, the spine helps prevent kidney displacement during puncture from the posterior axillary line, whereas in the prone position there is little to prevent the kidney from moving away from the needle when the needle is advanced (Fig. 17).

The procedure can be done as a single-needle or two-needle pass method. In the former, a large needle (18-gauge) is introduced into the renal pelvis after ultrasound localization to allow passage of guide wires, dilators, and catheters. In the latter method, a thin 22-gauge or 20-gauge needle is introduced into the renal pelvis for urine sample and opacification of the renal pelvis, and once opacified the larger needle system is introduced into the collecting system under fluoroscopic control for placement of the drainage catheter.

FIG. 13. Nephrostomy tube placement in benign conditions. **Top:** Infected urinoma secondary to ureteric fistula from passage of a calculus in a quadriplegic patient. **Top left,** IVP; **Top right,** nephrostomy tube antegrade pyelogram. **Bottom:** Management of pregnancy in a patient who developed total ureteropelvic junction obstruction by a large calculus. A nephrostomy tube was placed in the renal pelvis with ultrasound guidance only. Fluoroscopy was not utilized because of pregnancy. **Bottom left,** IVP. In **Bottom right,** retrograde after delivery of healthy baby shows complete ureteropelvic junction obstruction by calculus.

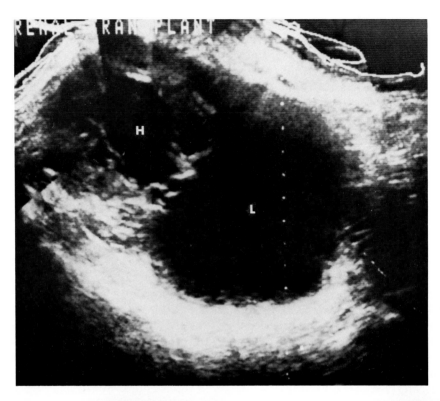

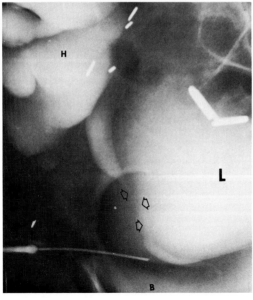

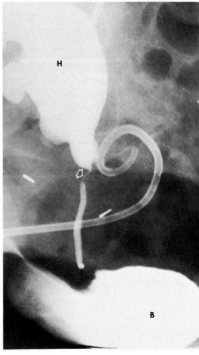

Often the choice is determined by the position of the kidney relative to the patient's ribs. The larger needles and nephrostomy tube should enter below the 12th rib to avoid a penetration of the lower posterior costophrenic angle and pneumothorax (Fig. 18A and B). If ultrasound imaging shows that a great deal of angulation of the larger needle is required to have the entry site remain below the 12th rib and enter the collecting system, I prefer to opacify the renal pelvis with a more vertical and intercostal skin entry site using a thin needle, and perform a second puncture with the larger needle under fluoroscopy.

If a portable real-time or A-mode transducer is available, initial localization can be performed with the patient on the fluoroscopy table. A real-time needle guide attachment permits direct single-needle-pass placement of the large needle into the renal pelvis in the fluoroscopy suite with placement of dilators and guide wire under fluoroscopic control. This is particularly helpful when only minimal hydronephrosis is present (Fig. 9). If no portable ultrasound equipment is available, the collecting system is localized with a conventional B-scan and puncture is performed in the ultrasound laboratory. Either a single-needle pass or double-needle pass approach can be used but I would recommend two precautions because the patient must be moved to the fluoroscopy room for placement of guide wires, dilators, and drainage catheter. First, as soon as urine is aspirated in the ultrasound room, contrast agent should be injected into the collecting system. Should the puncture needle be dislodged, as frequently occurs during transfer of the patient from the stretcher to the fluoroscopy table, repuncture can be performed under fluoroscopy. Second, I frequently use a teflon-sheathed needle (20-gauge) for the two-needle pass method (18-gauge for the single-pass method), and when urine is obtained and contrast injected, I advance the sheath further into the collecting system while holding the metal needle parts fixed. The metal needle is removed leaving only the soft plastic sheath within the collecting system. This reduces the risk of traumatizing the patient during his transfer to fluoroscopy and helps prevent dislodgement of the sheath from the collecting system. When either a guide wire or trocar system is used one should be certain that both puncture needles, guide wires, and dilators are long enough. Occasionally with a very obese patient, one must use longer puncture needles, stiffer guide wires (Ring or Lunderquist), and longer angiographic (vanAndel) dilators.

Premedication for nephrostomy tube placement is indicated. Vasovagal hypotension can occur and atropine 0.6 mg to 1 mg i.m. or i.v. is recom-

FIG.13. continued. Transplant kidney with hydronephrosis and a fluid collection. **Top:** Transverse ultrasound scan. B, Bladder; H, hydronephrosis; L, lymphocele. **Bottom left:** Nephrostomy tube has been placed in transplant kidney, and fluid collection also opacified. Is obstruction secondary to pressure from fluid collection? A pigtail catheter was also placed to drain the fluid collection subsequently shown to represent a lymphocele. **Bottom right:** Two days later a repeat antegrade pyelogram performed by nephrostomy tube injection demonstrates a residual short fixed stenosis caused by a fibrous band that was lysed at subsequent surgery.

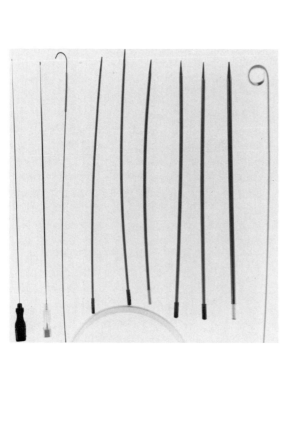

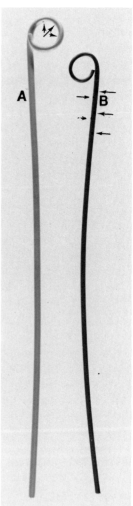

FIG. 14. Nephrostomy systems. **Left:** Vance nephrostomy system is available with 8-French or 10-French pigtail nephrostomy catheters and includes fine needle (22-gauge) for opacification, 18-gauge needle for puncture, guide wire introduction, and set of dilators. A Malecot catheter system is also available from Vance. **Right:** Nephrostomy pigtail (A) has sideholes only in the pigtail portion so that no sideholes can lie outside the renal cortex when properly positioned. Sideholes (*arrows*) lie only in the inner curve; a guide wire is unlikely to become caught in a sidehole since the tip of a guide wire usually seeks the outer curvature when introduced into the catheter. Nephrostomy catheter is unsuitable for angiographic injection because it would tend to uncoil. Angiographic pigtail (B) is unsuitable for urinary drainage because sideholes (*arrows*) lie proximal to the pigtail and would result in leakage of urine into the perinephric tissues.

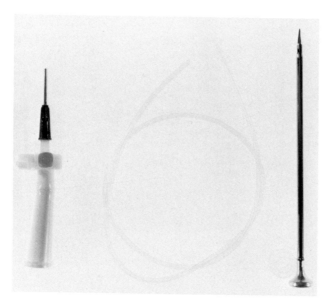

FIG. 15. Cystocath system, originally devised for suprapubic tube placement, can also be used. Since a guide wire and dilators are not used, fluoroscopy is not required. The trocar and cannula assembly (**left**) is advanced into the renal collecting system. When urine is obtained the trocar is removed, and the soft nephrostomy tube (**middle**) is advanced through the larger cannula to coil in the collecting system before the cannula is removed. The drainage port adaptor (**right**) is then attached to the external end of soft nephrostomy tube for drainage.

mended for prophylaxis or treatment (4). Strong-acting analgesics such as intravenous Sublimaze (fentanyl) or intramuscular or intravenous Demerol should be administered to minimize patient discomfort and improve patient cooperation. Intravenous antibiotics are administered prior to the procedure if there is any suggestion of infection in the urinary tract.

Complications of needle or catheter intubations of the kidney are relatively few, with an overall incidence of major complications of 4% for nephrostomy tube placement (50). If an obstructed collecting system is found on antegrade pyelography, we proceed with nephrostomy tube drainage to avoid potential infection or urine leakage. When an obstructed or partially obstructed collecting system may be infected, it is essential that the patient receive full-dose intravenous antibiotics because injection and needle or catheter manipulation into the collecting system are frequently accompanied by transient septicemia. Transient hematuria is almost an invariable immediate result of nephrostomy tube placement, but this usually clears within 1 or 2 days and is not to be regarded as a complication. Tube blockage may occur as a result of blood clot in the renal pelvis but this can be cleared with flushing of the catheter.

Small perinephric hematomas are found in 8% of cases, and usually resorb unless they become infected (50). Perforations of the collecting system can

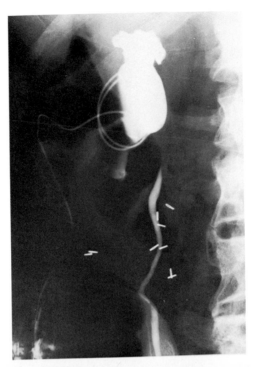

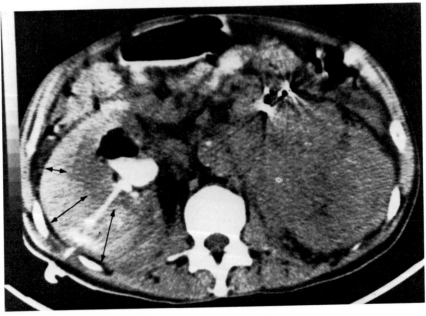

FIG. 16. Perirenal hematoma in a patient with thrombocytopenia after chemotherapy for lymphoma. The left kidney has been replaced by a mass of lymphomatous infiltration. **Top:** Nephrostomy tube using cystocath system. Patient received platelet transfusion during procedure. **Bottom:** CT scan with injection cystocath nephrostomy tube. A large dense halo of blood and some contrast (*arrows*) outlines the (less dense) right kidney. The cystocath system predisposes to perinephric hematoma formation in patients with clotting abnormalities, since the final drainage tube is smaller than the caliber of the trocar-cannula introduction system.

occur during the procedure but they will usually heal spontaneously provided urine drainage is established. Urinomas have been reported as a complication of nephrostomy placement. The primary contraindication is bleeding diathesis (Fig. 16), and deaths have been reported that were caused by hemorrhage following nephrostomy tube placement in patients with coagulopathies (34).

Percutaneous ureteral stents are available as either indwelling double pigtail stints (Vance Co.) (Fig. 19) or as single-pigtail stent with external injection

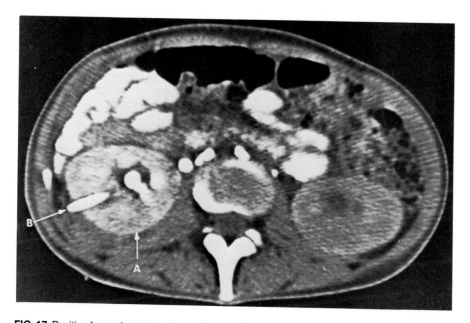

FIG. 17. Position for nephrostomy tube position. **A:** Direct posterior approach is more likely to injure the large vessels as they enter the renal hilum. There is less cortex to prevent urine extravasation around a catheter placed from this position. **B:** Posterior lateral (posterior axillary) placement is better not only for patient comfort, but also because the renal pelvis often presents a larger target in this position; the spine helps reduce kidney displacement during puncture, and there is less chance of injuring any major vessels. In addition, puncture is aligned with the renal pelvis and ureter when antegrade stent placement is planned.

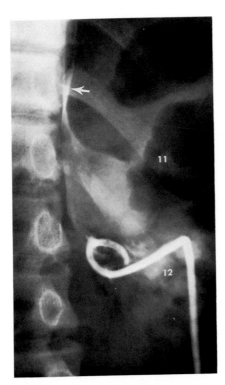
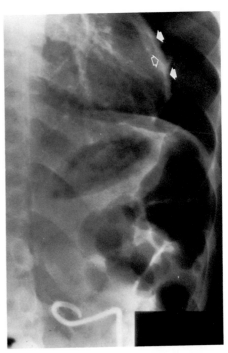

FIG. 18. Complication of improper nephrostomy entry site. The nephrectomy tube can enter any position of the renal collecting system, but the skin entry site should be below the 12th rib to avoid traversing the posterior costophrenic angle, which extends down as far as the upper half of the 12th rib. **Left:** Misdirected nephrostomy tube lay in the retroperitoneal soft tissue and entered between the 11th and 12th rib interspace. Injection shows contrast lining the posterior costophrenic angle (*arrow*). **Right:** Further injection showed contrast outlining the left heart border (*arrows*), indicating tracking of contrast into the mediastinum or pericardium.

or access port. Introduction of the indwelling double-pigtail system is well described by Mazer (33). Once a guide wire is manipulated through the renal pelvis, down the ureter, and into the bladder, the double pigtail is passed over the wire and advanced using a pusher catheter to push the proximal end into the renal pelvis. The pusher catheter is held in place as the guide wire is removed, and finally the pusher catheter is removed. We recommend an additional step to measure the length of stent that is needed. A tapered catheter is provided in the double-pigtail stent set. After the guide wire is passed down into the bladder, the tapered catheter is advanced all the way until the tapered tip is in the bladder. The guide wire is slowly withdrawn under fluoroscopy until the tip is just in the bladder and a Halstead clamp is placed on the wire where it enters at the hub of the catheter. The guide wire is then further withdrawn until the tip lies in the renal pelvis and a second clamp

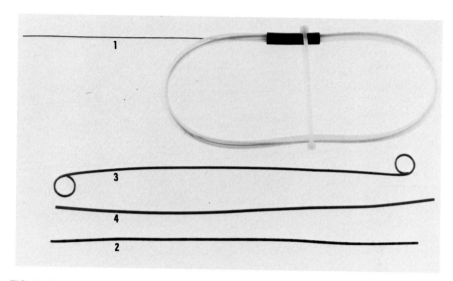

FIG. 19. Double-pigtail internal ureteral stent set available from Vance division of Cook Catheter Co. (1) Guide wire; (2) straight endhole exchange catheter (shortened for photography); (3) double pigtail; (4) "pusher" catheter.

is placed on the wire where it enters the hub of the straight catheter. The distance between the two clamps represents the distance the tip of the guide wire covered while withdrawn from bladder to renal pelvis, and provides the length of stent required (Fig. 20). Because the pigtail stent and pusher catheter are difficult to visualize during placement under fluoroscopy it is also useful to use a guide wire to measure the length from skin entry site to renal pelvis. This measurement is transferred to the pusher using a sterile bandaid or tape as indicator, and later the pusher can be advanced to this depth to ensure that the proximal end of the internal stent is within the renal pelvis. Since indwelling double pigtail stents usually require cystoscopic replacement every 3 months or less, no indwelling stent should be placed without consultation with an urologist. Single-pigtail stents with external injection or access port can be made by shortening and introducing sideholes into an aortic angiographic catheter, or a 65-side hole Ring biliary drainage catheter can be used. Such stents are now also available from Vance Co. These stents have an advantage that an external port provides access for follow-up antegrade pyelography, for flushing, and for stent removal and changing. Indeed, the patient does not need an internal double-pigtail stent, if he can tolerate a percutaneous stent with an external access port. We have been using stents with external ports in patients where only temporary diversion is needed such as in the preoperative or postoperative patient undergoing ureteric surgery (Figs. 21 and 22). It is important that the most proximal sidehole lie within the collecting system but above the obstructing lesion (Fig. 23).

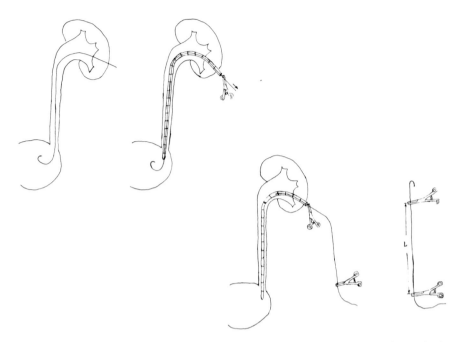

FIG. 20. Determination of proper stent length. A guide wire or catheter can be used as an *in vivo* ruler for selection of stent length. After a guide wire has negotiated the ureter and has passed antegrade down into the bladder (or ileal loop), the straight catheter is passed over the wire into the bladder to secure this position. The distal end of guide wire lies in the bladder and a clamp is placed over the wire where it enters the straight catheter. The wire is withdrawn slowly under fluoroscopy until the tip lies in the renal pelvis. At this point a second clamp is placed over the wire where it enters the catheter. The distance between the two clamps represents the distance the tip of the guide wire travelled from the bladder to its position in the renal pelvis and gives the proper (center of pigtail-to-pigtail stent length).

BILIARY APPLICATIONS

Ultrasound examination has become an important screening test in the jaundiced patient, because dilatation of either intrahepatic or extrahepatic biliary tree is easily detected. Although ultrasound is not as precise as transhepatic cholangiography in determining the actual cause of obstruction, useful correlations exist. If the obstruction lies at the level of the distal common bile duct, dilatation of the parts of the biliary tree appears to follow a characteristic sequence. The gallbladder will be distended first provided the wall has not been chronically inflamed, creating the so-called "Courvoisier" gallbladder. The common bile duct and extrahepatic duct will be the next most dilated structures because of the absence of surrounding parenchyma. The intrahepatic ducts will be next to dilate, but to a lesser degree than the extrahepatic ducts. When the ultrasonic appearance deviates from the above

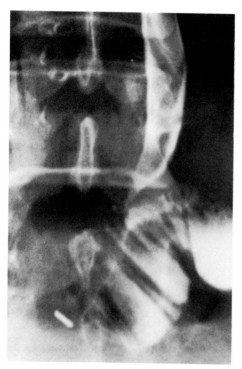 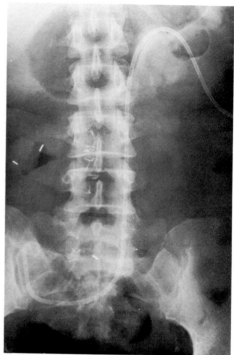

FIG. 21. Patient with solitary left kidney and ileal loop treated with ureteral stents. **Left:** Stenosis of anastomosis of left ureter to ileal loop. **Right:** A 8.3-French Ring biliary catheter with 65 sideholes was used as an antegrade ureteral stent with external injection or drainage port.

pattern one should suspect other lesions. For example, if the intrahepatic ducts appear more dilated than the extrahepatic ducts, or the common bile duct is normal in size, then an intrahepatic or high obstructing lesion in the porta hepatis should be suspected; cholangiography usually reveals a Klatzkin's tumor, metastasis, gallbladder carcinoma, or intrahepatic stone.

A second manner in which ultrasound provides important information is in planning a drainage procedure in patients with intrahepatic or high porta hepatis obstruction. One can assess the degree of duct dilatation in the right and left lobes of the liver. Occasionally, a patient presents with isolated dilatation of the left ducts and with nondilated right ducts. In such cases, the traditional approach of percutaneous transhepatic cholangiography (PTC) and drainage from the right side can be futile (Fig. 24). With knowledge of the ultrasound findings, one can anticipate and plan drainage procedure of the left biliary system from the start.

Since biliary duct dilatation is easily detected, it is natural to assume that ultrasound will be helpful in guiding the actual puncture for cholangiography

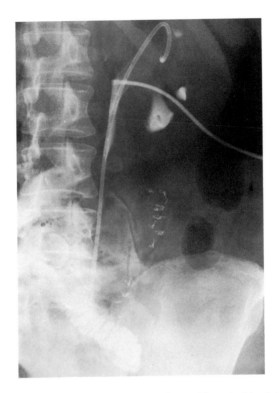

FIG. 21. continued. Subsequently, this catheter was changed for a double-pigtail internal ureteral stent. Prior to final removal of the "pusher," one can inject contrast through the pusher catheter to perform an antegrade pyelogram to verify proper stent function. If malfunction is seen on antegrade injection, or if there is significant blood clot, one passes a guide wire through the "pusher" to place a temporary nephrostomy tube, instead of removing the pusher. Subsequently, this double-pigtail stent has been changed by snaring the lower end of the pigtail, using a retrograde exchange for a new stent by passing the new catheters from ileal loop to renal pelvis, over a guide wire.

(37). In experience with 10 patients we have been able to puncture an intra-hepatic duct routinely on first pass if the patient has dilated intrahepatic ducts (Figs. 24 and 25). A real-time needle guide or aspiration transducer is an absolute necessity because even the largest dilated intrahepatic ducts are smaller than the respiratory excursion and are too small to localize well with A-mode transducers. A portable real-time unit is essential so that the procedure can be performed directly on the fluoroscopic table. Transfer of a patient from a stretcher to a fluoroscopy table will almost certainly result in dislodgement of the needle when one is dealing with such small structures. We have used needle guide attachments for the ATL and Hewlett-Packard portable ultrasound units.

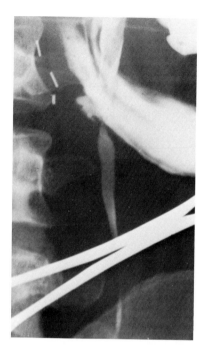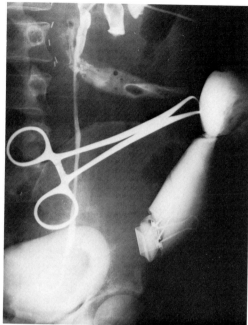

FIG. 22. Left: Treatment of ureteric fistulas. This patient had complete transection of the ureter during lymph node dissection for carcinoma of the colon. Although reanastomosis was performed a large amount of urine was drained to the skin as seen on this antegrade pyelogram. **Right:** Injection of external ureteral stent. An aortic arch angiographic catheter with multiple sideholes is passed antegrade down the ureter into the bladder to provide urine diversion and stent for healing. Note large amount of contrast extravasation to collection bag.

The left hepatic ducts are chosen for puncture under ultrasound guidance whenever they are dilated. This is because the left lobe is anterior and the left ducts are relatively superficial, and there are no ribs to impede visualization or puncture. Further, when contrast is injected into the left ducts, it naturally flows to fill the posterior right hepatic ducts and common bile duct unless there is intrahepatic biliary obstruction (Fig. 25). On the other hand, when puncture and injection of the right ducts are performed, one must turn the patient onto the left side to exclude intrahepatic obstructing lesions and to try to fill the left ducts, because the more anterior left ducts do not normally fill from injection of the right ducts in the supine position, unless a great deal of contrast is injected.

Because we are able routinely to puncture the dilated left hepatic duct on first pass, we use a 20-gauge needle rather than the fine 22-gauge needle because this slightly larger needle is easier to see and allows a bile aspiration as

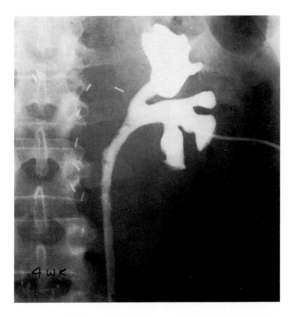

FIG. 22. continued. Four weeks later, an antegrade injection shows no evidence of fistula, and the temporary stent is withdrawn and replaced with a small nephrostomy tube for 2 or 3 days. Repeat injection showed no evidence of fistula, and nephrostomy tube was removed.

proof of successful puncture and for bile culture, and decompression prior to cholangiography. This can be particularly important in the patient with cholangitis.

Although it is possible to puncture the larger common hepatic duct under ultrasound guidance with the larger needle for placement of Ring drainage catheters in patients with low common bile duct obstructions, this portion of the biliary system may be extrahepatic, and, therefore, carries a risk of bile peritonitis. For this reason, we prefer to puncture a major branch of the common hepatic duct when drainage in such a patient is needed. Because scanning for puncture under ultrasound is restricted to scan planes parallel to intercostal spaces, only in a minority of cases (3 out of 10) can one actually trace a dilated branch of the common hepatic duct for suitable puncture under ultrasound. In the majority of cases, drainage is best performed under fluoroscopic control where the overall relationship of duct branches to the common hepatic duct is more easily assessed. In a few initial cases where punctures were attempted when ducts could not be traced to the common hepatic duct, we entered either a duct unsuitable for manipulation and advancement of guide wire or catheter, or a portal or hepatic vein.

However, in cases where there is isolated left hepatic duct obstruction as in Klatzkin's tumor or metastatic disease, puncture with the larger drainage

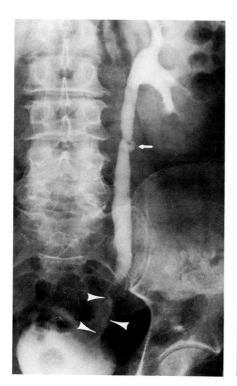 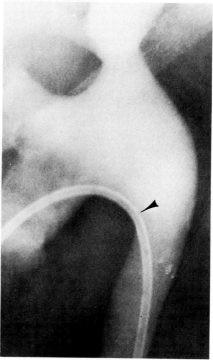

FIG. 23. External stent positioning. In both ureteric or biliary stent catheters, absence of opaque markers makes the sideholes difficult to see during fluoroscopic positioning. It is critical that the most proximal sidehole is proximal to the stenotic lesion, but yet still remain within the collecting (or biliary) system. Again a guide wire can be used as a ruler. When the stent catheter is first prepared, one advances a guide wire through the external injection port until the tip reaches the most proximal sidehole. Either a clamp or bend in the wire is made where it enters the injection hub, and this guide wire is set aside for reinsertion after the stent has been advanced into the patient. **Left:** Placement of an antegrade ureteral stent for carcinoma of colon with metastasis to pelvis near distal left ureter. In placement a sheathed needle was used, and the upper kink in the ureter (*arrow*) was negotiated with small J-shaped guide wire. The lower segment (*arrowheads*) was then negotiated with a straight guide wire. **Right:** Spot film after placement of antegrade stent. Patient prone, left kidney. The most proximal sidehole is barely seen on spot film (*arrowhead*) and undetectable under fluoroscopy.

needle is easier under ultrasound guidance than under fluoroscopy alone. Even in this instance we initially perform ultrasound-guided cholangiography with a 20-gauge needle to be certain that the duct seen on ultrasound examination will be suitable for placement of guide wires and catheters for subsequent drainage.

One further area of biliary tract disease lends itself to ultrasound-guided procedures: suspected gallbladder disease. It has been shown that trans-

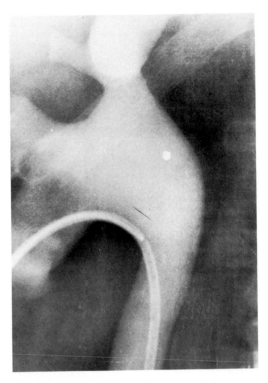

FIG. 23. continued. After insertion of premeasured guide wire, the position of the most proximal sidehole is clearly seen under fluoroscopy.

hepatic puncture of the gallbladder in animals is a safe procedure (21,26), and isolated cases have been reported in humans (42). We have performed a transhepatic gallbladder aspiration and cholecystocholangiogram in a 3-year-old Korean girl with elevated alkaline phosphates and occasional stool specimens suspicious for ova (Fig. 26). The cholangiogram demonstrated filling defects compatible with liver flukes, and ova was found in the bile. Following cholecystocholangiography, the gallbladder was completely evacuated under flouroscopic visualization before removal of the 20-gauge Longwell sheath used for the procedure as a precaution against bile leakage. As more experience is gained with this procedure, further applications and indications are likely to develop analogous to the use of percutaneous nephrostomy. The procedure will be safe in patients without cystic duct or common bile duct obstruction, but either immediate catheter or operative bile diversion will be required in any patients demonstrated by this procedure to have common duct obstruction.

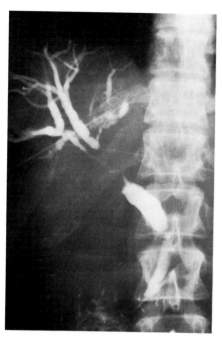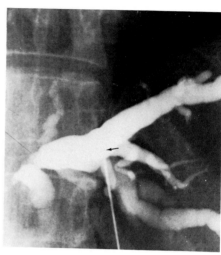

FIG. 24. Value of ultrasound in planning biliary drainage procedure. **Left:** Transhepatic cholangiogram under fluoroscopy with Chiba needle. The right biliary ducts are not dilated although partial obstruction is present. The left biliary ducts were not demonstrated on this or previous cholangiogram. Again, biliary drainage was attempted and a small catheter could be placed in a right bile duct after great difficulty. Only a small amount of bile drained daily and there was no improvement in jaundice. **Right:** Real-time ultrasound examination revealed the left biliary ducts were much more dilated than the right ducts, and a single-pass puncture of the left biliary system was performed using the ATL needle guide. Bile was aspirated and contrast was injected to obtain "clean" cholangiogram with no parenchymal contrast extravasation (contrast with **A**). AP view.

PERCUTANEOUS ABSCESS DRAINAGE

Percutaneous abscess drainage has been shown to be both safe and effective (15,18–20), and in many cases should be considered the treatment of choice. For example, when an abscess lies deeply embedded in a solid organ such as the liver, the lesion is detected and visualized much better with noninvasive imaging such as ultrasound or CT than at open laparotomy. For this reason, the drainage catheter can be placed more accurately into the abscess collection and with considerably less trauma to the patient.

Different catheter systems have been used for abscess drainage. While no system has been shown to be superior to another, larger-size drainage catheters are more effective than smaller-size drainage catheters. Most abscess collections can be drained by any catheter that is 8-French or larger in

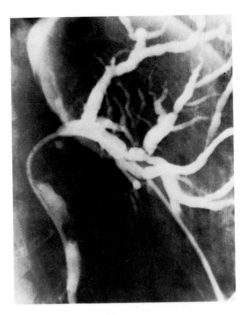

FIG. 24. continued. Ring guide wire and catheter was placed into left biliary ducts and through the obstruction produced by extension of gallbladder carcinoma, with rapid reduction in jaundice. Left posterior oblique view.

diameter. A popular system for abscess drainage is the guide wire-dilator-pigtail catheter combination. After an 18-gauge needle has been introduced into the abscess collection, a guide wire is passed into the collection for exchange for dilators and catheters in standard angiographic technique under fluoroscopic control. The Vance nephrostomy (8-French or 10-French) has been used with or without placement of additional sideholes in the catheter. Another system is the trocar-catheter. The trocar is considerably larger than the 18-gauge needle used in a guide wire and dilator system, so that the body is not likely to tolerate many unsuccessful passes with a trocar system. In theory, use of the trocar system does not require fluoroscopic control if the lesion can be clearly localized with ultrasound imaging. Use of the trocar system is quicker and often less painful if the lesion is superficial or very large and only one pass need be made. We limit the use of these systems to easily

FIG. 25. Single-pass real-time ultrasound-guided transhepatic cholangiography is routinely successful in patients with intrahepatic duct dilatation. Frequently, the left biliary ducts are dilated slightly more than right biliary ducts, even with common bile duct obstruction. The left biliary ducts are ducts most easily punctured under ultrasound. **Top:** A 20-gauge needle is advanced under direct real-time observation into a left biliary duct using the Hewlett-Packard needle guide (*arrow*, needle tip). **Bottom:** Injection of contrast into duct creates a burst of echoes that fills the bile duct (*arrows*).

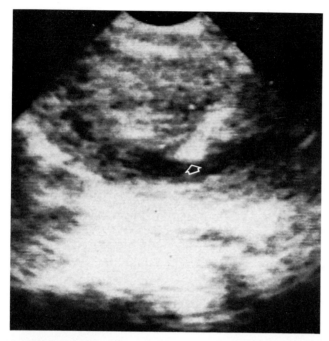

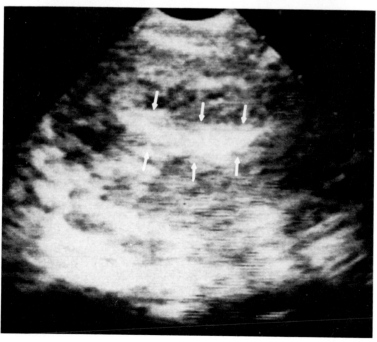

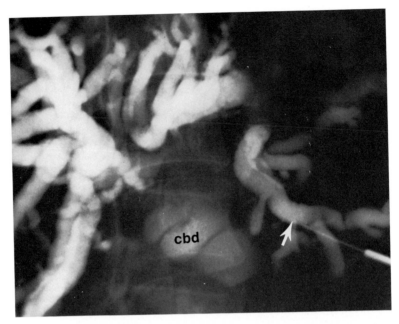

FIG. 25. continued. Cholangiogram obtained shows common bile duct obstruction due to carcinoma of the pancreas (*arrow*, tip of 20-gauge needle). Note cholangiogram is obtained without parenchymal staining or extrahepatic or subcapsular injections. Patient had Ring catheter biliary drainage catheter placed.

accessible abscesses where there is no danger of puncturing adjacent organs and the chance of a repeat puncture being required is very low. In other cases, softer catheter systems are sometimes needed. Either the cystocath system or soft rubber urinary catheters can be used with introduction over a guide wire and small catheter or dilator in a coaxial fashion.

Although some abscesses are detected on plain film because of the presence of an abnormal gas collection, many abscesses do not contain sufficient gas to be detectable with conventional radiography. In the majority of cases, the abscess is detected and localized for puncture with either ultrasound or CT (7,15,18–20). With the exception of abscess collections totally obscured from ultrasound detection by surrounding bowel gas, CT or ultrasound can be used as guidance for puncture. The choice is primarily one of operator experience and equipment availability. Because of the development of portable real-time imaging equipment, ultrasound has an advantage especially when fluoros-

FIG. 26. A Korean child presented with elevated alkaline phosphatase with possible parasitic ova seen in occasional stool specimens. **Top:** Ultrasound showed some debris (*arrow*) in the gallbladder. Minimal intrahepatic duct dilatation was present. **Bottom:** Percutaneous transhepatic cholecystogram with introduction of a 20-gauge long-dwell Teflon sheath into gallbladder (*arrows*) for aspiration and injection.

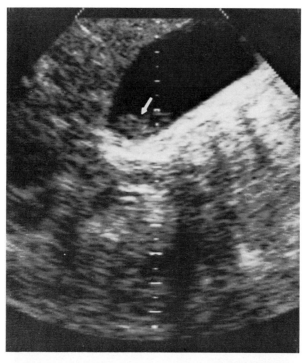

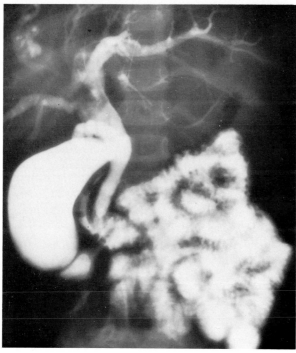

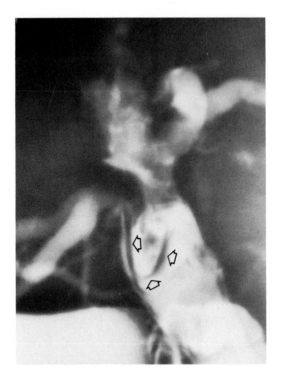

FIG. 26. continued. Patient is turned to fill intrahepatic ducts. An "ash leaf"-like filling defect (*arrows*) represents liver flukes (*Fasciola hepatica*). Bile aspiration revealed ova of the liver fluke. The gallbladder was completely decompressed by aspirating all contrast (and bile) under fluoroscopic observation, prior to removal of the sheath catheter.

copy is also required, e.g., whenever a guide wire-dilator-catheter system is used for abscess drainage (Fig. 27). Even with trocar systems, fluoroscopy can be helpful in optimizing final placement of the drainage catheter, and the importance of fluoroscopy in placement of any catheter system, whether for abscess or urinary drainage, is often understated in relation to ultrasound or CT (7). Abscess drainage frequently requires fluoroscopy even with puncture under CT. When this is true, there is always ample opportunity for dislodgement of the needle or trocar when transferring the patient to the fluoroscopy table. With portable real-time ultrasound, however, the entire procedure can be performed with the patient on the fluoroscopy table with no need for

FIG. 27. Upper pole renal abscess in a 7-year-old girl with congenital solitary right kidney. **Top:** Gallium scan, posterior view, demonstrated intense activity in right flank in this patient who presented with fever of unknown origin. **Bottom:** Longitudinal scan shows large dumbbell-shaped collection displacing remainder of kidney inferiorly. Urine cultures were negative.

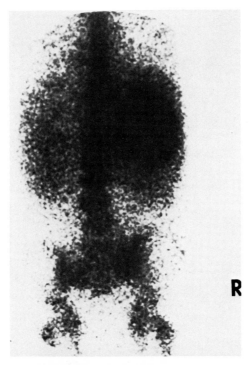

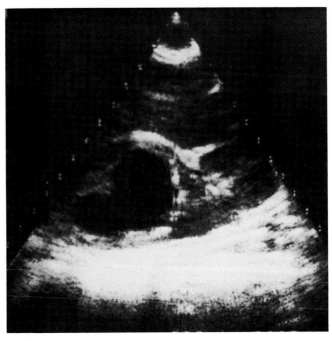

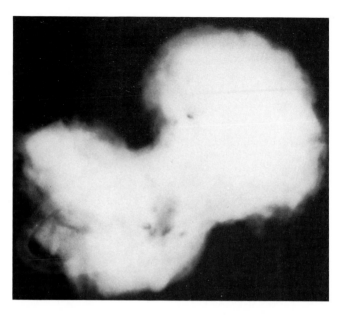

FIG. 27. continued. Sinogram after placement of pigtail drainage catheter (lateral view). There is no communication with collecting system. Abscess collection grew Proteus. Study performed with ATL needle guide in fluoroscopy room.

transfer of the patient from one area of the radiology department to another. Secondly, when an unusual oblique-angled puncture is required, the angle for puncture must be computed from simple trigonometric relationships when performed with CT. While this can also be done with ultrasound, it is frequently unnecessary. If a gray-scale compound scanner is used, the best angle for puncture is determined from the scan and the transducer can be held at the proper angle with the puncture performed advancing the needle parallel to the transducer. Alternatively, if an appropriate skin entry site is selected, puncture can be performed by placing an A-mode transducer over the site, and angling the transducer until the characteristic A-mode pattern of a fluid collection appears.

FIG. 28. Drainage of abscesses located near bowel loops and solid organs. **Top:** Following gastric resection, this patient became febrile and a large left upper quadrant collection containing a debris-fluid level was found. Longitudinal scan of left side. A, Abscess; B, ? bowel; L, liver; K, left kidney. Catheter drainage would be easy with a collection of this size, but unfortunately only needle aspiration was performed with apparent complete evacuation of (900 ml) collection. **Bottom:** Two days later, patient was still febrile and repeat (longitudinal) scan shows partial reaccumulation. It is more difficult to perform catheter drainage of this small collection without injuring adjacent liver or bowel loops.

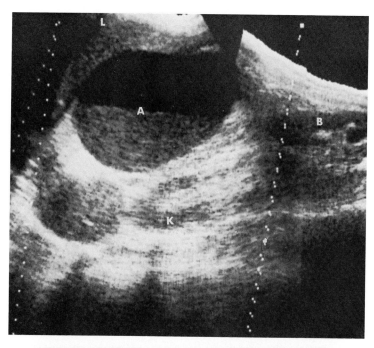

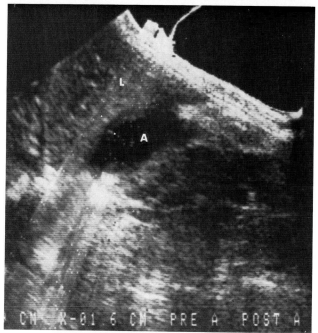

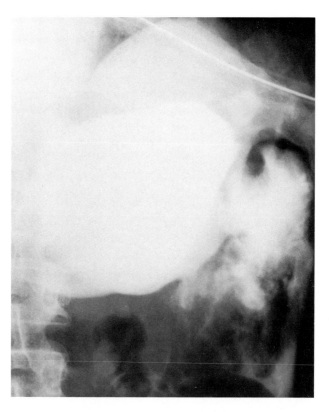

FIG. 28. continued. Under ultrasound localization, a fine 22-gauge needle is inserted trans-hepatically into collection to inject contrast to opacify and partially redistend this collection. The liver edge is marked on the skin and the angle of the inferior surface of the liver is determined.

A systematic approach to abscess drainage requires consideration of the relationship of the abscess to adjacent vital structures (Fig. 28). When an abscess lies totally enclosed in a solid organ such as liver or kidney, puncture can be performed under ultrasound guidance alone, with the skin entry site chosen low enough to avoid traversal of the costophrenic angles and pneumothorax (Fig. 18A and B). Puncture is made directly with the larger needle or trocar, unless there is some uncertainty concerning the nature of the fluid collection, in which case a diagnostic aspiration with a 22-gauge fine needle is performed.

When an abscess is intimately related to nearby bowel loops, successful drainage is usually still possible because any large abscess will cause sufficient displacement of these bowel loops to allow a safe puncture path. We take advantage of the fact that ultrasound and fluoroscopy are complementary imaging modalities and perform the drainage under combined ultrasound and fluoroscopy. Whereas ultrasound is exquisite in delineating the borders of

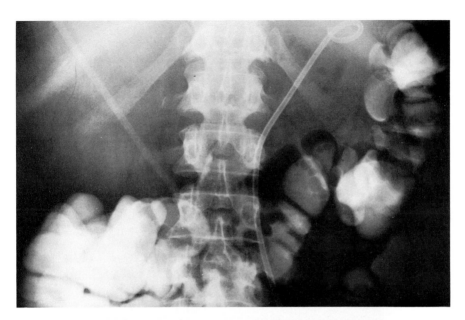

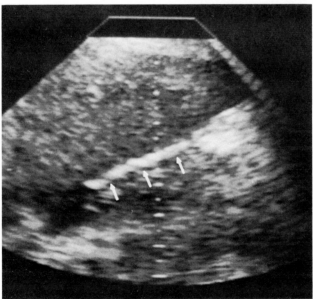

FIG. 28. continued. **Top:** The patient was transferred to fluoroscopy, where a contrast enema was performed to locate the large bowel and a safe puncture site was found between the liver edge (as determined by ultrasound) and the bowel edge (as determined under fluoroscopy). The puncture needle was advanced at the angle predicted by ultrasound to avoid injuring the liver. AP view after placement of catheter shows that contrast in collection has already been drained by the catheter. **Bottom:** Follow-up longitudinal scan shows collection has markedly collapsed and catheter (*arrows*) lies parallel to inferior surface of liver.

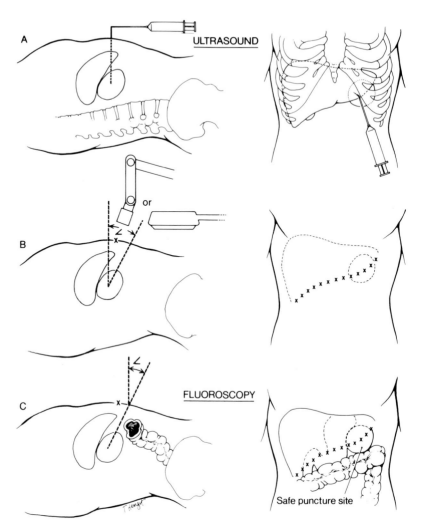

FIG. 28. continued. Summary diagram showing how ultrasound is used to demarcate solid organs and how, with introduction of contrast agents, fluoroscopy can be used to evaluate the relationship of the bowel loops to an abscess collection. **A:** Skinny needle aspiration and opacification abscess cavity with injection of contrast. **B:** Localization of adjacent solid organs and puncture angle determination with ultrasound. **C:** Opacification of adjacent bowel loops under fluoroscopy and drainage under fluoroscopic control.

solid organs and vascular structures in the region of the abscess, it is particularly poor in defining the relationship of adjacent bowel loops. With the introduction of air or positive contrast (barium or water-soluble contrast), adequate visualization of adjacent bowel loops is provided by fluoroscopy, although fluoroscopy is particulary poor in determining the margins of adjacent solid organs. When a fluid collection is found in a region where adjacent

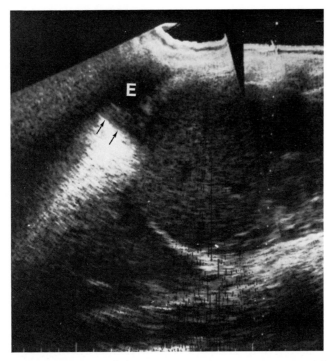

FIG. 29. Empyema drainage. Longitudinal scan over right lower chest and upper abdomen reveals anterior loculated empyema (*arrows*) in anterior right chest in this patient with pneumococcal pneumonia. A 10-French nephrostomy type pigtail catheter was placed into empyema. E, Empyema.

bowel loops may be present, the margins of adjacent solid organs are mapped out on the patient's skin, and the angle for a safe puncture (to avoid injury to solid or vascular structures) is determined from the ultrasound scan. A fine-needle aspiration is performed for diagnosis, and if pus is obtained water-soluble contrast is injected so that the abscess can also be seen under fluoroscopy. The patient is then transferred to fluoroscopy, where the bowel is visualized by introduction of either air or positive contrast. Frequently, the colon contains air and can easily be identified, but the small bowel almost always requires positive contrast for visualization. Using metallic markers under fluoroscopy, one can then determine whether a suitable skin entry site can be found that avoids injury to both solid organ and hollow viscus. A dermatotomy is made and the larger needle or trocar is advanced into the opacified collection at the angle determined by ultrasound imaging (Fig. 28).

If the abscess is surrounded only by bowel loops, ultrasound is used only for detection, diagnostic aspiration, and opacification of the collection, and the collection is drained under fluoroscopy where positive contrast can be introduced to identify the adjacent bowel loops. In this systematic way safe

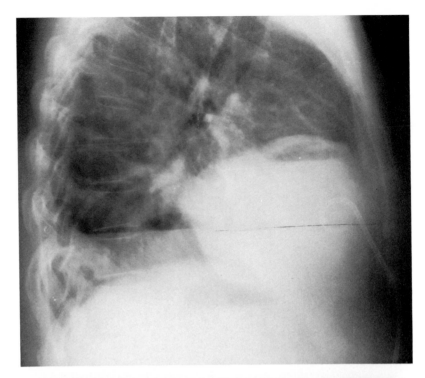

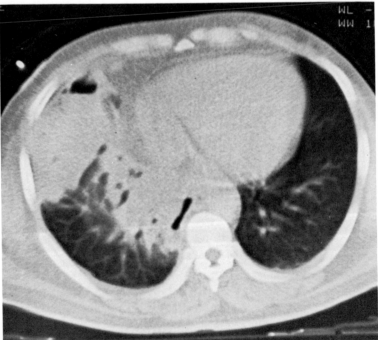

FIG. 29. continued. **Top:** Lateral chest X-ray after tube placement. Air was introduced by tube. **Bottom:** CT scan shortly after tube placement shows large collection and infiltrate in medial portion of right lower lung.

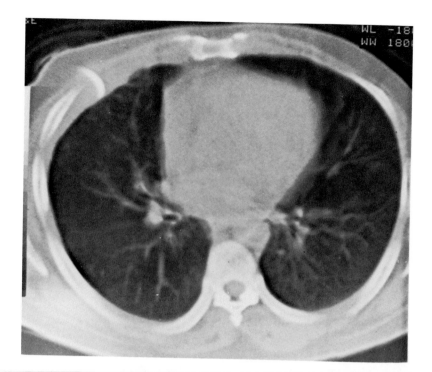

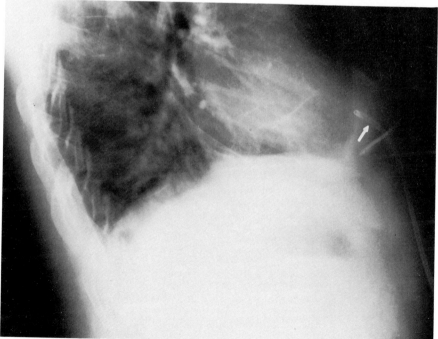

FIG. 29. continued. **Top:** Follow-up CT scan 2 weeks later shows collection has collapsed around catheter. **Bottom:** The patient was discharged, and later when he came in as an outpatient, the catheter (*arrow*) was gradually withdrawn. Lateral chest X-ray at 3 weeks showing only pleural thickening near catheter. The tube was then removed.

puncture and drainage of abdominal abscesses are performed in areas previously reserved for operative exploration and incision and drainage.

In the chest, ultrasound provides a simple method for localization and drainage of the loculated empyema (Fig. 29).

With the exception of mediastinal and intracranial abscesses or abscesses surrounded by the bony pelvis, there are few abscesses that cannot be drained by the radiologist who makes judicious use of current modern imaging techniques.

REFERENCES

1. Alken, P., Hutschenreiter, G., Gunther, R., and Marberger, M. Percutaneous stone manipulation. *J. Urol.*, 125:463–466, 1981.
2. Bigongiari, L. R. The seldinger approach to percutaneous nephrostomy and ureteral stent placement. *Urol. Radiol.*, 2:141–145, 1981.
3. Bigongiari, L. R., Kyo Rak, L., Moffat, R., Mebust, W. K., Foret, J., and Weigel, J. Percutaneous ureteral stent placement for structure management and internal urinary drainage. *AJR*, 133:865–868, 1979.
4. Bigongiari, L. R., Linshaw, M. A., Stapleton, F. B., and Weigel, J. W. Vagal hypotension after percutaneous biopsy; possible confusion with hypovolemic shock. *Urol. Radiol.*, 1:217–220, 1980.
5. Bolton, W. K., Tully, R. J., Lewis, E. J., and Ranniger, K. Localization of the kidney for percutaneous biopsy. *Ann. Intern. Med.*, 81:159–164, 1974.
6. Bush, W. H., Jr., Burnett, L. L., and Gibbons, R. P. Needle tract seeding of renal cell carcinoma. *AJR*, 129:725–727, 1977.
7. Dixon, G. D. Combined CT and fluoroscopic guidance for liver abscess drainage. *AJR*, 135:397–399, 1980.
8. Dretler, S. P., Pfister, R. C., and Newhouse, J. H. Renal stone dissolution via percutaneous nephrostomy. *N. Engl. J. Med.*, 300:341–343, 1979.
9. Ennis, M. G., and MacErlean, D. P. Percutaneous aspiration biopsy of abdomen and retroperitoneum. *Clin. Radiol.*, 31:611–616, 1980.
10. Er-Lin Yeh, Kronenetter, C., Monde, R. C., and Ruetz, P. P. Technical considerations in B-mode scanning with an aspiration transducer. *Radiology*, 129:527–530, 1978.
11. Evans, W. K., Ho Chia-Sing, M. B., McLaughlin, M. J., and Tao Liung-Che. Fatal necrotizing pancreatitis following fine needle aspiration biopsy of the pancreas. *Radiology*, 141:61–62, 1981.
12. Fernstrom, I., and Johansson, B. Percutaneous pyelolithotomy: A new extraction technique. *Scand. J. Urol. Nephrol.*, 10:257–259, 1976.
13. Ferrucci, J. T., Wittenberg, J., Margolies, M. N., and Carey, R. W. Malignant seeding of needle tract after thin needle aspiration biopsy. *Radiology*, 130:345–346, 1979.
14. Ferrucci, J. T., Jr., Wittenberg, J., Mueller, P. R., Simeone, J. F., Harbin, W. P., Kirkpatrick, R. H., and Taft, P. D. Diagnosis of abdominal malignancy by radiologic fine-needle aspiration biopsy. *AJR*, 134:323–330, 1980.
15. Gerzof, S. G., Birkett, D. H., Pugatch, R. D., Robbins, A. H., Johnson, W. C., and Vincent, M. E. Percutaneous catheter drainage of abdominal abscesses guided by ultrasound and computed tomography. *AJR*, 133:1–8, 1979.
16. Gittes, R. F., and Varady, S. Nephroscopy in chronic unilateral hematuria. *J. Urol.*, 126:297–300, 1981.
16a. Goldberg, B. B., Pollack, H. M., and Kellerman, E. Ultrasonic localization for renal biopsy. *J. Radiol.*, 115:167–170, 1975.
17. Goldin, A. R. Percutaneous ureteral splinting. *Urology*, 10:165–168, 1977.
18. Gronvall, J., Hegedus, V., and Gronvall, S. Ultrasound-guided drainage of fluid-containing masses using angiographic catheterization techniques. *AJR*, 129:997–1002, 1977.

19. Haaga, J. R., Alfidi, R. J., Havrilla, T. R., Cooperman, A., Seidelmann, F. E., Reich, N. E., Weinstein, A. J., and Meaney, T. F. CT detection and aspiration of abdominal abscesses. *AJR*, 128:465–474, 1977.
20. Haaga, J. R., and Weinstein, A. J. CT-guided aspiration and drainage of abscesses. *AJR*, 135:1187–1194, 1980.
21. Hogan, M. T., Watne, A., Mossburg, W., and Castaneda, W. Direct injection into the gallbladder in dogs using ultrasound guidance. *Arch. Surg.*, 3:564–565, 1976.
22. Horton, J. A., Bank, W. O., and Kerber, C. W. Guiding the thin spinal needle. *AJR*, 134:845–846, 1980.
23. Isler, R. J., Ferrucci, J. T., Jr., Wittenberg, J., Mueller, P. R.. Simeone, J. F., VanSonnenberg, E., and Hall, D. A. Tissue core biopsy of abdominal tumors with a 22-gauge cutting needle. *AJR*, 136:725–728, 1981.
24. Jacobs, S. C., and Gittes, R. F. Dissolution of residual renal calculi with hemiacidrin. *J. Urol.*, 115:2–4, 1977.
25. Klang, E., Alexander, R., Barnett, T., Palomar, J., and Hamway, S. Brush biopsy of pyelocalyceal lesions via a percutaneous lumbar approach. *Radiology*, 129:623–627, 1978.
26. Klapdor, R., Scherer, K., Sepelur, H., and Kloppel, G. Ultrasonically guided puncture of the gallbladder in animals. *Endoscopy*, 9:166–169, 1977.
27. Lalli, A. F., McCormack, L. J., and Zelch, M. Aspiration biopsies of chest lesions. *Radiology*, 127:35–40, 1978.
28. Lang, E. K. Diagnosis and management of ureteral fistulas by percutaneous nephrostomy and antegrade stent catheter. *Radiology*, 138:311–317, 1981.
29. Lang, E. K., Lanasa, J. A., Garret, J., Stupling, J., and Palomar, J. The management of urinary fistulas and structures with percutaneous ureteral stent catheters. *J. Urol.*, 122:736–740, 1980.
30. Lundquist, A. Fine needle aspiration biopsy of the liver. *Acta Med. Scand.*, 520:1–28, 1971.
31. Marshall, V., and Whitaker, R. H. Ureteral pressure flow studies in difficult diagnostic problems. *J. Urol.*, 114:204–207, 1975.
32. Masahito, S., and Watanabe, H. Ultrasonically guided percutaneous pyeloscopy. *Urology*, 17:457–459, 1981.
33. Mazer, M. J., LeVeen, R. F., Call, J. E., Wolf, G., and Baltaxe, H. A. Permanent percutaneous antegrade ureteral stent placement without transurethral assistance. *Urology*, 14(4):413–419, 1979.
34. Newhouse, J. H., and Pfister, R. C. Percutaneous catherization of the kidney and perinephric space: Trocar technique. *Urol. Radiol.*, 2:157–164, 1981.
35. Newhouse, J. H., and Pfister, R. C. Therapy for renal calculi via percutaneous nephrostomy; dissolution and extraction. *Urol. Radiol.*, 2:165, 1981.
36. Newhouse, J. H., Pfister, R. C., Hedren, W. H., and Yoder, I. C. Whitaker-test after pyeloplasty: Establishment of normal ureteral perfusion pressure. *AJR*, 137:223–226, 1981.
37. Ohto, M., Karasawa, E., Tsuchiya, Y., Kimnra, K., Saisho, H., Ono, T., and Okuda, K. Ultrasonically guided percutaneous contrast medium injection and aspiration biopsy using a real-time puncture transducer. *Radiology*, 136:171–176, 1980.
38. Palestrant, A. M., Rad, F. F., Sacks, B. A., et al. Postoperative percutaneous kidney stone extraction. *Radiology*, 134:778–779, 1980.
39. Pereiras, R. V., Meiers, W., Kunhardt, B., Troner, M., Hutson, D., Barkin, J. S., and Viamonte, M. Fluoroscopically guided thin needle aspiration biopsy of the abdomen and retroperitoneum. *AJR*, 131:197–202, 1978.
40. Pesky, L., Hampel, N., and Kedea, K. Percutaneous nephrostomy and ureteral injury. *J. Urol.*, 125:298–300, 1981.
41. Pfister, R. C., and Newhouse, J. H. Interventional percutaneous pyeloureteral techniques II percutaneous nephrostomy and other procedures. *Radiol. Clin. North Am.*, 17:351, 1979.
42. Shaver, R. W., Hawkins, I. F., and Soong, J. Percutaneous cholecystostomy. AJR, 138:1133–1136, 1982.
43. Pingoud, E. G., Bagley, D. H., and Zeman, R. K. Percutaneous antegrade bilateral ureteral dilation and stent placement for internal drainage. *Radiology*, 134:780, 1980.
44. Reimer, D. E., Coleman, C., and Oswalt, J. R. Iatrogenic ureteral obstruction treated with balloon dilation. *J. Urol.*, 126:689–690, 1981.

45. Rohner, T. J., Jr., and Tuliszewski, R. M. Fungal cystitis: Awareness, diagnosis and treatment. *J. Urol.*, 124:142–144, 1980.
46. Rosenberger, A., and Adler, O. Fine needle aspiration biopsy in the diagnosis of mediastinal lesions. *AJR*, 131:239–242, 1978.
47. Smith, A. D., Lange, P. H., Miller, R. P., et al. Dissolution of cystine calculi by irrigation with acetylcysteine through percutanenous nephrostomy. *Urology*, 13:422–423, 1979.
48. Smith, A. D., Reinke, D. B., Miller, R. P., et al. Percutaneous nephrostomy in the management of ureteral and renal calculi. *Radiology*, 133:49–54, 1979.
49. Spataro, R. H., Linke, C. A., and Barbaric, Z. L. The use of percutaneous nephrostomy and urinary alkalinization in dissolution of obstructing uric acid stones. *Radiology*, 129:629–632, 1978.
50. Stables, D. P., Ginsberg, N. J., and Johnson, M. L. Percutaneous nephrostomy: A series and review of the literature. *AJR*, 130:75–82, 1978.
51. Stark, H., and Savir, A. Dissolution of cystine calculi by pelvio caliceal irrigation with D-penicillinune. *J. Urol.*, 124:895–898, 1980.
52. Whitaker, R. H. Methods of assessing obstruction in dilated ureters. *Br. J. Urol.*, 45:15–22, 1973.
53. Whitaker, R. H. Equivocal pelvi-ureteric obstruction. *Br. J. Urol.*, 47:771–779, 1976.
54. Whitaker, R. H. Percutaneous upper urinary tract dynamics in equivocal obstruction. *Urol. Radiol.*, 2:187–189, 1981.
55. Wilbur, H. J. The flexible choledochoscope: A welcome addition to the urologic armamentarium. *J. Urol.*, 126:380–381, 1981.
56. Zornoza, J., Cabanillas, F. F.. Altoff, T. M., Ordonez, N., and Cohen, M. A. Percutaneous needle biopsy in abdominal lymphoma. *AJR*, 136:97–103, 1981.
57. Zornoza, J, Jonsson, K., Wallace, S., and Luheman, J. M. Fine needle aspiration biopsy of retroperitoneal lymph nodes and abdominal masses: An updated report. *Radiology*, 125:87–88, 1977.
58. Review of endourology procedures. *Urol. Clin. North Am.*, 9:1, 1982.

Ultrasound Annual 1982,
edited by Roger C. Sanders.
Raven Press, New York © 1982.

Breast Ultrasonography

William Robert Lees

*Departments of Ultrasound and CT Scanning, The Middlesex Hospital,
London, U.K.*

DIAGNOSIS OF BREAST DISEASE: CLINICAL PROBLEMS

Two aspects of diagnosis of breast disease are often confused in the imaging literature: detection and proof of diagnosis. Despite 20 years experience with X-ray mammography it is still true that the vast majority of breast cancers are discovered by the patient herself or by a physician at a routine examination. Referral to a specialist is usually secondary. The typical pattern of such a specialist practice has been concisely summarized by Haagensen, who reports the diagnostic outcome of each 1,000 women presenting with symptoms at his breast clinic (Table 1).

It is readily apparent that the majority of women with breast symptoms require no treatment other than reassurance. The capacity of physical examination alone to exclude cancer in these patients is generally severely limited, although surgeons with great experience in breast disease will miss very few. Early experience with X-ray mammography demonstrated that at least 20% of the malignant lesions thus detected were impalpable, and in more recent screening programs almost half were detected by mammography alone (11).

Symptomatic Women

Clinical signs and symptoms that require definitive investigation and proof of diagnosis are (22): a dominant tumor, a marked increase in the size or firmness of one breast, retraction signs, redness or edema of the skin, a spontaneous nipple discharge (this sign alone carries an 11.5% incidence of cancer), and changes in the nipple epithelium. To this can be added characteristic mammographic evidence of breast disease. Until recently the only successful nonsurgical methods that have proved sufficiently accurate are needle biopsy and conventional mammography.

Given that definite proof of diagnosis can only be obtained from histopathology of the paraffin block of the entire volume of abnormal breast tissue, all the methods of preoperative diagnosis exist to avoid or expedite biopsy (3). Reliable diagnosis or exclusion of cancer at an early stage can greatly simplify the management.

TABLE 1. *Diagnostic outcome of 1,000 women presenting with symptoms at Haagensen's clinic[a]*

No care required	750
Abnormal physiology only	80
Infection	5
Benign	
Gross cysts	70
Adenosis	2
Fibrosis	5
Duct ectasia	3
Adenofibroma	25
Intraduct papilloma	5
Malignant	55

[a]From ref. 22.

In experienced hands, fine-needle aspiration cytology has yielded true-positive rates as high as 96%, with few or no false-positive results (51,53). The technique is practiced widely in Europe and in the U.S.A. by surgeons but is limited to the patient with a well-defined palpable breast lump. Techniques of biopsy guidance borrowed from general radiology and ultrasonography are now being introduced to extend the range and improve the precision of cytodiagnosis. Useful results have even been returned on the cytology of benign breast disease (16).

Conventional mammography is often of limited value in the symptomatic patient. It is generally accepted that there are four main roles for mammography in the symptomatic woman. The first is to improve the detection of cancer. Gravelle reports a detection rate of 93% using xeromammography, compared with a clinical accuracy of 85%. However, the two combined detect 98% of malignant lesions (19). The second is to detect synchronous malignancy. This may be present in as many as 5% of women with cancer (14,17). The third function is to stage the lesion. The more conservative methods of surgical management require precise staging. Finally, mammography is important in the diagnosis of benign disease.

Although occult carcinomata can be excluded with much greater confidence than by palpation alone (13,36), the capacity of mammography to differentiate benign from malignant processes is poor. In a recent retrospective review of the mammograms of 607 patients with cancer conventional film/screen mammography yielded a definite diagnosis of malignancy in only 483 (80%) (42). The most common causes of failure were difficulty in seeing tumor when it was obscured by dense breast tissue, or failure to observe slight asymmetry in density between the two breasts.

The limitations of these conventional studies make it clear that breast ultrasonography has value in several ways:

(a) As an additional method to reduce the risk of missing a cancer where it is not clinically obvious

(b) In the young dense breast, where the conventional mammographic study is likely to be limited

(c) In the study of dysplasia, particularly fibrocystic disease and its long-term management

(d) As an adjunct to mammography in patients with suggestive but non-specific symptoms such as nipple discharge or retraction

(e) As a guidance method to improve the yield of aspiration biopsy.

Detection of Early Cancer

Screening for breast cancer is a separate issue. There are numerous classification systems that seek to define those at high risk, and screening protocols for these women and for the general population have been proposed (19).

The results of screening programs vary widely. This is largely attributable to variations in methods of patient selection. The gold standard is still the HIP study (46). This was a controlled study of 60,000 women with follow-up now complete to the 12th year. Two-thirds of the cancers detected were found on physical examination and only one-third were found by mammography alone. Many later series have been less controlled and have probably included women at higher risk than the general population (11). In the Cincinnati BCDDP study to 1976 the yield of mammography alone was almost 50%. Despite the strongest criticism of such programs (1,2), it is not disputed that cancers detected at an early stage carry a more favorable prognosis (4,5,15,18,21,39,48,54). Much controversy has been raised about the risk/benefit equations governing screening with conventional mammography (1,2,7,23,37,38). Even though radiation doses have been greatly reduced the potential risk has stimulated considerable research into breast ultrasonography in attempts to develop it to a point where it could be used as a screening tool, either alone or as an adjunct to conventional mammography.

ULTRASOUND MAMMOGRAPHIC TECHNIQUES

The breast presents several problems in ultrasonographic visualization that are unique to that organ. The anatomy of the breast is simple, with well-defined anatomical compartments. However, the constituent tissues vary greatly in their acoustic properties and significant problems are created by reflective and refractive artifacts. This inhomogeneity is probably the limiting factor in resolution.

The conical shape of the breast makes it difficult to maintain uniform path lengths, consistent focusing, or constant angles of beam incidence to the skin surface. The deformability of the breast makes contact B-scanning almost useless. Three different methods have been developed to circumvent these problems, and no single technique is entirely satisfactory. These are (a) scanning the patient prone with the breasts dependent in a water bath, (b) compressing the breast from above against the chest wall with a water bath, (c) scanning the breast with a real-time hand-held instrument, usually with a small built in water stand-off.

Each technique has advantages and disadvantages. Table 2 is an attempt to summarize their relative merits.

It can be seen from Table 2 that no single method can satisfy all the imaging requirements. Direct signs relating to the acoustic properties of the tumor itself are equally well seen with all these methods. It is in the display of secondary signs and the ease of imaging different parts of the breast that there are significant differences.

The Dependent Breast

Automated scanning of the dependent breast in a water bath is the method that has found favor in Australia and in the U.S.A., and is well suited to the predominant practice in these countries of ultrasonographer-based services. Intensive development of these machines has taken place during the past few years, and many of the physical problems listed in Table 2 have been solved.

TABLE 2. *Comparison of scanning methods*

	Dependent breast	Compressed breast	Real time
Costs	Very high	Moderate	Low
Availability	Dedicated	Dedicated	General purpose
Automation	High level	High level	Low level
Scan time	30 Min	30 Min	10 Min
Operator	Sonographer	Sonographer	Physician
Real time	Low frame rate	No	High frame rate
Compound scans	Yes	Limited	Not yet
Arc scans	No	Yes	Not needed
Linear scans	No	Yes	Yes
Sector scans	Yes	No	Yes
Freedom of scan plane	Yes	Limited	Unlimited
Different transducer frequencies	No	No	Yes
Path length (single scan)	Variable	Constant	Constant
Visualization			
Behind nipple	Poor	Poor	Excellent
Deep structures	Poor	Good	Excellent
Axillary tail	Poor	Good	Good
Axilla	No	Poor	Good
Skin signs	Excellent	Good	Poor
Distortion of suspensory ligaments	Excellent	Good	Poor
Fixation/tethering	Poor	Poor	Excellent
Compression signs	No	No	Excellent
Comparisons			
Between breasts	Excellent	Good	Moderate
With physical signs	Poor	Poor	Excellent
With X-ray mammography	Poor	Poor	Moderate
With guided biopsy	Poor	Poor	Excellent

To overcome the problems of focusing at variable depths, two different techniques have emerged. One employs large-aperture transducers, often in the form of annular arrays (45). This can give good focusing over a long focal zone. The other method uses multiple transducers that are strongly focused at different depths, the final image being laid down in strips from each focal zone (40).

Great freedom of scan plane can be achieved with water baths of suitable design (43), and nearly all types of scanning mode are possible including C-scans.

This method gives the best visualization of the overall shape of the breast. It is very sensitive to distortion and allows precise comparisons to be made between breasts, a factor of proven importance in X-ray mammography. The major disadvantages are (a) considerable shadowing from the nipple, (b) poor visualization of the deep structures of the breast, and (c) poor correlation with the findings of clinical examination. It is not possible to palpate the breast at the same time as the ultrasonic examination.

The Compressed Breast

Automated scanning of the breast compressed from above has been widely used in Japan and Europe. By means of this form of compression the breast tissue to be penetrated is reduced to no more than a few centimeters in thickness, and the tissue planes are aligned perpendicular to the beam. The geometry is straightened out, which gives much more uniform transmission and reduces degradation by artifact. The reduced penetration needed allows the use of higher frequency trasducers that can be very strongly focused.

Large breasts must be scanned by quadrants with frequent changes in position. Comparisons between the right and left breast are then inexact, and there is always the danger of an incomplete scan. Problems with acoustical shadowing from the nipple are reduced.

The information given by dynamic tissue compression is also not available. This method works extremely well with small breasts.

Real-Time Studies

Hand-held real-time scanning has only very recently become feasible with the advent of instruments capable of B-scan quality images. Table 2 shows this technique to have most of the advantages. A large scanning head is required to avoid the limitation of a narrow field of view. This can be provided with either a linear array or a sector scanner with a small attached water bath. The resolution of linear array real-time instruments with synthetic focusing is now comparable to that of the single-crystal devices.

Both the advantages and disadvantages of the real-time approach stem from the relative mobility of breast and transducer. This allows one the ability to scan and palpate simultaneously, with instantaneous correlation with the clinical findings—a method of obvious attraction to the clinician. However, without careful technique large areas of the breast may not be adequately scanned, and it is difficult to make exact comparisons between the breasts. A logical method of approach has been proposed by Ching (7a). The transducer is used in the hand of a clock with one end kept constantly at the nipple.

Variations in breast size present no problem. The very large breast can be examined with transducers of different frequencies and in different positions. In the erect posture the large breast is considerably flattened and is easily scanned in many different planes. Dynamic compression studies are a routine part of the examination, and can add significantly to the number of signs available for diagnosis (10,36). The rigidity of the tissue around a mass can be estimated. Small fatty inclusions can often simulate lesions, particularly in the dysplastic breast. Very firm compression will induce shape changes that are not seen in true neoplasms.

Fine-needle biopsy is greatly facilitated with real-time scanning; by biopsying under direct vision, samples can be taken from any part of a tumor mass.

The best information is currently gained by a combination of automated water bath scanning and a high-resolution hand-held real-time scanner. When used by the radiologist, the latter is the most direct tool for clinical problem-solving, regardless of whether the problems become evident on physical examination, X-ray mammography, or automated ultrasound mammography. In the absence of examination with either a contact or real-time scanner, mass lesions can be missed with an automated system. In one study 8 of 68 lesions were missed (47a).

Recording and display of the very large number of individual scans that make up a thorough study is a much more difficult problem than in abdominal or other small parts scanning. Simple photographic recording is really adequate only where the diagnostician has performed or witnessed the original scan. The simplest option is to record the information in video format such that the images can be sequenced through at varying frame rates. This is possible by using either videodiscs or time-lapse videotape recorders. The corresponding sections of the two breasts can be presented simultaneously with appropriate recording techniques. This allows precise and continuous comparisons.

The greatest problems are encountered in recording the real-time study. A time-lapse videotape recorder acquiring 2 to 10 frames per second interspersed with short bursts of relevant real-time loses little of the original data and can be reviewed at high speed.

DIAGNOSTIC CRITERIA IN ULTRASOUND MAMMOGRAPHY

By the time that they are first seen by a physician most breast cancers are obvious with all the diagnostic methods so far described. Strict criteria are most needed for evaluating the subtle signs, which may be the only indicators of malignancy in the less obvious group or in those with very early cancer.

To establish and validate diagnostic criteria was the major goal of most of the basic research in this field, and valuable comparisons of ultrasonographic findings with X-ray mammography, clinical findings, and histopathology have been made. The most precise method has used the whole-organ thin-sectioning technique (17). With specific stains this can give a 1:1 map relating the ultrasound image to the underlying tissue components (27). Clinical research has produced large data bases, and tested empirical criteria in a clinical environment (Table 3).

As a result of this work over the last 20 years, diagnostic criteria for breast diseases are more soundly based on experimental research and carry a better clinical consensus than criteria in any other field of ultrasound imaging (9,35).

The analysis of the ultrasonographic features is clarified by dividing them into two groups, direct and indirect.

Indirect Signs

These are features resulting from the effect of the tumor on the surrounding tissue and on the anatomy of the breast as a whole. These findings are minimal in early or circumscribed cancers, and most obvious in scirrhous lesions (29).

Many of these signs are subtle and require careful comparisons with control tissues. The breast is a radially symmetrical organ, and comparisons are best made along radii between opposing quadrants within the breast and along the same radius in the other breast. A simple record of the radial angle along which the scan was made will suffice for comparisons to be made in subsequent scans. The signs include:

(a) Skin thickening. With suitable water baths and sufficient resolution skin thickening can be reliably demonstrated at low gain by a slight separation of the two echoes comprising the skin line (Fig. 1) (34).

(b) Skin retraction.

TABLE 3. *Diagnostic features of breast cancer[a]*

	53 Carcinomas		
	Medullary (24 cases) (%)	Papillary (15 cases) (%)	Scirrhous (14 cases) (%)
Diagnostic accuracy	83	87	100
Malignant features			
Loss of distal limit of tumor echo	70	80	78
Irregular boundary sign	95	93	78
Acoustic middle shadow sign	75	73	100
Benign features			
Bilateral loss of distal limit of tumor echo	17	6	0
Tadpole tail sign	17	13	0
Lateral shadow sign	17	13	0

[a]Data from ref. 3.

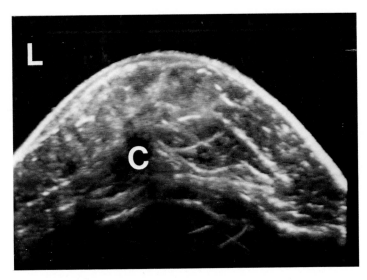

FIG. 1. A 2-cm scirrhous carcinoma (compressed breast). There is marked skin thickening in the lateral quadrant (L) overlying the carcinoma. C, Carcinoma.

(c) Loss of the subcutaneous fat space, or focal changes in echo amplitudes.

(d) Anatomical distortion within the breast.

(e) Differences in echo density and amplitude between quadrants, and between breasts (Fig. 2).

(f) Loss of or increased echo amplitudes in the retromammary fat space.

(g) Asymmetry of intraglandular fat within or between breasts.

(h) Nonuniform shadowing, and poor visualization of the chest wall.

(i) Differences in size between the two breasts.

(j) Differences in compressibility.

Direct Signs

These signs relate to the morphology and acoustic properties of the tumor mass itself.

Features of malignancy are:

(a) The presence of a well-defined mass lesion.

(b) Signs relating to the margins of a lesion: (1) sharp definition; (2) contour: irregular, spiculated, lobulated. Irregularity of part or all of the margin of a lesion is the single most significant sign (Table 3). (3) Increased echogenicity of the surrounding reactive breast tissue (Fig. 3); (4) lateral shadowing (26,33); (5) local changes of marginal definition; (6) duct ectasia associated with a mass lesion.

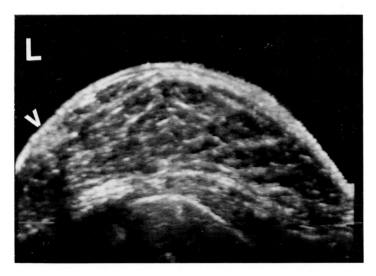

FIG. 2. Multicentric carcinoma (compressed breast). The only demonstrable ultrasound features are slight skin thickening anteriorly, loss of the subcutaneous fat space laterally, and a general increase in echo amplitudes (*arrowhead*) in the lower lateral quadrant (L).

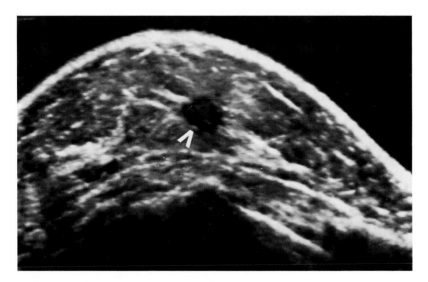

FIG. 3. An 8-mm scirrhous carcinoma (compressed breast). The tumor is centrally placed (*arrowhead*) with sharp margins, internal heterogeneity, increase in the surrounding echo amplitudes, and loss of the subcutaneous fat space.

(c) Internal echo pattern: (1) heterogeneity. In scirrhous carcinomas this is usually extreme; in circumscribed carcinomas a high degree of order may be apparent (Figs. 4 and 5). (2) Differences in echo pattern between different parts of the lesion.

(d) Internal echo amplitudes. (1) Usually reduced, but depends on tissue type, and particularly on connective tissue content (33). (2) Echo amplitudes approximating those found in the normal glandular structures of the breast are more often found in benign lesions (Fig. 6) (20,31,35).

(e) Acoustic shadowing. (1) Virtually all scirrhous carcinomas produce shadowing that is usually irregular across the lesion. Shadowing alone without a well-defined lesion can be seen in up to 5% of detectable cancers (41). (2) Shadowing is minimal distal to well-defined lesions of the circumscribed type. The importance of minor degrees of shadowing behind specific parts of a lesion cannot be overemphasized (41). There is a strong relationship between connective tissue content and shadowing (28).

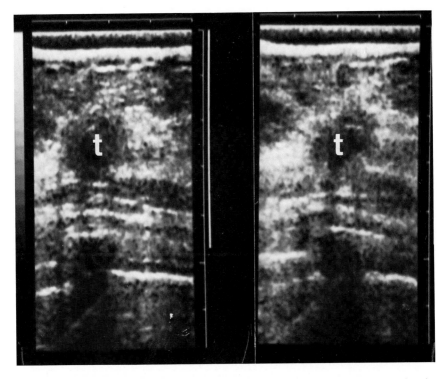

FIG. 4. A 12-mm scirrhous carcinoma (t) (hand-held real-time). The internal echo amplitudes show considerable variation. (Microcalcification is present on the radiograph.) Note the irregular shadowing distal to the lesion and distortion of the subcutaneous fat space.

(f) Local reaction. (1) The most specific sign is indrawing of the ligaments of Astley-Cooper (Fig. 7). (2) The transition of echo amplitudes is almost invariably from low amplitude to high amplitude leaving the tumor. This sign is often visible as an echogenic crown (Fig. 7).

(g) Others. (1) TM scanning. Dale has described a method of graphically displaying the rigidity of the tissues surrounding a tumor using the TM mode with a contact B-scanner. A slow M-mode sweep is combined with repeated compression by the transducer as it is scanned across the area of interest (10). Similar information can be gained by combining real-time scanning with manual palpation and repeated compression by the transducer (36). (2) Velocity changes. These can only be adequately demonstrated by computed tomographic image reconstruction. (3) Changes induced by tissue heating. Wagai has shown that by insonating breast lesions at energy levels sufficient to cause temperature rise there are changes in acoustic properties and boundary characteristics that enhance a lesion from its background (50). (4) Doppler studies. Abnormal blood flow signals can be demonstrated in at least 85% of malignant tumors that have presented as palpable lumps (52). This method is entirely independent of imaging but it is likely that those tumors that give the strongest Doppler signals are the aggressive scirrhous carcinomas that are so well diagnosed by all the other methods. The typical cancer is poorly echogenic, markedly heterogeneous, sharply marginated with an irregular con-

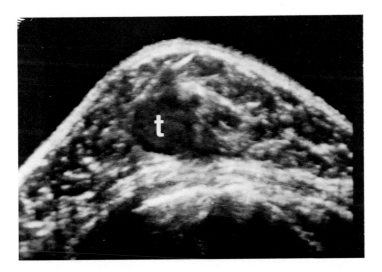

FIG. 5. Circumscribed carcinoma (compressed breast). The central lesion (t) is well defined with uniform low-amplitude internal echoes. The only definite indicators of malignancy are irregularity of the posteromedial contour and associated slight acoustic shadowing.

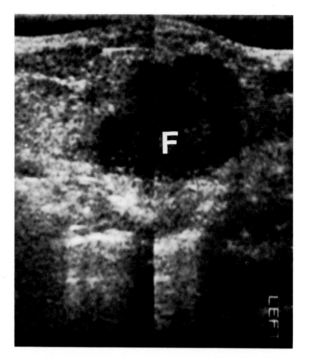

FIG. 6. Fibroadenoma (hand-held real-time). This is a well-defined lobulated lesion (F). The internal echo pattern is uniform and of low amplitude. There is retrotumoral shadowing. The mass was freely mobile and incompressible.

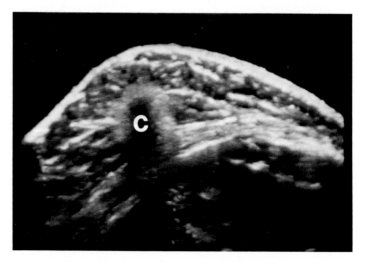

FIG. 7. A 20-mm carcinoma (C) (compressed breast). Note the marked acoustic shadowing and dense echogenic crown anteriorly.

tour, and produces nonuniform distal shadowing, and considerable distortion and reaction in the surrounding tissue. Any significant degree of marginal irregularity, internal heterogeneity, acoustic shadowing, or strong local reaction alone should be regarded as an indicator of malignancy. Coupled with one or more indirect signs, malignancy becomes much more likely.

The use of criteria derived from this long list of ultrasonographic features allows a specific or highly probable diagnosis to be made in 80 to 90% of patients with symptoms (Table 4).

Severe difficulties are encountered in trying to establish reliable criteria of malignancy in three groups of patients. These are:

(a) Those with microcalcification on mammography but with no palpable lump.

(b) Those with preexisting dysplasia, fibrocystic disease, or sclerosing adenosis.

(c) Those with chronic inflammatory conditions, particularly when related to synthetic implants.

These are just those groups that all diagnostic methods find difficult. If ultrasound mammography is to offer information not otherwise available, then clear and useful criteria must be laid down in these areas. This remains to be done.

TABLE 4. *Accuracy of ultrasound mammography*

Reference	Ultrasonography		X-Ray Mammography	
	True positive	False positive	True positive	False positive
Harper, 1981 (24) 1,000 patients	33/35 (95%)	11/865 (1.3%)		
Pluygers, 1981 (44) 12,500 patients	337/352 (96%)		315/352 (89%)	
Cole-Beuglet, 1981 (8) 278 patients	55/79 (69%; studied blind) 62/79 (79%; with clinical data)		(74%)	
Wagai, 1979 (49) Nongray scale 1966–1974	249/281 (88.6%)	32/694 (5%)		
Gray scale 1974–	90/101 (89.1%)			
Kobayoshi, 1974 (49) 618 patients	51/57 (89%)		(83%)	
Dale, 1981 (10) 20,000 patients	791/942 (84%)		(84%)	

ACCURACY OF ULTRASOUND MAMMOGRAPHY

Virtually all the large published series on the accuracy of breast ultra-sonography have been comprised of women with symptoms. The results of most of the major studies are summarized in Table 4.

Most of these large series have been performed with prototype instruments, the precursors of those now offered commercially. Only limited results have been presented from these new machines, but the preliminary evidence does not suggest any obvious improvement in either detection rates or in speci-ficity. Two factors probably account for this. The first is the patient selection bias that changes as a new technique becomes more widely used. The early series is often weighted to those patients with obvious disease, each receiving the detailed attention of a research worker. As experience accumulates, a higher percentage of both normals and patients with minimal disease become included. The second factor is the law of diminishing returns. This is primarily a function of the pathology of the breast itself. Early work has been encour-aged by the fact that the first 80% of tumors are easy to detect. To then halve the error rate to 10% requires a twofold improvement in performance. The extra lesions then detected are smaller and their features more subtle. To reduce further the error rate to 5% (which is probably the level that the clinician requires before he has confidence in the method) involves an improvement of the same magnitude. The physical limitations of ultra-sonography are soon reached.

Most of the reported series have used X-ray mammography as the imaging control. The available data (Table 4) show no significant difference between the two modalities in the populations studied. Although many individual series report better results than these, a summation of the world literature is as follows. For X-ray mammography Dodd summed 65,000 reported cases between 1955 and 1975 to obtain a detection rate of 87.2% with 14.9% false-positives (11). A similar compilation by Kobayoshi to 1976 for ultrasound gave a detection rate of 85.2% (1,180 cancers) (30).

Other studies have examined accuracy relative to tumor size (Table 5). There is an apparent decline in accuracy with the very large cancers. This is related to both their diffuse spread within the breast and the low referral rate from the clinician when a very obvious lesion is present.

Diagnostic accuracy varies significantly with tumor type. Typical results are given by Wagai (50): scirrhous carcinoma 83% (45/54), medullary 78% (54/70), and papillary 29% (27/88).

Kasumi has considered circumscribed cancers as a separate group and described clusters of signs assisting in diagnosis (26). Of 616 cancers studied 91 were classified by histopathology as circumscribed. Of these, ultrasonogra-phy gave a definite diagnosis of malignancy in 68.1%. Seventy-one percent showed suspicious features, and 24.2% were ostensibly benign.

Benign breast tumors show a similar low specific accuracy (20).

TABLE 5. *Diagnostic accuracy (%) and tumor size*

Reference	T0	T1	T2	T3	T4
Dale, 1981 (10) 942 cancers	33	62	85	93	87
Kobayoshi, 1978 (30) 112 cancers		78	90	93	
Wagai, 1977 (49) 101 cancers		80	95	85	

SCREENING

The quantity of literature on breast cancer screening is immense. Most of the arguments for and against are based on different interpretations of the available statistics. A sensible discussion of the benefits versus costs of any screening program can be based only on a mathematical model. The more elaborate the model the more difficult it is to understand it, with most doctors getting lost in the mathematical arguments (12,47).

The consensus view is that X-ray mammographic screening is of definite benefit in selected populations: (a) In all women over 55; (b) in those at high risk—strong family history ($\times 2$), previous cancer in other breast ($\times 7$), gross cystic disease ($\times 4$), lobular neoplasia ($\times 6$), and multiple intraduct papillomata ($\times 7$).

The classic HIP study was based on a combination of mammography and physical examination. It was shown that if X-ray mammography had been excluded from the program the yield of screening would have been reduced by only one-third. Over half the cancers discovered during the screening period presented in the intervals between screening examinations (46).

Later studies have been more encouraging. Lester, reporting the BCDDP experience, defined three groups; those at high risk, the true screening population, and those women who present themselves for screening. This latter group have a higher incidence of breast cancer than the population at large (37). In this study the overall incidence was 3.5 per 1,000 women. There was also a higher proportion of cancers in young women than would have been expected (25% of cancers in women under 50). The yield of mammography was much greater than that of physical examination: 91% and 56%, respectively.

The cancers detected by such programs tend to be smaller than those in women presenting with symptoms. There have been many attempts to quantify the effect of early detection on survival. Fisher's study of 2,000 patients predicted that if all tumors over 2 cm in size had been detected before they reached that stage, then the 5-year recurrence rate might have been reduced by 10 to 18%, and the overall survival rate increased by 11 to 20% (15).

The biology of the tumor and the host response govern survival. The key functions are growth rate and the point in the natural history at which metastases develop (4,18,21,48). Experimental studies in rats have shown that 1 g of tumor tissue is capable of shedding $3-4 \times 10^6$ neoplastic cells into the circulation daily (6).

This is the background against which any proposed ultrasonographic screening program must be judged. Proponents of X-ray mammographic screening have not had all their own way in the arguments (1), yet the ability of this modality to display tumors of subcentimeter size cannot be approached by current ultrasonographic techniques. The only significant attempt to set up an ultrasonographic screening program has been by Wagai (49), and it will be some time before statistically significant data are collected.

There is no doubt that the extensive injection of commercial resources into the development of new machines is predicated on the premise that if ultrasonography is effective in screening, then the lack of a radiation hazard makes the potential market enormous. Despite this there are few serious advocates for ultrasound screening, even in the context of a controlled study.

BENIGN BREAST DISEASE

As has been shown, benign diseases comprise the greatest proportion of the clinical case load. A benign neoplasm almost always require excision for proof of diagnosis. The only exception is an obvious fibroadenoma in women under 25 (22). Cystic disease and the various dysplasias are better followed by ultrasonography than by X-ray mammography, and the effectiveness of hormonal therapy is readily monitored by longitudinal studies (25).

Most women with benign breast diseases have a much higher incidence of malignancy than the normal population. Radiography carries a significant radiation burden and images this group relatively poorly. Effective management is to use a combination of the two methods for long-term follow up with radiography elucidating ultrasonographic abnormalities. The features of dysplasia are readily recognized, but vary widely in any single patient with age and menstrual status (Fig. 8; Table 6).

ROLE OF ULTRASONOGRAPHY IN CLINICAL PRACTICE

In symptomatic women ultrasonography can be as accurate as radiography; as it is based on a different physical principle it is complementary rather than competitive. It is likely to prove of greater value than mammography in women with benign breast disease, and of lesser value in those without well-defined clinical signs or symptoms. A combination of ultrasonography, radiography, and fine-needle aspiration biopsy used judiciously promises to increase the specificity of nonoperative diagnosis to the point where the clinical management can be significantly influenced.

In screening programs the only valid role for ultrasound is to evaluate further radiographic abnormalities and thus reduce the numbers coming to

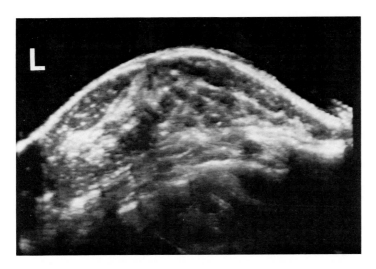

FIG. 8. Sclerosing adenosis (compressed breast). There is an increase in echo amplitudes in the stroma of the lateral aspect (L) of the breast. There is no loss of the subcutaneous fat space or distortion of the glandular cone.

biopsy unnecessarily. A useful subsidiary role is to assist in the localization of impalpable lesions shown radiographically.

Ultrasonographic mammography has an evolving role in the elucidation of clinical problems. The majority of such problems can be resolved by the radiologist using a relatively cheap high-resolution real-time instrument. These machines have possessed adequate resolution for this task only during the past year or so, and no comparative studies on their effectiveness are yet available, but given their ability to demonstrate the relevant signs, it would be surprising if they were significantly inferior to automated systems given the demonstrated variable accuracy of the latter instruments. The demands on the radiologist's time that this technique makes are of little importance when clinical problems are being solved.

TABLE 6. *Ultrasound features of dysplasia*

	Benign mammary dysplasia (37 cases) (%)	Normal (44 cases) (%)
Duct prominence	57	16
Cysts	46	—
Glandular tissue		
High echo amplitudes	51	20
Normal echo amplitudes	22	36
Poor definition of anatomical compartments	27	29
Involution	—	16

Data from ref. 35.

REFERENCES

1. Bailar, J. C. Mammography: A contrary view. *Ann. Intern. Med.*, 84:77–84, 1976.
2. Bailar, J. C. Screening for breast cancer: Pros and cons. *Cancer*, 39:2783–2795, 1977.
3. Bauermeister, D. E. The role and limitations of frozen section and needle aspiration biopsy in breast cancer diagnosis. *Cancer*, 46:947–949, 1980.
4. Bloom, H. J. G., Richardson, W. W., and Harries, E. J. Natural history of untreated cancer (1805–1933). Comparison of untreated and treated cases according to histological grade of malignancy. *Br. Med. J.*, 2:213–221, 1962.
5. Brady, L. W., and Croll, M. N. Clinical uses of bone scanning. *Skeletal Radiol.*, 1:161–167, 1977.
6. Butler, T. P., and Gullino, P. M. Quantitation of cell shedding into efferent blood of mammary adenocarcinoma. *Cancer Res.*, 35:512–516, 1975.
7. Chiacchierini, R. P., and Lundin, F. E. Benefit/risk ratio of mammography. In: *Breast Carcinoma*, edited by W. W. Logan, 1977. John Wiley, New York.
7a. Ching, J. C., Martin, J. F., Leinbach, L., and Wooten, E. Real-time ultrasound of 277 breasts using 7 MHz focused linear array transducers. *Paper No. 334 presented at AIUM, New Orleans, 1980.*
8. Cole-Beuglet, C., Golberg, B. B., Kurtz, A. B., Rubin, C. S., Patchefsky, A. S., and Shaber, G. S. Ultrasound mammography: A comparison with radiographic mammography. *Radiology*, 139:693–698, 1981.
9. Cole-Beuglet, C., Soriano, R., and Golberg, B. B. Ultrasound mammograms: Diagnostic criteria of malignant breast disease correlated with histopathology. *Paper presented at the Second International Conference on the Ultrasonic Examination of the Breast, London, 1981.*
10. Dale, G., and Gros, Ch. Tissue characterization in breast echography. In: *Ultrasonic Tissue Characterization*, edited by J. M. Thijssen, pp. 16–25, 1980. Stafleu, Brussels.
11. Dodd, G. D. Present status of thermography, ultrasound and mammography in breast cancer detection. *Cancer*, 39:2796–2805, 1977.
12. Eddy, D. M. *Screening for Cancer.* Prentice-Hall, Englewood Cliffs, New Jersey, 1980.
13. Egan, R. L., Goldstein, G. T., and McSweeney, M. M. Conventional mammography, physical examination, thermography, and xeroradiography in the detection of breast cancer. *Cancer*, 39:1984–1992, 1977.
14. Egan, R. L., and McSweeney, M. B. Mammographic parenchymal patterns and risk of breast cancer. *Radiology*, 133:65–70, 1979.
15. Fisher, B., Slack, N. H., and Bross, I. D. J. Cancer of the breast: Size of neoplasm and prognosis. *Cancer*, 24:1071–1080, 1969.
16. Frazier, T. G. The value of aspiration cytology in the evaluation of bysplastic breasts. *Cancer*, 45(II):2878–2879, 1980.
17. Gallagher, H. S., and Martin, J. E. The study of mammary carcinoma by mammography and whole organ sectioning. *Cancer*, 24:855–873, 1969.
18. Gallagher, H. S., and Martin, J. E. Early phases in the development of breast cancer. *Cancer*, 24:1170–1177, 1969.
19. Gravelle, I. H., Bulstrode, J. C., Wang, D. Y., and Hayward, J. L. The relation between radiographic features and determinants of risk of breast cancer. *Br. J. Radiol.*, 53:107–113, 1980.
20. Griffiths, K. Ultrasonic appearances of fibroadenoma—The elusive breast mouse. *Paper presented at the Second International Conference on the Ultrasonic Examination of the Breast, London, 1981.*
21. Gullino, P. M. Natural history of breast cancer. *Cancer*, 39:2697–2793, 1977.
22. Haagenesen, C. D. *Diseases of the Breast*, W. B. Saunders, Philadelphia, 1971.
23. Haenszel, W. Screening—Does it apply to breast cancer. *Cancer*, 46:957–960, 1980.
24. Harper, P., and Kelly-Fry, E. Ultrasound visualization of the breast in symptomatic patients. *Radiology*, 137:465–469, 1980.
25. Jellins, J., Kossoff, G., and Reeve, T. S. Detection and classification of liquid filled masses in the breast by gray scale echography. *Radiology*, 125:205–212, 1977.

26. Kasumi, F., Hori, M., and Fukami, A. Characteristic features of circumscribed cancer. *Paper presented at the Second International Conference on the Ultrasonic Examination of the Breast, London, 1981.*
27. Kelly-Fry, E., and Harper, A. P. Clinical and research factors associated with accurate diagnosis of breast masses by ultrasound imaging. *Paper presented at the Second International Conference on the Ultrasonic Examination of the Breast, London, 1981.*
28. Kelly-Fry, E. Breast imaging. In: *Diagnostic Ultrasound in Obstetrics and Gynaecology*, edited by R. E. Sabbagha, pp. 327–350, 1980. Harper & Row, Hagerstown, Maryland.
29. Kobayoshi, T. Gray-scale echography for breast cancer. *Radiology*, 122:207–214, 1977.
30. Kobayoshi, T. *Clinical Ultrasound of the Breast*. Pitman Medical, 1978.
31. Kobayoshi, T. Current status of sonography for the early diagnosis of breast cancer and its tissue character. *Paper presented at the Second International Conference on the Ultrasonic Examination of the Breast, London, 1981.*
32. Kobayoshi, T. Gray-scale echography for breast cancer. *Radiology*, 122:207–214, 1977.
33. Kobayoshi, T. Diagnostic ultrasound in breast cancer: Analysis of retrotumourous echo patterns correlated with sonic attenuation by cancerous connective tissue. *J. Clin. Ultrasound*, 7:421–429, 1979.
34. Kopans, D. B., Meyer, J. E., and Proppe, K. H. The double line of skin thickening on sonograms of the breast. *Radiology*, 141(2):485–487, 1981.
35. Lees, W. R. Tissue characterization in the breast. In: *Ultrasonic Tissue Characterization*, edited by J. M. Thijssen. Stafleu, 1980.
36. Lees, W. R. Real-time compression studies: Applications in symptomatic women and in screening. *Paper presented at the Second International Conference on the Ultrasonic Examination of the Breast, London, 1981.*
37. Lester, R. G. A radiologist's view of the benefit/risk ratio of mammography. In: *Breast Carcinoma*, edited by W. W. Logan. John Wiley, New York, 1977.
38. Mahoney, L. J., Bird, B. L., and Cooke, G. M. Annual clinical examination: The best method available screening test for breast cancer. *N. Engl. J. Med.*, 301:315–316, 1979.
39. Martin, J. E., Moskovitz, M., and Milbraith, J. Breast cancer missed by mammography. *AJR*, 132:737–739, 1979.
40. Maturo, V., Zosmer, N. R., Gilson, A. J., Smoak, W. M., Janowitz, W. R., Bear, B. E., Goddard, J., and Dick, D. E. Ultrasound of the whole breast utilizing a dedicated automated breast scanner. *Radiology*, 137(2):457–463, 1980.
41. Meyer, J. E., Kopans, D., and Steinbock, T. The appearance of breast carcinoma on whole breast water bath ultrasound: A spectrum of sonographic findings. *Paper presented at the Second International Conference on the Ultrasonic Examination of the Breast, London, 1981.*
42. Parsons, C. A. Accuracy and limitations of breast cancer diagnosis using X-rays. *Paper presented at the Second International Conference on the Ultrasonic Examination of the Breast, London, 1981.*
43. Piggins, J. M., McDicken, W. N., and Nicoll, J. J. A versatile scanner for ultrasonic examination of the breast. *Paper presented at the Second International Conference on the Ultrasonic Examination of the Breast, London, 1981.*
44. Pluygers, E., Rombaut, M., and Dormal, C. Present status in the ultrasonic diagnosis of breast disease: The Jolimont experience. *Paper presented at the Second International Conference on the Ultrasonic Examination of the Breast, London, 1981.*
45. Reeve, T. S., Jellins, J., Kossoff, G., and Barraclough, B. H. Ultrasonic visualisation of breast cancer. *Aust. NZ J. Surg.*, 48(3):278–281, 1978.
46. Shapiro, S. Evidence on screening for breast cancer from a randomized trial. *Cancer*, 39:2772–2782, 1977.
47. Slack, N. H., Blumenson, I. E., and Bross, I. D. J. Therapeutic implications from a mathematical model characterizing the course of breast cancer. *Cancer*, 24:960–971, 1969.
47a. Teubner, J., vanKaick, G., Schmidt, W., and Kubli, F. Echomammographie with different scanning units. *Paper presented at the Second International Conference on the Ultrasonic Examination of the Breast, London, 1981.*
48. Underwood, J. C. E. A morphometric analysis of human breast carcinoma. *Br. J. Cancer*, 26:234–237, 1972.

49. Wagai, T., Tsutsumi, M., and Takeuchi, H. Diagnostic ultrasound in breast diseases. In: *Breast Carcinoma*, edited by W. W. Logan. John Wiley, New York, 1977.
50. Wagai, T. Echographic differentiation of histological structure of breast cancer and enhancement of breast cancer images by ultrasound irradiation. *Paper presented at the Second International Conference on the Ultrasonic Examination of the Breast, London, 1981.*
51. Webb, A. J. Fine needle aspiration biopsy in the diagnosis of breast lesions. *Paper presented at the Second International Conference on the Ultrasonic Examination of the Breast, London, 1981.*
52. Wells, P. N. T., Burns, P. N., Cole, S. E., Halliwell, M., Turner, G. M., and Webb, A. J. Ultrasonic Doppler studies of the breast. *Paper presented at the First International Congress on the Ultrasonic Examination of the Breast, Philadelphia, 1979.*
53. Zajicek, J. Fine needle aspiration biopsy of palpable breast lesions. In: *Breast Carcinoma*, edited by W. W. Logan. John Wiley, New York, 1977.
54. Zelen, M. A hypothesis for the natural time history of breast cancer. *Cancer Res.*, 28:207–216, 1968.

Ultrasound Annual 1982,
edited by Roger C. Sanders.
Raven Press, New York © 1982.

Fetal Echocardiography

Charles S. Kleinman

*Department of Pediatrics, Yale University School of Medicine,
New Haven, Connecticut 06510*

While ultrasonic imaging techniques have been applied to the evaluation of the human fetus extensively during the past several years, expertise in evaluation of fetal cardiac structure and function has only recently begun to be developed. The use of static B-mode scanning, although applicable for the assessment of fetal growth patterns and the diagnosis of certain congenital somatic malformations (19), is of limited use for evaluation of the fetal heart because it is a moving structure. The earliest studies of fetal cardiac integrity involved the use of M-(time-motion) mode echocardiographic techniques. Such studies provided early insights into measurement of the fetal heart and intracardiac structures (3–6,8,11–14,20,22,25,28,30,36,37,40,41,43,44,46). The M-mode echocardiogram has been used by some for confirming fetal life during the first and early second trimesters. In 1972, Winsberg (44) combined B-mode and M-mode imaging to measure ventricular dimensions in fetal hearts and to estimate left ventricular output. Winsberg commented at that time that it did not seem feasible to diagnose congenital cardiac disease *in utero* utilizing this technique. M-mode echocardiography has the obvious disadvantage of lacking spatial orientation, especially in the presence of the frequently changing intrauterine position of the fetus. The more recent application of high-resolution linear, phased-array, and mechanically driven real-time two-dimensional imaging techniques has allowed study of both fetal cardiac dynamics and cardiac structure (1,2,15,16,18,21–24,31,35,38,42, 45,47–49). The application of the newest high-resolution dynamic imagers in tandem with the newer techniques of pulsed Doppler determination of cardiac flows have lately stimulated interest in the study of the developing fetal heart (32,42,46–48).

RISK FACTORS WARRANTING
FETAL ECHOCARDIOGRAPHIC STUDY

The application of these specialized imaging techniques to the average normal pregnancy would require a commitment of time and resources that would be prohibitive. If such techniques are to be cost-effective, screening for fetal heart disease should be restricted to certain groups of pregnant women who are defined to be at "higher than normal risk" for fetal cardiac abnormalities. In our laboratory at the Yale-New Haven Medical Center we have found it convenient to restrict such studies to pregnancies falling into "risk factor" groups that we have defined as fetal risk factors, maternal risk factors, and familial risk factors (Table 1) (21,22).

Fetal Risk Factors

Included among fetal risk factors is intrauterine growth retardation. This group has accounted for approximately 16% of the patients scanned. McCallum (31) has noted that approximately 25% of all intrauterine growth-retarded fetuses are symmetrically growth retarded, and of these 20 to 30% will have a chromosomal abnormality. Of these abnormalities the most common are trisomies 21, 18, and 13, all of which impart an exceptionally high risk (50 to 99%) of structural cardiac abnormalities.

Almost 10% of the fetuses studied in our group have had fetal cardiac arrhythmias. Most arrhythmias have been of no hemodynamic significance to the fetus. The presence of sustained tachyarrhythmia, however, has been associated with intrauterine congestive heart failure, and in two fetuses

TABLE 1. *Risk factors warranting fetal echocardiographic study*

Fetal risk factors
  Intrauterine growth retardation
  Fetal cardiac dysrhythmia
  Abnormal amniocentesis (trisomies)
  Somatic anomalies (ultrasound, fetoscopy)

Maternal risk factors
  Heart disease
    Congenital
    Acquired
  Drug ingestions
    Alcohol
    Narcotics
    Amphetamines
    Anticonvulsants
    Lithium
    Birth control pills
    Other sex hormones
  Polyhydramnios
  Oligohydramnios
  Rh sensitization
  Diabetes mellitus
  Preeclampsia
  Collagen vascular disease

Familial risk factors
  Congential heart disease
  Genetic syndromes

arrhythmias were associated with complex structural cardiac disease, imparting, ultimately, a poor prognosis for survival (23).

Fetuses with nonimmune hydrops, presenting on occasion as sudden-onset maternal polyhydramnios, should be carefully screened for structural or functional causes of fetal "right heart failure" (23). Other fetal risk factors indicating a need for cardiac imaging include abnormal chromosomal studies on amniocentesis. As previously noted, patients with the common trisomies have an exceptionally high risk of significant congenital abnormalities, as do patients with XO Turner's syndrome (35%) and 5p-(Cri-du-chat) (approximately 20%). Fetuses diagnosed at ultrasound or fetoscopy to have certain somatic abnormalities may also be at high risk for associated major cardiac structural abnormalities. For example, Ellis-von-Creveld (association of polydactyly and single atrium, 50%); Holt-Oram (association of radial dysplasia with ventricular and/or atrial septal defects, 50%), or Ivemark's syndrome (asplenia with complex cardiovascular abnormalities, 100%) (Table 2).

Maternal Risk Factors

Certain maternal diseases impart a risk for structural or functional heart disease in the developing fetus. Included are maternal diabetes mellitus, which may be associated with hypertrophic cardiomyopathy of the develop-

TABLE 2. *Selected syndromes and associated cardiac defects*

Syndrome	Cardiac lesion
Di George	Conotruncal malformations Interrupted aortic arch
Ellis von-Creveld	Single atrium
Holt-Oram	Atrial (ASD) and Ventricular septal defects (VSD)
Hurler	Cardiomyopathy
Noonan	Pulmonary stenosis ASD Hypertrophic cardiomyopathy
Trisomy 13–15	VSD Tetralogy of Fallot Double outlet RV (DORV)
Trisomy 16–18	VSD, PDA, DORV
Trisomy 21	A-V canal defect VSD
Turner (XO)	Aortic coarctation Bicuspid aortic valve
5p-(Cri-du-chat)	VSD, PDA, ASD

ing fetus. Initial studies suggest that poor maternal diabetic control may impart a stimulus for the development of the hypertrophic changes, and therefore that the fetal echocardiogram may be of some utility in assessing the "adequacy" of maternal diabetic therapy.

Mothers with collagen vascular disease, in particular systemic lupus erythematosis, have been found to have a significant risk of abnormalities of the atrioventricular conduction system in the developing fetus with a significant risk for the development of congenital complete heart block. One such case has been diagnosed in our laboratory (22).

Maternal ingestion of certain medications may impart a risk for the development of congenital cardiac disease detectable by *in utero* scanning. Diphenylhydantoin is a cardiovascular teratogen with a 2 to 3% incidence of pulmonary stenosis, aortic stenosis, aortic coarctation, and patent ductus arteriosus. Among the anticonvulsant medications, trimethadione appears to be significantly associated with a 15 to 30% incidence of complex fetal heart disease including great arterial transposition, tetralogy of Fallot, and left heart hypoplasia (31). The ingestion of sex hormones has been associated with significant heart disease in 2 to 4% of cases.

There has been strong evidence to associate lithium with abnormalities of the tricuspid valve (Ebstein's malformation or tricuspid atresia) and to associate excessive maternal alcohol ingestion with left-to-right shunt lesions (ventricular and atrial septal defects) (31).

As noted above, the presence of maternal hydramnios should be used as an indication for study, in an effort to detect the presence of structural or functional disease of the fetal heart resulting in nonimmune fetal hydrops.

Familial Risk Factors

Much of the data currently used to assist in genetic counselling of families in whom one or more parent or other siblings has had congenital heart disease has been based on the work of Dr. James J. Nora of the University of Colorado (32,33). Dr. Nora has commented that in comparing recurrence risks that have been found in different series in the world literature, there are some similarities and other striking dissimilarities (Table 3). This is related to the small size of each single series, making it impossible to overcome biases introduced by outlying observations. A recent review of the experience at Yale-New Haven Medical Center (R. Whittemore, *personal communication*) agrees with Nora's recent observation that the risk appears to be generally higher in the offspring than for the sibling of an affected individual. It should be noted that Nora reports familial recurrence of congenital cardiovascular disease *whether the lesions are concordant or are apparently discordant*, feeling that the meaningful figure in counselling is the recurrence of congenital cardiovascular disease and not necessarily the recurrence of a specific defect. Of the major birth defects (accounting for approximately 27 per 1,000 live

TABLE 3. *Recurrence risks for congenital heart disease given one affected sibling*

Anomaly	Suggested risk[a] (%)	Suggested risk[b] (%)
Ventricular septal defect	3	1.2
Patent ductus arteriosus	3	2.2
Atrial septal defect	2.5	2.0
Tetralogy of Fallot	2.5	1.5
Pulmonic stenosis	2	2.5
Coarctation	2	0.6
Aortic stenosis	2	3.1
Transposition of great arteries	2	2.5
Endocardial cushion defect	2	
Endocardial fibroelastosis	4	6.1[c]
Tricuspid atresia	1	1.1
Ebstein	1	
Pulmonary atresia	1	1.3
Hypoplastic left heart	2	0[d]
Truncus arteriosus	1	1.2

[a]Data from refs. 32, 33.
[b]Data from ref. 26.
[c]Endocardial fibroelastosis only; stillbirth rate increased.
[d]Hypoplastic left heart only; 1.8% for all CHD.

births) congenital cardiac disease accounts for 8 per 1,000 live births (0.8%). Nora's collected data suggest recurrence risks given one sibling who has a cardiovascular abnormality varying from 1% to 4% (the latter the recurrence risk of endocardial fibroelastosis cited by Chen et al. in 1971). The recurrence risk cited given a sibling with one of the most common cardiovascular abnormalities (ventricular septal defect, atrial septal defect, patent ductus arteriosus, tetralogy of Fallot) varies from 2.5 to 3%. Similar data given one parent with a congenital heart defect suggest that for the common defects listed above the recurrence risk ranges from 2.5% (atrial septal defect) to 4% (VSD, PDA, tetralogy of Fallot) (Table 4).

TABLE 4. *Recurrence risks for congenital heart disease given one affected parent*

Anomaly	Suggested risk[a] (%)	Suggested risk[b] (%)	Suggested risk[c,d] (%)
Ventricular septal defect	4	3.6	22
Atrial septal defect	2.5	1.9	11
Patent ductus arteriosus	4	3.6	11
Tetralogy of Fallot	4		
Pulmonary stenosis	3.5	2.8	19
Coarctation of the aorta	2	2.6	} 26
Aortic stenosis	4		

[a]Data from refs. 32, 33.
[b]Data from ref. 26.
[c]Data from R. Whittemore, *personal communication.*
[d]Does not exclude genetic syndromes.

The information obtained from fetal echocardiographic study may be invaluable for counselling prospective parents at risk for bearing children with heart disease and may allow the obstetric and pediatric team to lay plans for the psychologic and medical management of the remainder of the pregnancy, delivery, and the neonatal period. In particular, in cases of the most severe forms of congenital heart disease previously encountered in families, fetal echocardiography may provide information of great import to the genetic counselor. We have found fetal scans to be of great use in allaying the fears of parents who have previously encountered the hypoplastic left heart syndrome, a lesion that Nora states imparts a recurrence risk for all congenital heart lesions in siblings exceeding the general population expectation with, however, only a 25% risk of such recurrences being hypoplastic left heart. We have encountered one case of recurrent hypoplastic left heart syndrome in 35 cases scanned for this diagnosis.

THE FETAL ECHOCARDIOGRAPHIC EXAMINATION

B-mode and/or real-time scans are performed to determine the position of the fetus within the uterus and amniotic sac and the position of the fetal heart with respect to the maternal abdominal wall. Thereafter, a two-dimensional echocardiographic examination is performed, with an effort made to simulate the standard precordial views that are usually recorded in postnatal studies (21). Occasionally the fetus may be positioned with limbs or spine interposed between the anterior abdominal wall and the fetal heart. In this situation it is often helpful to have the mother reposition herself on the examining couch or have her take a short walk. An alternative approach is to have the obstetrician gently apply external version technique. Utilizing such methods, we have been successful in imaging the fetal heart in 90 to 95% of cases. Visualization is facilitated by the fact that the fetus is immersed in the fluid-filled amniotic sac with fluid-filled lungs surrounding the heart. Limiting factors have been maternal obesity, a persistent unfavorable fetal lie, and oligohydramnios.

Utilizing echocardiographic systems that have the capacity for either simultaneous or sequential M-mode recording during two-dimensional echocardiographic scanning, it is possible to obtain hard-copy M-mode studies that are spatially oriented with reference to the accompanying two-dimensional scan (Fig. 1). The M-mode study provides an excellent means to analyze cardiac motion against time, and therefore the M-mode scan is useful for analysis of disturbances in cardiac rhythm (Fig. 2). The M-mode study may also be used for measurement of intracardiac chamber and wall dimensions (2). Lange et al. in Tucson found that four-chamber and short-axis views at the arterial level were most useful for evaluation of two-dimensional anatomy (24). They were successful in obtaining such images in 95 to 96% of cases studied. The ascending and descending aorta and arch vessels were visualized in 87% of examinations and superior and inferior vena cavae in 76% of the 88 cases studied. Lateral resolution at 6 cm down was 1.5 to 2 mm and axial

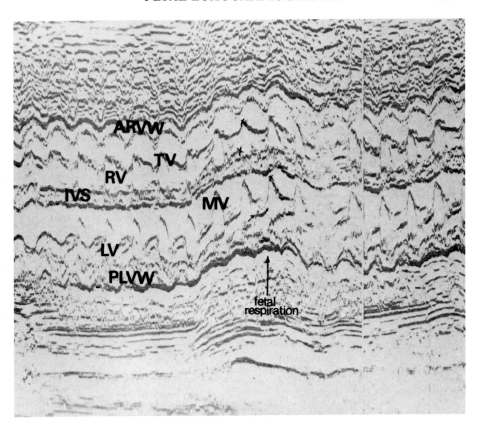

FIG. 1. M-mode fetal echocardiogram performed at midventricular level in a 33-week fetus. Note the effect of fetal respiration on the echocardiogram. ARVW, Anterior right ventricular wall; IVS, interventricular septum; LV, left ventricular cavity; MV, mitral valve; RV, right ventricular cavity; TV, tricuspid valve. (From ref. 22, with permission.)

resolution 0.2 mm at a depth of 6 to 10 cm from the transducer. They did not choose to use their M-mode capability "due to the absence of electrocardiographic availability." Allan et al. in Great Britain performed elegant anatomic sections of the human fetus and standardized the fetal echocardiographic examination (1). They noted the more horizontal position of the fetal than of the neonatal heart and the large fetal liver extending to the left abdominal wall. This results in cranial displacement of the cardiac apex. Allan described eight standard planes of examination, finding the long-axis view to be the most difficult to obtain.

Longitudinal Planes

The most consistently obtained longitudinal plane scanned two-dimensionally in the fetus was described by Allan et al. (Fig. 3) as the "tricuspid-pulmonary" plane, showing the pulmonary valve anterior and cranial to the

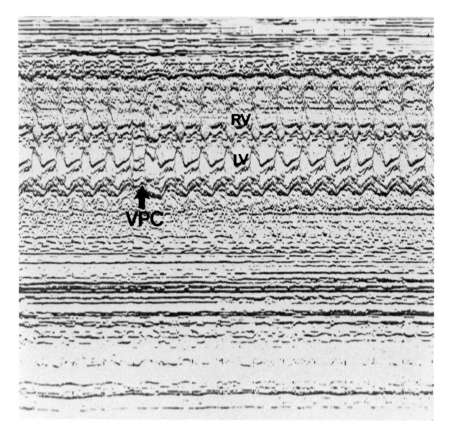

FIG. 2. M-mode echocardiogram at midventricular level. Note that the regularity of posterior left ventricular wall motion is interrupted by the presence of a premature beat. There is no evidence of a presystolic atrial wave on the mitral valve. The premature beat is therefore a ventricular premature contraction (VPC).

aortic root (1). This could be identified in 90 to 96% of cases studied. By changing the obliquity from the sagittal plane, the pulmonary trunk is continuous (Fig. 4) with the patent ductus arteriosus and enters the descending aorta. This plane is referred to as the "ductus plane." The head and neck vessels are seen originating from the aortic arch in this view. By orienting the transducer somewhat more to the left of the sternum, the left ventricle is seen in short axis with the right ventricular infundibulum ascending anteriorly. This is referred to as the "left ventricular short axis plane" (Fig. 5) (1).

Transverse Planes

With the fetus scanned in transverse view, the most consistently obtained view is a four-chamber view of the heart (Fig. 6). In this plane, the right ventricular cavity appears to have a coarser trabecular pattern and the cavity

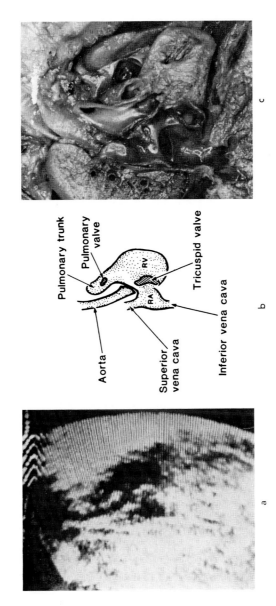

a

Aorta

Pulmonary trunk

Pulmonary valve

RV

Superior vena cava

RA

Tricuspid valve

Inferior vena cava

b

c

FIG. 3. The longitudinal echocardiographic findings (**A**) in the tricuspid-pulmonary plane. The structure as visualized is shown in (**B**), and (**C**) shows the simulated plane in a dissected human abortus. (From ref. 1, with permission.)

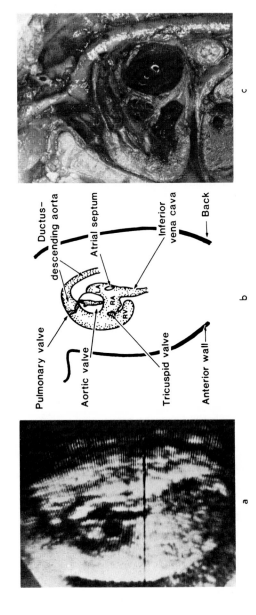

FIG. 4. Similar composite of echocardiographic findings (**A**), drawing (**B**), and anatomical dissection (**C**) illustrating the ductus longitudinal plane. (From ref. 1, with permission.)

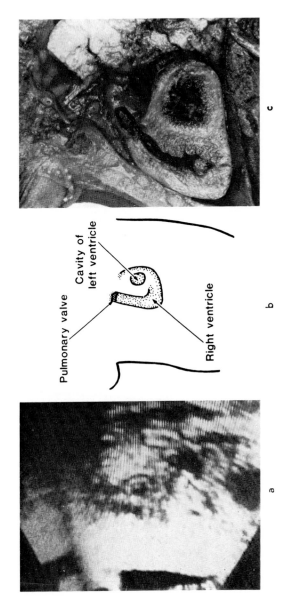

FIG. 5. A composite showing the left ventricular short axis plane. (From ref. 1, with permission.)

Pulmonary valve

Cavity of
left ventricle

Right ventricle

a

b

c

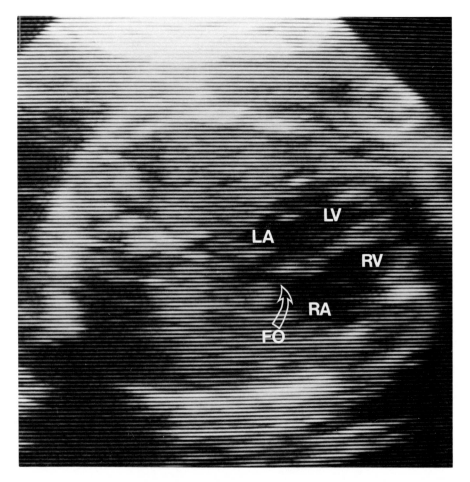

FIG. 6. Four-chamber two-dimensional echocardiographic section in a 22-week fetus. Note the prominent foramen ovale (FO, *arrow*) between the right (RA) and left (LA) atria. The flap of the foramen ovale is on the left atrial side of the interatrial septum.

often appears foreshortened as a result of the moderator band near the apex. In this view the atria are visualized and the flap of the foramen ovale may often be seen moving with the cardiac cycle. This is a useful "marker" for "left sidedness," the foramenal flap normally being located within the left atrial cavity. In addition, the vena cavae may be traced to their entry into the sinus venosus portion of the right atrial cavity (Fig. 7). By orienting the transverse plane toward the right shoulder, the left ventricular long axis view is seen (Fig. 8). In this view, septal continuity with the anterior aortic wall and mitral valve continuity with the posterior aortic wall are normally visualized. The

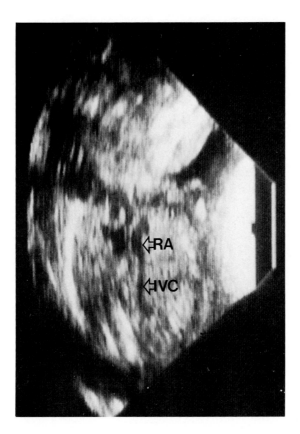

FIG. 7. Two-dimensional scan in the tricuspid-pulmonary plane. The transducer is oriented to follow the insertion of the inferior vena cava (IVC) into the right atrial cavity. (From ref. 1, with permission.)

four-chamber plane was obtained in virtually 100% of cases whereas the long axis view of the left ventricle was found to be more difficult to obtain, with success attained in 41% of the second 100 patients scanned.

Abnormalities Noted

Utilizing the standardized two-dimensional views, normal anatomic inter-relationships between intracardiac structures should be easily demonstrable in greater than 90% of cases studied. In particular, demonstration of the presence of a normal complement of intracardiac valves and septae may be quite useful in ruling out the presence of major forms of cardiac pathology. Reports are becoming abundant in the literature of major structural abnormalities detected by *in utero* scans (Table 5) (2,15,18,21–23,35). The applica-

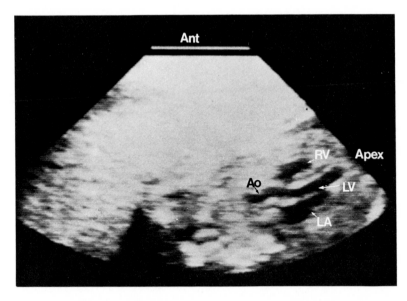

FIG. 8. Two-dimensional echocardiographic scan in the left ventricular long axis view. The ascending aorta (AO) arises from the left ventricular cavity (LV). In this view, septal-to-aortic and mitral-to-aortic fibrous continuity can be appreciated.

tion of echocardiographic imaging techniques in a prospective fashion for study of "high-risk" pregnancies has yielded a significant number of positive studies (see Table 5). As predicted in earlier studies, success has been encountered in anticipating the presence of hypoplastic right heart syndrome (Fig. 9), hypoplastic left heart syndrome (Fig. 10), univentricular heart (Fig. 11), malalignment defects such as in tetralogy of Fallot (Fig. 12), persistent truncus arteriosus, atrioventricular canal defects (Fig. 13), and, quite significantly, the British group has recently successfully diagnosed a hypoplastic aortic arch *in utero* (Fig. 14).

Fetal Echocardiographic Evaluation of Nonimmune Hydrops

We have recently applied fetal echocardiographic imaging for the evaluation of 13 fetuses with nonimmune hydrops (23). Ten fetuses had cardiovascular abnormalities resulting in acute right-sided venous hypertension (Table 6). In three cases hydrops was related to supraventricular tachycardia. One of these fetuses responded to cardioversion at birth and another responded to transplacental digoxin therapy. There were no cases of "idiopathic" hydrops. Our results suggest that fetal echocardiography may be useful for determining cardiac causes of *in utero* heart failure resulting in hydrops fetalis. The fetal echocardiogram was also useful in monitoring transplacental therapy of heart failure secondary to supraventricular tachyarrhythmias.

TABLE 5. *Major structural abnormalities detected by in utero scans*

Source of report	Lesion	Gestational age (weeks)
Yale's experience	Pulmonary atresia Intact interventricular septum Hypoplastic right ventricle	34
Yale's experience	Tricuspid atresia Hypoplastic right ventricle	34
Yale's experience	Univentricular heart	28
Yale's experience	Tetralogy of Fallot	28
Yale's experience	Septal rhabdomyoma	32
Yale's experience	Polysplenia syndrome A-V canal defect	28
Yale's experience	Ventricular septal defect	32
Yale's experience	Acardia (twin pregnancy)	36
Yale's experience	Acardia (twin pregnancy)	28
Yale's experience	Absent pulmonary valve syndrome	32
Yale's experience	Asplenia syndrome A-V canal defect	32
Yale's experience	Myocardial infarction	33
Yale's experience	Hypoplastic left heart	32
Yale's experience	Hypoplastic left heart	38
Allan et al.	Ostium primum and secundum atrial septal defect	16
Allan et al.	Aortic isthmal hypoplasia	18

Roelands
  (cited by Allan)
	Pulmonary atresia and ventricular
	  septal defect
		(Middle trimester)

Hackeloer and Hansmann	Enlarged ascending aorta	28
Hackeloer and Hansmann	Enlarged left ventricle	33
Redel	Prematurely closed foramen ovale	35
Henrion and Aubry	Polysplenia syndrome Single ventricle	24

Echocardiographic Evaluation of Fetal Cardiac Arrhythmias

Fetal cardiac rhythm disturbances are usually first suspected on the basis of auscultatory findings. Due to the inability of transabdominal fetal electrocardiography to demonstrate atrial depolarization this technique is of limited value in the analysis of cardiac rhythm disturbances *in utero*. Initial studies in our laboratory have been performed utilizing real-time directed M-mode echocardiography for evaluation of cardiac motion against time (23). Due to its rapid sampling rate and its capacity to obtain a hard copy of the tracing demonstrating cardiac motion as a function of time, M-mode echocardiogra-

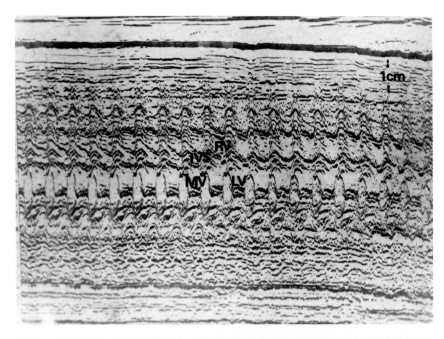

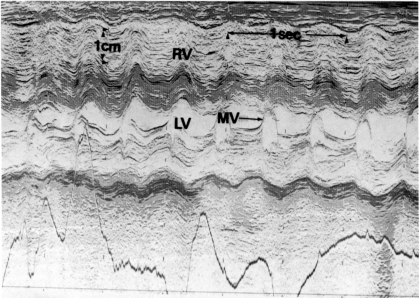

FIG. 9. Top: M-mode echocardiogram at midventricular level in a 34-week fetus with intrauterine growth retardation. Only one atrioventricular valve could be identified, within a large (18–22 mm) ventricular chamber. The anterior ventricular cavity is hypoplastic (6 mm diameter). **Bottom:** M-mode echocardiogram at midventricular level in the same patient. Study performed at one hour of age. Note the similarity between studies. IVS, Interventricular septum; LV, left ventricular cavity; MV, mitral valve; RV, right ventricular cavity. (From ref. 22, with permission.)

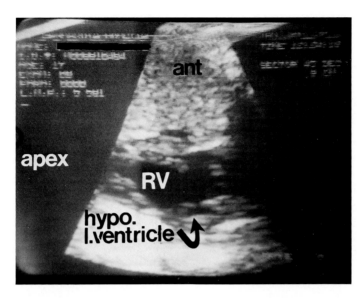

FIG. 10. Two-dimensional echocardiographic scan at 38 weeks gestation. A hypoplastic left ventricular cavity (hypo. 1 ventricle) (*curved arrow*) is seen in the posterior aspect of this longitudinal scan. This cavity is dwarfed by the enlarged anterior right ventricular cavity. (Courtesy of Department of Radiology, LaGuardia Hospital, New York, New York.)

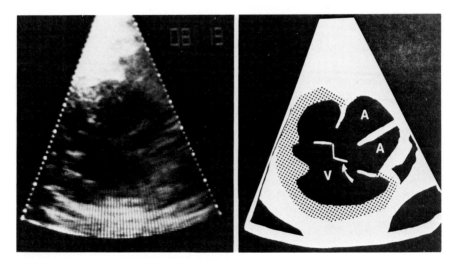

FIG. 11. Left: Single frame of video tape recording of real-time two-dimensional study in a fetus with a univentricular heart. Frame is a four-chamber view in early diastole. **Right:** Diagram representing frame on left. Atrial septum separates the two atrial chambers **(A)**. The common anterior valve leaflet (*arrow*) bridges the large atrioventricular canal-type ventricular defect and opens into the large single ventricular chamber (V). This has been photographed as one figure. (From ref. 22, with permission.)

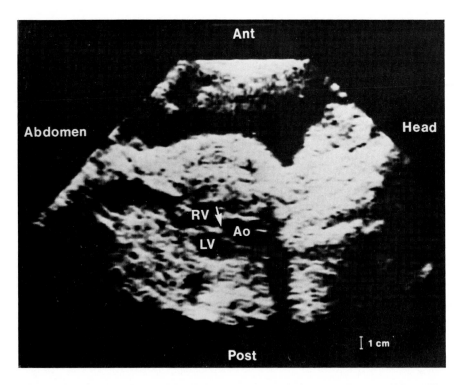

FIG. 12. Long axis two-dimensional echocardiographic scan in a 28-week hydropic fetus. The aorta "overrides" a large interventricular septal defect (*arrow*). The *in utero* diagnosis of tetralogy of Fallot was confirmed at cardiac catheterization shortly after birth. (From ref. 23, with permission.)

phy is well suited for the evaluation of cardiac rhythm disturbances. By allowing evaluation of the atrioventricular contraction sequence, septal motion pattern, ventricular wall contraction, postectopic pauses, and atrioventricular valve motion, the site of origin of premature contractions may be suggested (3,7,9,15,17,18,21–23). For example, premature supraventricular beats are usually preceded by demonstrable "A" waves prior to premature mitral valve closure whereas ventricular premature contractions frequently precede atrial depolarization, thereby aborting the mitral "A" waves during that cardiac cycle. Our initial studies are in agreement with previous workers who have suggested that isolated ectopy in the fetus is usually supraventricular in focus and is almost always self-limited, carrying a favorable prognosis. We are, however, in agreement with others who have suggested that further study during pregnancy is warranted because of the possibility that isolated ectopic beats could trigger reentrant tachyarrhythmias. Sustained arrhythmias may carry a more ominous prognosis, particularly if they are associated with structural heart disease and/or *in utero* congestive heart

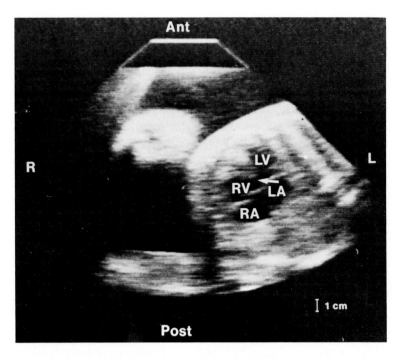

FIG. 13. Four-chamber echocardiographic scan in a 32-week fetus with hydrops fetalis. A large interventricular septal defect is seen (*arrow*). The atrioventricular valves arose within the same plane and this suggests the diagnosis of atrioventricular canal defect. This was confirmed at catheterization shortly following birth. (From ref. 23, with permission.)

failure. We have encountered two cases with frequent supraventricular premature beats who subsequently were found to have supraventricular tachycardia (23). Both cases responded well to transplacental administration of therapeutic doses of digoxin administered orally to the mother. The fetal echocardiogram was therefore useful for diagnosis as well as for monitoring of the therapy of these potentially lethal arrhythmias.

QUANTITATIVE STUDIES

Initial attempts at measurement of the fetal heart and its component parts were made from M-mode echocardiograms in 1970. Winsberg (44) refined the studies in a small series in 1972 in which measurements of left ventricular end-diastolic and end-systolic dimension were made and simple formulae applied in an effort to estimate left ventricular output. More recently, Sahn et al. (38) performed two-dimensional echocardiographic studies on 69 pregnancies in healthy mothers referred for routine fetal cardiac scanning. Normal fetal cardiac development was studied quantitatively through the second half of

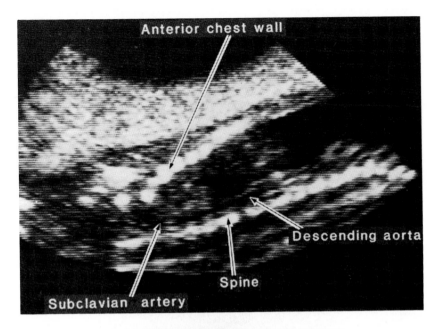

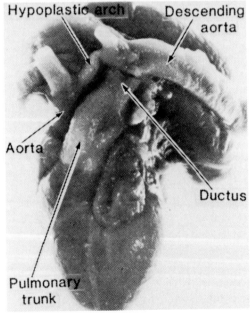

FIG. 14. Top: Echocardiogram showing the descending aorta in an 18-week fetus with hypoplasia of aortic arch. **Bottom:** The anatomical specimen from the 18-week fetus shown in **A,** illustrating gross tubular hypoplasia of the aortic arch between the left common carotid and the left subclavian artery. (From ref. 2, with permission.)

pregnancy by this group. Statistical analysis of total fetal cardiac dimension, right ventricular and left ventricular dimension, and great artery size showed good correlation with the calculated fetal weight and growth curves for these cardiac sturctures were derived. The right ventricular/left ventricular dimensional ratio was 1.18 ± 0.01 prior to birth decreasing to 0.99 ± 0.03 ($p < 0.001$) within 36 hr of birth. These findings are useful for evaluation of cardiac growth and are important if studies of cardiac abnormalities are to be performed antenatally. In addition, such studies have provided insights into the major adjustments of the circulatory system for extrauterine life and may be useful in confirming the applicability of previous reports in humans and animals. More recently, Wladimiroff has measured internal cardiac structures from two-dimensional images (47).

Flow Studies

Wladimiroff et al. have applied combined compound B-scanning or real-time scanning and pulse Doppler studies for study of umbilical or descending aortic blood flow (47,48). A number of problems have been encountered regarding the accuracy and reproducibility as a result of the need to know the location and orientation of the vessel, measurement of the arterial lumen, and the flow velocity within the vessel. An error of only five degrees in angulation will change flow calculations by as much as 10%. An error of "E" in measurement of the arterial lumen alters flow calculation by a factor E^2. These workers have therefore measured the diameters of vessels using either electronic calipers applied to the two-dimensional image or measurements made from time-motion recordings. There has been a discrepancy noted between the calculated flows using the maximum versus the minimum diameter as calculated from the T-M studies. This has been accounted for by a 10 to 15% change in the diameter of the vessel resulting from the diastolic/systolic changes. Because the fetal aorta is 6 to 10 cm from the transducer, a low-frequency Doppler source is needed. Wladimiroff has used a 2 MHz source. On the basis of the mean profiles obtained, the blood flow in the intraabdominal portion of the umbilical cord has been estimated at between 100 and 110 ml/kg/min, which is in agreement with the data of Assali et al. Wladimiroff has estimated the descending aortic flow at 168 to 179 ml/kg/min (47,48).

Fetal Systolic Time Intervals

Wladimiroff has measured the fetal systolic preejection period utilizing transabdominal fetal electrocardiography and the Doppler signal (47,48). A significant positive relationship was found between the preejection period and gestational age. A diverse relationship existed with fetal heart rate. Utilizing M-mode echocardiographic imaging of semilunar valve motion in association with simultaneous transabdominal fetal electrocardiography, Donnerstein in

TABLE 6. *Evaluation of heart failure in utero. Summary of clinical data in 13 consecutive fetuses with nonimmune hydrops*[a]

Case no.	Gestation (week)	Presenting problem	Prenatal diagnosis	Outcome and pathologic findings
1	32	Polyhydramnios	Hydrops fetalis Intracardiac tumor	Stillbirth (33 weeks) Septal rhabdomyoma
2	28	Polyhydramnios	Hydrops fetalis Isolated levocardia Situs inversus Complete heart block Atrioventricular-canal defect	Premature birth (34 weeks) Complete heart block Death at 10 min of life Isolated levocardia, situs inversus, poly- splenia, atrioventricular-canal defect
3	28	Irregular fetal cardiac rhythm (Polyhydramnios)	Hydrops fetalis Complex dysrhythmia Tetralogy of Fallot Congestive cardiomyopathy	Cesarean delivery (31 weeks) Tetralogy of Fallot Ventricular dysrhythmia Congestive cardiomyopathy Hyaline-membrane disease Death at 5 days
4	32	Polyhydramnios	Placental hydrops Right ventricular and right atrial dilatation Pulmonary stenosis ?Pulmonary insufficiency	Premature birth (33 weeks) "Absent-pulmonary-valve" syndrome; interventricular septum intact Hyaline-membrane disease Death at 1 day
5[b]	38	Polyhydramnios	Hydrops fetalis Atrial flutter with 2:1 atrio- ventricular block	Cesarean delivery (38 weeks) Atrial flutter with 2:1 atrioventricular block Cardioversion (direct current) Survival in good condition
6	34	Polyhydramnios	Hydrops fetalis Atrial flutter with 5:1 and 6:1 atrioventricular block	Stillbirth (34 weeks) Dilated, anatomically normal heart
7	27	Irregular fetal cardiac rhythm	Pericardial effusion Supraventricular ectopy leading to paroxysmal atrial tachycardia	Transplacental digoxin therapy leading to sinus rhythm Delivery at 39 weeks Chaotic atrial rhythm treated with digoxin and propranolol Survival in good condition

8	Polyhydramnios	36	Twin pregnancy Twin A—"acardiac" monster Twin B—hydrops fetalis	Premature delivery Twin A—stillbirth and dysmorphism Twin B—hydrops, hyaline-membrane disease, disseminated intravascular coagulation; death at 3 days
9	Polyhydramnios	32	Hydrops fetalis Normal heart	Stillbirth (33 weeks) Cystic adenomatoid malformation of right lung
10	Polyhydramnios	29	Hydrops fetalis Mediastinal mass Normal heart	Stillbirth (29 weeks) Extralobar sequestration Normal heart
11	Polyhydramnios	22	Hydrops fetalis Normal heart Abnormal chromosome No. 11	Premature birth (28 weeks) Respiratory distress Death at 30 min of life Multiple anomalies Diaphragmatic hernia Hypoplastic lungs Bilateral hydroureter
12	Polyhydramnios	32	Hydrops fetalis Large ventricular septal defect	Premature delivery Asplenia, atrial isomerism, atrioventricular-canal defect, transposition of great arteries, pulmonary atresia, anomalous pulmonary venous drainage Death at 3 days
13	Polyhydramnios	33	Hydrops fetalis Dilated atrium and ventricle	Premature delivery Death at 30 min of life Dilated left atrium and ventricle Coronary-artery embolus with massive myocardial infarction

aData from Kleinman et al. (23).
bThis case has been previously described.

our laboratory performed calculated fetal systolic time intervals including right and left ventricular preejection period and ventricular ejection times. Preliminary data suggest that *in utero* myocardial dysfunction influences these intervals in a manner similar to the alteration seen postnatally. These studies may be useful for evaluating the effects of cardioactive medications (such as the betamimetic agents currently being used to arrest premature labor) on fetal myocardial function.

SUMMARY

Qualitative and quantitative evaluation of high-resolution images of the developing fetal heart during the second and third trimesters of pregnancy may provide structural information concerning the developing heart that may be of use for counselling prospective parents and may also allow the medical team to make plans for the medical management of the remainder of the pregnancy, delivery, and neonatal period of these offspring. The application of quantitative studies of cardiac output, umbilical blood flow, relative ventricular sizes, and systolic time intervals may provide information to improve our understanding of cardiac dynamics in the developing human fetus and adaptations of the cardiovascular system to the changes occurring at birth in the presence of normal and abnormal cardiac function. The M-mode echocardiogram may be used to analyze the rhythm of the fetal heart and may be used to monitor transplacental therapy of fetal cardiac arrhythmias.

REFERENCES

1. Allan, L. D., Tynan, M. J., Campbell, S., et al. Echocardiographic and anatomical correlates in the fetus. *Br. Heart J.*, 44:444–451, 1980.
2. Allan, L. D., Tynan, M., Campbell, S., and Anderson, R. H. Identification of congenital cardiac malformations by echocardiography in midtrimester fetus. *Br. Heart J.*, 46:358–362, 1981.
3. Baars, A. M., and Merkus, J. M. W. M. Fetal echocardiography: A new approach to the study of the dynamics of the fetal heart and its component parts. *Eur. J. Obstet. Gynecol. Reprod. Biol.*, 7:91–100, 1977.
4. Bang, J., and Holm, H. H. Ultrasonics in the demonstration of fetal heart movements. *Am. J. Obstet. Gynecol.*, 102:956–960, 1968.
5. Bessis, R., Schneider, L., Gortchakoff, M., and Zeitoun, M. Fetal echocardiography. In: *Abstracts, 3rd European Congress on Ultrasonics in Medicine*, p. 427, 1978.
6. Birnholz, J. C. Confirmation of fetal life. *Res. Staff Physician*, 105–108, 1975.
7. Blandon, R., Leandro, I., and Acuna, E. L. Ecocardiografia fetal en el 2o y 3er. trimestre de gestacion. *Separ. Revis. Med. Panama*, 2:1–15, 1978.
8. Bonicelli, L., Orsini, L. F., Rizzo, P., et al. Diagnostic and prognostic value of fetal heart activity in early abnormal pregnancy. In: *Abstracts, 3rd European Congress on Ultrasonics in Medicine*, p. 433, 1978.
9. Crowley, D. Combined (2-D) and M-mode echocardiographic evaluation of fetal arrhythmia. *Pediatr. Res.*, 14:442, 1980 (Abstr.).
10. Deitz, J. D. Fetal cardiac anatomy as imaged by ultrasonic methods. *Med. Ultrasound (USA)*, 2:15–18, 1978.
11. DeLuca, I., Ianniruberto, A., and Colonna, F. Aspetti ecografici del cuore fetale. *G. Ital. Cardiol.*, 8:776–780, 1978.

12. Demidov, V. N. On possibility of using echography in determining sizes of the heart, amplitude of the mitral valves, width of the myocardium and the stroke volume of the heart in the fetus. *Akush. Ginekol. (Mosk.)*, 9:31–33, 1979.
13. Dewbury, K. C., and Meire, H. B. Fetal mitral valve movement recording by pulsed ultrasound. *Clin. Radiol.*, 29:1–4, 1978.
14. Egeblad, H., Bang, J., and Northeved, A. Ultrasonic identification and examination of fetal heart structures. *J. Clin. Ultrasound*, 3:95–105, 1975.
15. Hackeloer, B. J. The value of combined real-time and compound scanning in the detection of fetal heart disease. *Contr. Gynecol. Obstet.*, 6:115–118, 1979.
16. Hahta, J. C., Hagler, D. J., Bailey, P. B., and Fish, C. R. Wide-angle two-dimensional echocardiographic assessment of fetal cardiac anatomy. *Am. J. Cardiol.*, 45:466, 1980 (Abstr.).
17. Harrison, N. A. "Missed beats" in a fetus—A case report. *S. Afr. Med. J.*, 57:330–332, 1980.
18. Henrion, R., and Aubry, J. P. Fetal cardiac abnormality and real-time ultrasound study: A case of Ivemark syndrome. *Contr. Gynecol. Obstet.*, 6:119–122, 1979.
19. Hobbins, J. C., and Winsberg, F. *Ultrasonography in Obstetrics and Gynecology*, Williams & Wilkins, Baltimore, 1977.
20. Ianniruberto, A., and DeLuca, F. Fetal echocardiography. In: *Abstracts 3rd European Congress on Ultrasonics in Medicine*, p. 429, 1978.
21. Kelley, M. J., Jaffe, C. C., and Kleinman, C. S. *Cardiac Imaging in Infants and Children*, W. B. Saunders, Philadelphia, 1982.
22. Kleinman, C. S., Hobbins, J. C., Jaffe, C. C., Lynch, D. C., and Talner, N. S. Echocardiographic studies of the human fetus: Prenatal diagnosis of congenital heart disease and cardiac dysrhythmias. *Pediatrics*, 65:1059–1067, 1980.
23. Kleinman, C. S., Donnerstein, R. L., DeVore, G. R., et al. Fetal echocardiography or evaluation of *in utero* congestive heart failure: A technique for study of nonimmune fetal hydrops. *N. Engl. J. Med.*, 306:568–575, 1982.
24. Lange, L. W., Sahn, D. J., Allen, H. D., et al. Qualitative real-time cross-sectional echocardiographic imaging of the human fetus during the second half of pregnancy. *Circulation*, 62:799–806, 1980.
25. Lee, F. Y. L., Batson, H. W. K., Alleman, N., and Yamaguchi, O. T. Fetal cardiac structure: Identification and recognition. A preliminary report. *Am. J. Obstet. Gynecol.*, 129:503–511, 1977.
26. Lenz, W. Aetiology, incidence and genetics of congenital heart disease. In: *Heart Disease in Infants and Children*, edited by G. Graham and E. Rossi, pp. 33–34, 1980. Yearbook Medical Publishers, Chicago.
27. Madison, J., Sukhum, P., Williamson, D., and Campion, B. Echocardiography and fetal heart sounds in the diagnosis of fetal heart block. *Am. Heart J.*, 98:505–509, 1979.
28. Merkur, H. Normal and abnormal antenatal ultrasonic cardiographic patterns. *Br. J. Obstet. Gynecol.*, 86:533–539, 1979.
29. Murata, Y., Martin, C. B., Jr., Ikinoue, T., and Lu, P. S. Antepartum evaluation of the preejection period of the fetal cardiac cycle. *Am. J. Obstet. Gynecol.*, 132:278–284, 1978.
30. Murata, Y., Takemura, H., and Karachi, K. Observation of fetal cardiac motion by M-mode ultrasonic cardiography. *Am. J. Obstet. Gynecol.*, 111:287–294, 1971.
31. McCallum, W. D. Fetal cardiac anatomy and vascular dynamics. *Clin. Obstet. Gynecol.*, 24:837–849, 1980.
32. Nora, J. J. Update on the etiology of congenital heart disease and genetic counselling. In: *Etiology and Morphogenesis of Congenital Heart Disease*, edited by R. VanProagh and A. Takao, pp. 21–39, 1980. Futura Publishing, Mount Kisko, New York.
33. Nora, J. J., and Nora, A. H. The evolution of specific genetic and environmental counseling in congenital heart diseases. *Circulation*, 57:205–213, 1978.
34. Powell Phillips, W. D., Wittmann, B. D., and Davison, B. M. Real-time ultrasound B-scan as an aid to antepartum fetal heart rate monitoring in multiple pregnancy. Case report. *Br. J. Obstet. Gynecol.*, 86:666–667, 1979.
35. Redel, D. A., and Hansmann, M. Fetal obstruction of the foramen ovale detected by two-dimensional Doppler echocardiography. In: *Echocardiology*, edited by H. Rijsterborgh, pp. 425–429, 1981. Martinus Nijhoff Publishers, The Hague.
36. Robinson, H. P. Detection of fetal heart movement in first trimester of pregnancy using pulsed ultrasound. *Br. Med. J.*, 4:466–468, 1972.

37. Roczen, R. S. Fetal echocardiography: Present and future applications. *J. Clin. Ultrasound,* 9:223–229, 1981.
38. Sahn, D. J., Lange, L. W., Allen, H. D., et al. Qualitative real-time cross-sectional echocardiography in the developing normal human fetus and newborn. *Circulation,* 62:588–597, 1980.
39. Silver, T. M., Wicks, J. D., Spooner, E. W., and Cohen, S. M. Prenatal ultrasonic detection of congenital heart block. *Am. J. Roentgenol.,* 133:546–547, 1979.
40. Stygar, A. M. Comparative evaluation of the TM and Doppler's method in detection of the fetal cardiac activity during the first trimester of pregnancy. *Vopr. Okhr. Materin Det.,* 25:64–66, 1980.
41. Suzuki, K., Minei, L. J., and Schnitzer, L. E. Ultrasonographic measurement of fetal heart volume for estimation of birth weight. *Obstet. Gynecol.,* 43:867–871, 1974.
42. Vosters, R., Wladimiroff, J. W., Vletter, W., and Roelands, J. Assessment of fetal and neonatal cardiac dynamics by means of real-time ultrasound. In: *Abstracts, 3rd European Congress on Ultrasonics in Medicine,* p. 431, 1978.
43. Wagner, G., Klanbe, A., and Kunze, G. Echokardiographische Darstillung des fetalen Herzens (TM-Bild-Darstellung) *Z. Kardiol.,* 68:497–498, 1979.
44. Winsberg, F. Echocardiography of the fetal and newborn heart. *Invest. Radiol.,* 7:152–158, 1972.
45. Wladimiroff, J. W. Real-time assessment of fetal dynamics. In: *Echocardiology,* edited by N. Bom, pp. 135–140, 1977. Martinus Nijhoff Publishers, The Hague.
46. Wladimiroff, J. W., Vastus, R., and Vletter, W. Ultrasonic measurement of fetal and neonatal ventricular dimensions. *Contr. Gynecol. Obstet.,* 6:109–114, 1979.
47. Wladimiroff, J. W. The present state of art in the study on human fetal cardiovascular performance. In: *Progress in Medical Ultrasound: Reviews and Comments, Vol. 2,* edited by A. Kurjak pp. 43–58, 1981. Medical Examination Publishing Co.
48. Wladimiroff, J. W., McGhie, J., and Vastus, R. Fetal cardiac activity. In: *Echocardiology,* edited by H. Rijsterborgh, pp. 417–423, 1981. Martinus Nijhoff Publishers, The Hague.
49. Yamaguchi, D. T., and Lee, F. Y. L. Ultrasonic evaluation of the fetal heart. A report of experience and anatomic correlation. *Am. J. Obstet. Gynecol.,* 134:422–430, 1979.

Subject Index

Subject Index